"Throughout Susan McCarthy's *Bec*... amazed at her long succession of animal anecdotes and made uneasy by the feeling that we humans are only guessing at the private lives of other species. Readers will long for a conversation with the otter that searches for the perfect stone with which to bash its dinner, the chimp that fashions a tool to probe a mound of dirt for a termite lunch, the gazelle that learns about cheetahs by mobbing them—safety in numbers—to know how they look and smell, and the little songbird that expands its brain every year to learn new springtime tunes." —*San Diego Union-Tribune*

"*Becoming a Tiger* takes a fascinating area of exploration—how we all, as animals, learn—and brings it to life in all its richness. The individual stories are both intimate and surprising, and the book as a whole illuminates the ways all of earth's learning creatures share such important commonalities . . . and differences. Susan McCarthy's deep appreciation of the process of learning and the ways that learning joins us sings from each page."
—Dawn Prince-Hughes, Ph.D.,
author of *Gorillas Among Us: A Primate
Ethnographer's Book of Days* and *Songs of
the Gorilla Nation: My Journey Through Autism*

"Susan McCarthy knows more about animals than anyone I have ever met. She also writes beautifully, is unwilling to say anything she is not absolutely certain to be correct, and carries out meticulous research. But beyond all that she has an eye for the unusual, for what other writers have passed over, which means that she is never dull or conventional, no matter what she is writing about." —Jeffrey Moussaieff Masson, Ph.D.,
coauthor of *When Elephants Weep:
The Emotional Lives of Animals*

Brady Lea

About the Author

SUSAN McCARTHY is the coauthor (with Jeffrey Moussaieff Masson) of the *New York Times* bestseller *When Elephants Weep*. She holds degrees in biology and journalism, writes regularly for Salon.com, and has contributed to *The Best American Science Writing*. She lives in San Francisco.

ALSO BY SUSAN McCARTHY

When Elephants Weep: The Emotional Lives of Animals
(coauthor with Jeffrey Moussaieff Masson)

BECOMING A
TIGER

How Baby Animals Learn
to Live in the Wild

SUSAN McCARTHY

HARPER
PERENNIAL

HARPER ● PERENNIAL

A hardcover edition of this book was published in 2004 by HarperCollins Publishers.

BECOMING A TIGER. Copyright © 2004 by Susan McCarthy. All rights reserved. Printed in the United States of America. No part of this book may be used or reproduced in any manner whatsoever without written permission except in the case of brief quotations embodied in critical articles and reviews. For information address HarperCollins Publishers, 10 East 53rd Street, New York, NY 10022.

HarperCollins books may be purchased for educational, business, or sales promotional use. For information please write: Special Markets Department, HarperCollins Publishers, 10 East 53rd Street, New York, NY 10022.

FIRST HARPER PERENNIAL EDITION PUBLISHED 2005.

Designed by Nancy B. Field

The Library of Congress has catalogued the hardcover edition as follows:

McCarthy, Susan.
   Becoming a tiger : how baby animals learn to live in the wild / Susan McCarthy—1st ed.
     p. cm.
   Includes bibliographical references.
   ISBN 0-06-620924-2
   1. Learning in animals. 2. Animal behavior. 3. Animals—Infancy.
   I. Title.
   QL785.M45   2004
   591.5'14—dc22   2003067553

ISBN-10: 0-06-093484-0 (pbk.)
ISBN-13: 978-0-06-093484-2 (pbk.)

05 06 07 08 09 ❖/RRD 10 9 8 7 6 5 4 3 2 1

For Mary Susan Kuhn

With love, gratitude, and respect

Contents

Introduction: Why Learning?   ix

ONE   How to Do or Know Something New:
        Ways of Learning   1

TWO   Learning the Basics: How to Crawl,
        Walk, Climb, Swim, and Fly   31

THREE   Learning Your Species   59

FOUR   How to Get Your Point Across:
        Being Vocal, Being Verbal, and
        Otherwise Communicating   91

FIVE   How to Make a Living   141

SIX   How Not to Be Eaten   183

SEVEN   Invention, Innovation, and Tools:
        How to Do Something New, Possibly
        with a Stick   209

EIGHT   How to Get Cultured   245

NINE   Parenting and Teaching: How to Pass It On   273

TEN   What Learning Tells Us About Intelligence   311

        Conclusion: Secrets of a Tiger's Success   347

        Notes   351

        Bibliography   375

        Acknowledgments   399

        Index   401

Introduction:
Why Learning?

SOMEHOW A FUZZY, STUMBLING tiger kitten becomes a monstrously efficient killer. Somehow a big-footed fool of a raven fledgling becomes an aerial acrobat and a masterful survivor in the north woods. Somehow a wide-eyed spindle-limbed squirrel monkey infant becomes a wily adult who eats a wholesome caterpillar and avoids a poisonous one. Somehow a panda finds love, a spear-nosed bat joins a sorority, and a bear who successfully holds a territory passes her hard-won gains on to her ignorant cubs.

How do baby animals become competent adults? While part of the answer is that cubs and kittens and chicks mature and come into their powers, another part is that they learn what they can do and how to do it.

Learning is the ultimate combination of nature and nurture, in which a growing animal applies its powers of intelligence, curiosity, perception, and memory to the world around it, again and again, and ends up with knowledge and skills it did not have before. No newborn animal is a blank slate and no newborn animal has a complete instruction manual.

Learning and intelligence are connected, but are not the same thing. We often ask questions like "How intelligent is a chimpanzee?" "How smart are pigs?" or "How dumb is my sister's cat?" This book looks instead at what a chimpanzee, a pig, or a cat can learn. This kind of inquiry acknowledges change, examines the interplay between nature and nurture, and lends itself to narrative. In the end, it is a sneaky way of starting to answer the questions above.

Learning has an odd status in our esteem. We're impressed with people who know a thing without learning, who grasp it instantly. It's not unknown for animal behaviorists to dismiss an impressive performance by an animal as "only learned behavior"—as opposed to intelligent insight. Yet nothing seems dumber than being unable to learn.

(If you do much reading about learning and intelligence in animals, you will come across references to "Einsteins among the herons," "a raven Einstein," "an Einstein among macaques," or "an Einstein of a Herring Gull." I hope not to succumb to such temptation.)

Learning is a process, not the static image provided by an intelligence test. It's an intrinsically hopeful process of improvement. As an animal, I am also perpetually beguiled by the bumbling folly of baby animals, while also understanding that what I see is not stupidity, but an early stage of a journey toward grace, competence, and comprehension.

As an optimist I am all in favor of learning as much as possible. I am idealistic enough to think we're better off learning how sausage is made, though I admit there are times when I regret having learned that demonic form of solitaire my father plays.*

This is not a book about what the study of young animals can teach us about child-rearing. But since we are animals ourselves, a certain amount of illumination is unavoidable.

IN RESEARCHING *WHEN ELEPHANTS WEEP*, my coauthor Jeffrey Moussaieff Masson and I had to wrestle with a widespread scientific reluctance to write of emotion. Researching this book didn't present that problem. That animals feel emotion is still anathema in some circles; that animals learn is not. *How* animals learn is often the controversial part. Do animals imitate each other? (Is that true imitation?) Do they pass on learning in the form of culture? (Is that really culture?) Do they teach their children? (What counts as teaching?) I sifted the research for animal stories that illustrate different kinds of learning.

Many stories in this book come from scientific journals and books, but some come from wildlife rehabilitators. In wildlife rehabilitation there are many cases of baby animals and birds who didn't have natural

* "Used to play," he says.

childhoods in the wild. As a result, what sometimes seems like the effortless and neatly programmed progress of an animal from birth to life as a grown animal is discovered not to be inevitable. Seams show. Strange gaps appear. Unnatural liaisons are suggested. Wildlife rehabilitation is an occupation that seeks, among other things, to discover what an animal needs to be exposed to and to learn in order to have a normal life.

The reintroduction of endangered species, under the supervision of scientists, draws on the skills of wildlife rehabilitation. The pitfalls in reintroducing black-footed ferrets are not the same as the pitfalls in reintroducing whooping cranes, but all throw light on the nature of these animals. To save the species it may be vital to teach ferret kits not to spend so much time on the surface of the prairie dog colony; to ensure that crane chicks don't get wrong ideas about romance and family; or perhaps, someday, to provide mentors for tiger cubs.

The ability to learn is an adaptation of tremendous power, one which has taken our own species a long way. We even go to the extreme of learning about learning, as in this book.

ONE

How to Do or
Know Something New:
Ways of Learning

Studying killer whales off the Canadian Pacific coast, researcher Alexandra Morton spotted an eaglet learning that not all birds are alike. The nest he had hatched in was in a fir tree by the shore. Gazing keenly about, the eaglet saw a great blue heron standing on floating kelp. Deciding to do likewise, the eaglet flew down to alight on the kelp. Instead of standing on it with splayed toes as the heron did, he gripped it like a bough. The seaweed sank when the young bird landed on it, and he plunged in. He struggled free, but didn't give up the project. "Again and again the eagle alighted and sank to his breast, flapping wildly to avoid drowning," writes Morton. It took all morning before the eaglet abandoned hope of becoming the Terror of the Kelp.

AN EXPERIMENTAL PROCESS of trial and error persuaded the young bald eagle that at least one perching place wasn't for him. Some baby animals are born with most of the skills they will use in their lives, and need to gather very little information to make their way through life. Others must learn many skills and collect a great deal of information if they are to survive.

There are many ways to learn. An orphaned fox cub at a rehabilitation center who stops panicking every time someone puts a food pan in its cage is becoming habituated. An owlet flapping its wings and trying to fly to a higher branch is practicing. Wrestling puppies come to understand social relationships by playing. A tiger cub with newly opened eyes is learning to see by forming neural connections. A day-old chick that begins foraging by pecking indiscriminately at seeds, bugs, and its own toes (and switches to pecking only at seeds and bugs) is using trial and error. When a young raven, cawing fiercely, joins with the rest of the flock in mobbing a creature it has never seen before, the fledgling is undergoing social conditioning. These are all forms of learning.

Researchers have carefully examined many of the ways animals learn, and have focused a lot of attention on which ways to learn are available to all kinds of animals and which are special to only a few top-flight species, particularly the human species. But that animals of every kind learn in at least some ways is undisputed.

Why learn?

A young rabbit being closely pursued by a predator zigzags crazily from side to side. The rabbit covers less ground this way, but because it's nimbler than a big wolf or bobcat or eagle, it stands a chance of getting away by dodging. This is an excellent strategy that might not occur to me if a monster were suddenly hot on my heels. The young rabbit doesn't have to learn zigzagging—a rabbit's life is short, and if it had to learn its evasive tactics would be even shorter. This is great, unless the rabbit stupidly jumps out in front of your car, realizes that it's in trouble, and starts zigzagging down the road in front of the car. If rabbits lived a long time, and were protected from predators by their parents while they slowly learned about the world, we might hope they would come up with a better way not to be hit by cars. (How about not jumping in front of them in the first place, pal?) Not having time, rabbits are born ready to zigzag.

Being prepared ahead of time or thinking on your feet?

Why go to the trouble of learning if you can just be born knowing how to dodge? You may dodge when it's inappropriate, like the rabbit in front of the car. You may be unable to learn new strategies, like crossing the road after the car comes, not before. Species vary tremendously in how much of their behavioral repertoire is learned. The zoologist Ernst Mayr proposed the metaphor of closed and open programs. Mayr defines a closed program as one which does not allow appreciable modifications, and an open program as one "which allows for additional input during the lifespan of its owner."*

An animal with a closed program recognizes mates without having to learn what their species looks like, usually by one or two key features or a ritualized display. An animal with an open program learns what prospective mates should look like, often by observing its own family. A frog doesn't need to learn what a suitable frog mate looks like, but an owl must learn to spot a suitable partner. Many animals have closed programs for some aspects of their behavior and open for other aspects.

I once raised two small Virginia opossums whose mother had been hit by a car, and they operated largely on closed programs. Things they simply knew included how to hiss (showing 50 teeth), how to curl up in a ball, how to hang by their tails, how to beat up a cat that jumped them, and how to catch fledgling birds. They knew that fledgling birds and rotten apples were good to eat (they thought almost everything was good to eat). They knew they should waddle along, sniffing, until they smelled food. They knew that climbing upward was a good way to be safe.

They were open to some new information. They learned that I was a friend who would protect them, and so when I took them to the woods, thinking they'd like to explore, they whirled in alarm and clambered up my legs—because they hadn't learned to know the woods. They learned that I wouldn't really let them fall if they refused to hang by their tails.† They learned that my dog and my cat wouldn't bother them. And that was about all.

Species with short lifespans, like opossums, have little time to learn, so they are apt to have more closed than programs open. Animals whose parents take care of them for a long time have a chance to learn while protected, so they are apt to be able to afford more open programs.

Whenever there's usually one right thing to do in a clearly recognizable situation, a closed program is perfect. If there's something in

*In the 1970s students of animal behavior got a collective crush on computer metaphors, which shows no sign of letting up.

†They learned to trust me so well that I could never scare them into "playing possum."

your eye, you should blink, not ponder. If a cat attacks you, show all 50 teeth right away.

But if the choices you face are more complicated, and the world you live in keeps changing, open programs might be more successful. "The great selective advantage of a capacity for learning is . . . that it permits storing far more experiences, far more detailed information about the environment, than can be transmitted in the DNA of the fertilized zygote," writes Mayr. "If it is to survive in a constantly changing environment, a bird cannot rely exclusively on the genome. There are far too many gaps in this network of inborn information," writes Jürgen Nicolai, a scholar of birdsong.

Then there's the thrift issue. Getting information by learning (as opposed to having it stored in the genes) saves space in the genome. Of course, then you need more space in the brain, and brains are expensive to run, metabolically speaking.

The nature of nurture, and nurturing nature

Closed programs and open programs refer to nature and nurture, two sources of behavior that are pitted against each other in many arguments about why humans do what they do. Are we born the way we are, or do our environment and upbringing make us the way we are? Are there fewer women in politics because women are less competitive or because cultures derail their ambitions? Are criminals bad to the bone, or were they simply raised wrong? Do nations make war because our species can't resist, or is it just that we made the mistake of getting into agriculture?

Scientists regularly become discontented with terms like "instinct" and "innate" because they are so imprecise. But try not to use them, or come up with synonyms, and the concepts keep coming back. It's too hard to do without them. Many books on behavior, especially human behavior, begin with an obligatory passage about how both nature and nurture are important—but then often go on to stress only one of the two. The struggle is probably inescapable. If we could prove that some aspect of human behavior, for example, is 50 percent innate and 50 percent learned, battles would erupt about whether it isn't 51 percent and 49 percent or the other way around. Since the subject here is learning, the nurture/slate-with-a-little-room-left-to-write-on/open program camp will be well represented, but it's really true, just as the obligatory disclaimers say, that nature and nurture are incredibly entwined.* So there'll be lots of examples of closed programs, although they may get embroidered by an animal's experience and intelligence.

An innate behavior is often modified by learning. No one had to teach you to sneeze, but one hopes that someone taught you to grab for a handkerchief or tissue when sneezing. Even the most elaborate innate behaviors are assembled in a series of environments. Genes are transcribed and proteins are assembled in environments within the cell, embryos develop in environments within their mothers, and behavioral triggers are encountered in environments.

Imprinting, which will be discussed in scandalous detail later, is a classic example. The infant animal or bird has, in the middle of its closed program for many behaviors, a big blank spot that says only Your Parents' Names Here. The closed program of the gosling says to follow its mother, peeping, but who mother is must be learned.

Primatologist Hans Kummer has compared attempts to determine how much of a trait is genetic and how much is produced by the environment to an attempt to decide whether the sound of drumming is made by the drummer or the drum. But, as Frans de Waal puts it, if the sound changes, "we can legitimately ask whether the difference is due to another drummer or another drum."

I'm so mixed up

An interplay of instinct and learning can be seen in hybrid lovebirds. Lovebirds, who really are the cuddly and constant little parrots their name implies, come in different species with different habits. When building a nest, peach-faced lovebirds cut long strips from bark or leaves, and then tuck the ends of the strips under the feathers of their lower back and fly to the nest site. About half the strips fall out along the way, of course, but they get the job done.

Researchers crossbred peach-faced lovebirds with Fischer's lovebirds, who do the rational thing and carry strips of nesting material in their bills like most other birds. The young hybrids, nesting for the first time, "acted as though they were completely confused." They cut

*Chance also plays a role that is often overlooked.

lovely strips, and they seemed to have some vague but powerful idea that they should tuck something in their plumage, but they could almost never manage it. They occasionally seemed to feel that carrying strips in their bills might also be good, but first they needed to do some tucking. All the tucked strips fell out. After two months of this, the young hybrids carried about 40 percent of strips in their bills, but they still spent lots of time making tucking gestures before flying. It took them three years to more or less give up on the tucking thing.

Conditioning

For a large stretch of the twentieth century, the only kind of learning many animal behaviorists were willing to talk about was conditioning. Conditioning comes in two kinds, Pavlovian and operant conditioning, which do not always get kept neatly separate in life. Pavlovian conditioning is also called classical conditioning (because it got talked about first) or associative learning. It has been referred to as a correlation-learning device. Pavlov's dogs famously came to salivate when they heard a bell, because when the bell rang, they got fed. Drooling when they got fed wasn't learned, but associating the bell with food—and drooling—was.

Fish whistle

In the 1930s, when it was generally thought that fish could not hear, Karl von Frisch began whistling to a blind catfish in his laboratory before he put food in the tank. After a few days, when the fish heard whistling, it would come out of the drainpipe where it lurked and search for food. The proof that at least one fish could hear led to many other experiments on fish hearing.

Conditioning does not take a mighty brain. Fruit fly larvae (yes, tiny maggots) learned to form associations to the odor of either ethyl acetate or isoamyl acetate. When they were offered a choice between an odor they had smelled while given rich and delicious food full of wholesome brewer's yeast and an odor they had smelled while given nutrient-deficient food laced with quinine, 64 percent wriggled toward the side that reminded them of yeast. And when offered a choice between an odor that they had smelled when they were being harassed with a fine brush (to simulate predation) and an odor they had smelled when being left alone, 73 percent headed for peace and quiet. These were, of course, larvae from a strain of fruit fly renowned for its good test scores, but still, we're discussing the intellect of maggots.

Tiger, tiger, burning bright, on the roadways of the night

In the 1950s, Lieutenant Colonel Locke, of the Malayan Civil Service, in the state of Trengganu, had duties that included shooting problem tigers. Problem tigers, as Locke saw it, were tigers who ate people, tigers who ate cattle, tigers who ate dogs, and one tiger who had formed the habit of walking up to rubber tappers in the forest and growling. Although he only growled, this invariably caused the perturbed rubber tappers to take the rest of the day off, and the resulting financial losses to the local rubber industry spelled the tiger's doom.

Locke's shooting technique involved putting out bait, erecting a concealed platform in a nearby tree, and waiting there at night until he heard a tiger at the bait. Then he'd switch on a flashlight so he could aim, and shoot the tiger. One night Locke was after a cattle-killing tiger. This particular tiger was an elderly male who, Locke happened to know, had been in a car accident while crossing a road at night. The tiger had recovered from his injuries, but "retained an overwhelming dread of bright lights." There weren't many bright lights in Trengganu in those days.

On this evening, Locke finally heard the tiger come to the bait, a dead cow. Locke switched on his light, and the tiger immediately reared up and toppled over backward into some bushes, where he lay moaning dismally. The astonished Locke reports that the tiger was neither growling nor roaring, but moaning "as though the beast was in mental anguish. I was convinced that he thought another car was after him." The tiger sobbed for a while and then fell silent. After twenty minutes he got up and approached the carcass. Locke switched on the light again, and the tiger instantly bolted. The tiger did not return to the cow that night. In fact, he never touched cattle again and thereby escaped being shot.

The tiger seems to have learned, in one traumatic accident, to fear sudden bright lights in the night. Then it seems to have learned to associate eating cattle with the horrifying lights. Whether it actually thought, "If I touch a cow, a car will appear and attack me" is more speculative.

Associating the lights in the night with being struck by a car is an example of Pavlovian conditioning—the innate fear of being hurt was associated with the learned stimulus of lights. Associating messing with cattle with the dreaded lights is an example of operant conditioning— the tiger now connected his action of attacking cattle with the negative stimulus of the lights.

Operant conditioning

Operant conditioning is also called Skinnerian conditioning, after its famous and persuasive advocate, B. F. Skinner. The most common scientific example of basic operant conditioning is the white rat in a cage equipped with a lever and a food hopper, in which the rat learns that if it pushes the lever, it will be rewarded with a piece of rat chow tumbling into the hopper. (Such cages are called Skinner boxes.) One can go on from here to condition far more complex behaviors in any species you care to name, including the human. Rewards condition behavior, and so do unpleasant, negative things, such as being given an electric shock, very popular in the lab. (Or being hit by a car, as in the tiger's case.)

Sonja Yoerg describes one rat among a group of 20 that a colleague was training to press a lever to get a food reward, using automated Skinner boxes. When the colleague checked on their progress, 19 rats had become conditioned to press the lever with their paws to get a reward in the standard way. But apparently the twentieth rat had, at the beginning of the process, accidentally hit the lever with its head and gotten rewarded. As a result its technique involved facing away from the lever, rising on its hind legs, toppling over backward, and hitting the lever with its head. Repeatedly. Was this rat any stupider than the others, or just unluckier?

A vast array of behavior can be explained as the result of conditioning by pleasant and unpleasant experiences. Sometime conditioned behavior looks more intelligent than it is: the animal appears to understand what it is doing when in fact it has only learned, without knowing why, that if it does a certain thing, a certain good thing will result. (In essence, such actions are superstitions.) Not only is it an extremely effective way of training animals, it's the way many things are learned in the real world.

Skinnerians fell so deeply in love with this powerful way of explaining behavior that for a while they rejected explanations for learning other than conditioning, either operant or associative. Thus we find psychologist Irene Pepperberg grumbling that, "according to Skinner, . . . one needn't study a wide variety of animals, because none would react any differently from a pigeon or a rat: The rules of learning were universal."

While the basic concept of operant conditioning is valid, many exceptions and variations that were once thought to be impossible have turned up.

A rat is a pig is a dog is a boy

Beginning in the 1940s, two of Skinner's disciples, Keller and Marian Breland, used operant conditioning with great success to train performing animals. They published an eventually influential paper, "The Misbehavior of Organisms," on their findings about learning in different animals. Despite utterly standardized procedures, they reported, each animal put its own species' spin on what it was learning. They conditioned a chicken to stand on a platform, but the chicken couldn't stand still and kept scratching around on the platform, so instead they trained the chicken to "dance"—in other words, to scratch in a context that makes it look like dancing. In the final performance the chicken pulls a loop that starts a model jukebox, which plays while the chicken dances. Jitterbug mama!

The Brelands conditioned a raccoon to put money in a piggy bank. He quickly learned to pick up a coin and take it to the bank, but it was hard for him to let go of the coin. He'd start to put it into the slot only to pull it out at the last second and clutch it to him. When he finally mastered this, they tried him with two coins, but the raccoon couldn't bear to do it. "Not only could he not let go of the coins, but he spent seconds, even minutes, rubbing them together (in a most miserly fashion), and dipping them into the container. He carried on . . . to such an extent that the practical application we had in mind—a display featuring a raccoon putting money in a piggy bank—simply was not feasible." The more they tried to get him to bank his funds, the more tenaciously he rubbed and gloated.

The Brelands called this "a clear and utter failure of conditioning theory . . . the animal simply does not do what he has been conditioned to do." Chickens instinctively scratch for food, and raccoons instinctively handle or "wash" their food to do such things as peel a crayfish. Their behavior gradually drifted toward their natural inclinations, even when the result was less food for a hungry raccoon.

At an aquarium in Hawaii, trainers had a hard time conditioning river otters to do tricks. It wasn't that the otters didn't get it; it was that they got it right away, and then got over it. Trainer Karen Pryor began by training an otter to stand on a box. The moment she produced a box the otter rushed over, stood on the box, and was rewarded. Soon the otter understood that standing on the box earned a piece of fish. But instantly she began exploring the situation. What if she lay down on the box? Would she get fish for that? How about if she had three feet on the box— would that count? What if she hung upside down from the edge of the box or put her front paws on it and barked? When Pryor complained to some visiting behavioral psychologists, they said she must be mistaken. "If you reinforce a response, you strengthen the chance that the animal will repeat what it was doing when it was reinforced; you don't precipitate some kind of guessing game."

Pryor took the behaviorists to see the otters, and to back up her claim she tried to condition an otter to swim through a hoop. She put the hoop in the water, the otter swam through, and she gave it fish. The otter swam through again, and she rewarded it again. Very good. But, from the otter's point of view, already old news. The otter swam through the hoop—and stopped halfway through. And looked up for a reward. No reward. The otter swam through the hoop—but as it was almost through, it grabbed the hoop with its hind foot and tore it off. And looked up for a reward. No reward. Okay. The otter lay in the hoop, bit the hoop, and backed through the hoop, each time checking to see if that rated a prize. "See?" said Pryor. "Otters are natural experimenters." One bemused scientist replied that it took him four years to teach students to think like that.

Backward conditioning and latent learning

Another phenomenon once considered impossible is backward conditioning, in which, for example, an animal who has just had an unpleasant experience looks around for something or somebody to blame. Sure enough, animals as well as people do this.

Latent learning describes things an animal learns for no reward that may come in handy later. When a rat explores a maze even though it has never found food in a maze, that has been called latent learning. I suspect exploring is its own reward: rats like to poke around. E. C. Tolman, who discovered this phenomenon in the late 1940s, was ridiculed, since this kind of learning was not predicted by either classical or operant conditioning theory.

Animal trainers sometimes speak of the moment when the light goes on, the moment when something the animal has learned by rote is suddenly understood. Aha! I get it! Karen Pryor describes what she calls "the prelearning dip." Just as an animal is really starting to learn what's wanted, it stalls. "This can be most discouraging for the trainer. Here you have cleverly taught a chicken to dance, and now you want it to dance only when you raise your right hand. The chicken looks at your hand, but it doesn't dance. Or it may stand still when you give the signal and then dance furiously when the signal is not present," writes Pryor. "After that, however, if you persist, illumination strikes: Suddenly, from total failure, the subject leaps to responding very well indeed—you raise your hand, the chicken dances."*

Pryor argues that the chicken is unthinkingly responding to cues that mean it will get rewards. Gradually it gets better, and the trainer is pleased. Then suddenly the chicken "notices" the cue. It realizes that the cue has something to do with being rewarded, and starts paying attention to the cue instead of dancing. "When, by coincidence or the trainer's perseverance, it does once again offer the behavior in the presence of the cue, and it does get reinforced, the subject 'gets the picture.' From then on, it 'knows' what the cue means and responds correctly and with confidence."

James Gould describes something similar during concept learning in honeybees. An example of concept learning is when bees learn that a nectar reward will be marked by either a symmetric or an asymmetric marker. The marker changes, so the bee can't just learn which marker is the correct one, but has to learn that whichever marker is asymmetric is the correct one. "The learning curve is different from that of more

*Chickens in a Hawaiian tourist attraction have been taught to "hula."

standard tests in which bees are taught that a particular odor, color, or shape is always rewarded. During concept learning there is no evident improvement over chance performance until about the fifth or sixth tests, whereas in normal learning there is incremental improvement beginning with the first test. This delay is characteristic of what has been called 'learning how to learn,' which is interpreted as a kind of 'ah-ha' point at which the animal figures out the task." Bzzt!

Trial and error

Trial and error is experimenting to see what works. Strangely, this fine model of the scientific method is often spoken of scornfully by animal behaviorists. Perhaps they suspect animals' ability to formulate a hypothesis and follow up with further testing.

Young herring gulls, like adults, fly up and drop shellfish to break them open. When they start, they're not very good at it. They may not let the clam or mussel fall far enough, or they may drop it on a surface too soft to crack it. Joanna Burger describes young gulls on the New Jersey coast dropping mussels on sand. When the mussels don't open, they try dropping them from a greater height. If that doesn't work, they try dropping them on a dirt road. If that doesn't work, they'll try concrete—and that usually works. "Your enterprising gull will then figure out that he can break the shells open on a board. Should you, while beachcombing, come across a board surrounded by shells, you know you've seen the handiwork of an Einstein of a Herring Gull."

Trial and error is fine when you have time for it, like the gulls. Some things are more urgent. Young vervet monkeys are born ready to react when they hear alarm calls from other vervets. Very young babies dash for their mothers. Older infants learn what to do when they hear an eagle alarm call (hide in a bush) as opposed to when they hear a leopard alarm call (climb a tree). They learn this by seeing what other vervets do. As Frans de Waal points out, "It would be incredibly costly for them to do so by trial and error."

Getting to Carnegie Hall

One form of learning is practice. Practice is generally boring, but playing is fun, so it's handy that play can serve as practice. Two-month-old Inca terns on the Peruvian coast, who have just learned to fly and can't yet catch enough fish for themselves, have been seen "practicing hunting," which looks like playing. The young birds, hanging out on some rocks, take off, circle over the water, and then plunge down on an unsuspecting piece of seaweed. Bearing the seaweed off in triumph, a young tern will then drop it into the water and attack it again. Other juveniles, seeing this, either attack their own piece of seaweed or try to nab another bird's chosen victim. Other tern kids try the "contact dipping" approach of flying low over the water and dipping to snatch the coveted seaweed, or go into an aerobatic display of rapid twists and turns just above the surface. Grown-up terns don't do this. They have fish to catch.

Maturation

When an animal gets better at doing something or recognizing something, it's possible that it hasn't learned a thing. It's easy to mistake growth for learning. Newborn chicks peck zestfully at everything they see, but their aim is sloppy. If chicks born in a laboratory see a brass nailhead in a smooth field of clay, they peck at it. The pattern of peck marks they create in the clay around the nailhead is large and loose, and they often miss the nailhead by a lot. As they get older, their aim improves, and if they are tested four days later, the pecking pattern (since they still haven't learned not to try to eat nails) becomes smaller and clusters tightly around the nailhead.

A possible explanation is that their aim has improved because their better-directed pecks were rewarded by food, and so they learned through conditioning—target practice—to aim better. To see if this was so, Eckhard Hess fitted new-hatched Leghorn chicks with tiny rubber hoods which held goggles over their eyes.* The goggles displaced what the chicks were seeing to one side. As soon as they had been fitted with the goggles, the chicks were tested with the nailhead in clay. The pattern of the pecks was large and loose and displaced to one side, away from the nailhead. Then the chicks, still fitted with goggles, spent several days either in an environment where grains of food were loosely scattered or in which their food was spread thickly in

*These goggled hoods looked very much like World War I aviator helmets but were not accessorized with a long dashing scarf.

wide bowls so that they usually hit something to eat no matter how badly they missed.

Hess thought that the chicks who ate from wide bowls would not learn to correct for the goggles (because they still got food when they missed) and that the chicks whose food grains were scattered would learn to compensate for the goggles. He gave them the nailhead-in-clay test after four days, and both groups showed identical tightly clustered, precise pecking patterns—off to one side of the nailhead. They had all improved their pecking precision, not because they had learned but because they had gotten older.

Chicks do learn some things about pecking—don't peck your toes and don't peck chicken droppings. Chicks also seem to be born with the important knowledge that when you get a piece of food too big to eat in a few pecks, you should grab it and run like the wind. "I had always supposed, if I bothered to think about it at all, that when a hen picks up a particularly fat worm and immediately starts to run away the motive was an innate greed, an unwillingness to share with her fellows. Or else that it was a wisdom born of previous experience. The truth is otherwise," writes zoologist Maurice Burton. "A young chick, first able to run, will, on picking up a morsel of food that cannot be instantly swallowed, turn round and run, as if pursued by an imaginary host intent on stealing. It will do this even when there are no other chicks present."

Social learning

Animals influence each other's behavior in ways that researchers have tried desperately to pin down. One aspect of social learning is its direction, metaphorically speaking. In vertical learning, animals pass information down the generations. If your mother teaches you what fruits are safe to eat, or how to tie your shoes, that's vertical learning. It is conservative, in that information can be conserved and passed on indefinitely.

In horizontal learning, animals learn from their own generation. If you follow your sister or your friend to forage in the new Dumpster the humans set up by the construction site, or if your classmates take you to a great concert, that's horizontal learning. It is "rapid and ephemeral; hence most appropriate for the transmission of information pertaining to rapidly changing aspects of the environment," write Hilary Box and Kathleen Gibson.

In oblique learning, animals learn from unrelated individuals of another generation. If you follow an aged baboon to a hidden water hole (and he's not your father or grandfather), or you actually learn something a teacher tells you, that's oblique. Baby elephants reach into the mouths of other elephants to find out what they're eating, and this is not a liberty they only take with their mothers.

Social facilitation

Social facilitation is when the sight of someone else doing something which you already know how to do inspires you to do it. Everybody's dancing, and you feel like dancing too. You're not copying them, because you already know all the dances. Those kids have ice-cream cones, and suddenly you feel like having one. Or everyone else is throwing rocks and before you know it, you're throwing rocks too.

Claire Kipps raised Clarence, an abandoned house sparrow nestling, in her London home. Because one wing was slightly crippled, he could not be released. After Clarence had grown, neighbors brought her several sparrow fledglings they had wrested from cats. The sparrows feared her almost as much as they now feared cats, but Clarence gave them confidence. They followed him slavishly. If he chirped, they chirped. If he preened, they preened. If Clarence drank from a teaspoon held by Kipps, they hopped over and drank from the teaspoon. They already knew how to chirp and preen, and the sound and sight of Clarence chirping and preening was social facilitation to do likewise. As for drinking from a spoon, they already knew how to drink. But going to the spoon to drink is probably better categorized as stimulus enhancement.

Stimulus enhancement, local enhancement

Stimulus enhancement is a very common form of observational learning. An animal's attention is attracted to a place or an object by the actions of another animal. The observer doesn't copy what the first animal did, but the heightened interest created in the place or thing may eventually cause the observer to perform the same action.

Suppose I have somehow never run across a vending machine

before. I notice that you go to the machine and come back with brightly colored snack crackers, and I suddenly crave such crackers. Stimulus enhancement has kicked in. I go over to the machine—local enhancement—and look at it, notice the coin slot, maybe even read the directions, and eventually buy myself some crackers of unnatural appearance. This is not imitation. I didn't watch, understand, and copy what you did. My interest in the machine was enhanced by noticing your interaction with it, but I had to learn how to get crackers out of it by myself.

Similarly, juvenile sea bass were much more likely to figure out how to push a lever to get fish pellets if they'd seen other sea bass in an adjacent aquarium push a lever to get fish pellets. (They knew the other fish were getting pellets because pellets went into both tanks when the lever was pushed.).

In a large South African garden a bantam hen was persuaded to sit on five eggs abandoned by an Egyptian goose. She hatched them successfully and led them around the garden, scratching up bugs and seeds for them and looking on as they ate grass instead. When they were two weeks old she took her children to forage by the fishpond, and the goslings instantly realized that water was where they were meant to be. To the bantam's astonishment, they jumped in and swam gaily, uttering joyful minihonks. They ignored their mother's frantic calls to get out of the water. She flew back and forth, begging them to escape while they could. They dove and splashed. The rooster came to see what the fuss was about. "He couldn't believe his eyes either and stood gaping at the scene, shocked speechless," writes Kobie Krüger in *The Wilderness Family*.

The goslings insisted on going to the pool every day, and one day another bantam arrived with her chicks. She blinked in surprise. Meanwhile one of her chicks, duped by local enhancement, decided that the pond looked like fun, marched up to the brim, ready to step in, leaned over—and lost his nerve.

There is a well-known story of how tits in Britain learned to peck open milk-bottle tops and drink the cream at the top, and how this habit spread like wildfire across Britain as the birds copied each other.*

---

*The birds called tits in Britain are called chickadees in North America. One trusts in the refinement of the American reader in this regard.

This was once thought to be a case of imitation, but is now considered one of local and stimulus enhancement. A tit sees another on a milk bottle and thinks that milk bottles must be a good place to perch. A tit that sees another tit, even one that is just standing around, is also more apt to look for food, which involves pecking at things. A tit that finds a pecked-open milk bottle will learn to inspect milk bottles, and does not have to wonder what the best way to open milk bottles might be, since pecking is always a good bet. In the laboratory, chickadees* presented with a model of a milk bottle, and with no one to copy, came up with the idea of pecking through the foil to get the cream about a quarter of the time. So it is likely that tits in various parts of Britain came up with the idea independently, making the progress of the custom seem even faster than it was. Other birds, particularly house sparrows, also took it up.

Imitation

Debunking the milk-bottle imitation story produced its own wave of imitation, this time of scientists imitating other scientists by trying to prove that various cases—maybe all cases—of imitation by animals were no such thing.

Imitation used to be scorned. It was considered "a cheap trick that animals often use, which produces a spurious mimicry of real intelligence," writes Richard Byrne. "From this lowly status, imitation has recently been promoted to a sign of remarkable intellectual ability, one which involves a symbolic process—except when it is vocal imitation by birds, perhaps an anti-bird bias. Researchers studying human babies have called imitation 'an innate mechanism for learning from adults, a culture instinct.' And ironically it is now suggested that imitation can only be done by humans."

Frans de Waal writes that "increasingly the term 'imitation' is being reserved for cases in which a solution to a problem is copied with an understanding of both the problem and the model's intentions. This usage has turned 'imitation' into a small, cream-of-the-crop subset of social learning, one that may not apply to rats and cats, perhaps not even to monkeys and apes."

*They were in North America.

Just as being said to ape someone is not a compliment, being said to parrot someone is a criticism. If the anti-bird bias suggested by Richard Byrne means that vocal imitation by birds doesn't count, consider the case of Okíchoro. This African grey parrot was raised by psychologist Bruce Moore and imitated both words and gestures. Okíchoro learned to say "Ciao!" and wave one foot or wing. He'd say, "Look at my tongue," open his beak, and show his tongue. He'd say, "Turn" and turn on his perch. He had motions or gestures to accompany "Get back, you," "Back in your tree," "ready," "shake," "microphone," "heads up," and "jump." Okíchoro also imitated the sounds of Moore's footsteps while walking or marching in place. He imitated the sound of knocking on a door while making knocking movements in the air with his beak or foot.

Copycatting

Psychologist Edward Thorndike, a student of William James, rejected the notion that cats can imitate. Around 1900, Thorndike constructed a series of puzzle boxes, from which cats could escape by manipulating levers, treadles, latches, or strings. Cats who saw how other cats escaped did not escape any quicker. Cats got out by trial and error, and they showed no insight into the workings of the levers, treadles, latches, or strings. Many cats tried the mewing-to-be-let-out technique. This was ineffective with Thorndike, but it's an excellent strategy, which often works.

Thorndike was touchy about amazing animal stories. He complained that they focused on animal intelligence, not stupidity. "Thousands of cats on thousands of occasions sit helplessly yowling, and no one takes thought of it or writes to his friend, the professor; but let one cat claw at the knob of a door supposedly as a signal to be let out, and straightway this cat becomes the representative of the cat-mind in all the books."

Thorndike's low opinion of cats prevailed in scientific circles for some time, but he may have picked the wrong tests to get cats to imitate. In the 1960s, brain researchers doing electrophysiological studies became impatient with the lengthy process of conditioning cats and decided to see if cats could learn tasks (either jumping over a hurdle to avoid having their feet shocked or pushing a lever to get food) by imitation. They could, and they learned faster than cats conditioned by standard techniques.

Karen Pryor writes that cats are quite good at imitating. When a dog does what another dog does, it's usually because it's responding to the same thing as the first dog, she says. But if she teaches a trick to one cat in a household, the other cats will do it without being taught.

Pryor describes an incident in which her daughter spent an hour teaching her small poodle to jump into a child's rocking chair and then make it rock. She rewarded its efforts with bits of chopped ham. At the end of the lesson the poodle jumped down and a cat who had been watching jumped into the chair unbidden, set it rocking, and looked up for her ham.

Pryor doesn't take the view that the cat capacity for mimicking the actions of another animal shows that cats are smarter than dogs, only that they're better imitators.

Beak and claw

In recent years, some researchers have produced fairly clear-cut demonstrations of imitation by giving animals simple tasks that can be accomplished in more than one way. Only if they copy the precise way the task was done, as well as the end result, is it considered imitation.

If parakeets, also called budgies, saw a demonstrator budgie push a piece of cardboard off a cup of birdseed with its bill, they would push the cardboard off a cup of seed with their own bill. But if they saw a budgie remove the cardboard by gripping it in its bill and lifting, they too would grip and lift. And if they saw a budgie kick it off with its foot, why, then, they would boot it off too.

Challenging the view that apes can imitate but monkeys cannot, Bernhard Voelkl and Ludwig Huber worked with marmosets, small South American monkeys. They showed that if a marmoset sees another marmoset pry the lid off a film canister with its teeth and eat the scrumptious mealworm within, it will then pry the lids off film canisters with its teeth to retrieve mealworms. But if the marmoset sees another marmoset pry the lid off a film canister with its hands to win the mealworm, it will use its hands too.

Swim this way

Dolphins undeniably imitate. More than once an untrained dolphin in an aquarium who has witnessed another dolphin go through its act has turned out to be able to do the act perfectly without training. They don't limit themselves to imitating each other. Haig, a captive Indian Ocean bottlenose dolphin who appears to have been bored out of her mind, took to imitating Tommy, a fur seal in the same tank. She lay on the surface, keeping her tail still and rowing with her flippers as Tommy did. Unlike a dolphin, but just like Tommy, Haig rubbed herself with her flippers. She copied Tommy's habit of rising above the surface with open mouth. She even attempted Tommy's sleeping position, lying on her side with a flipper protruding stiffly. As Tommy did, she lay belly-up on the surface of the water with her flippers pressed against her body. This put her blowhole underwater, so from time to time she had to turn over to breathe. "The dolphin maintained the postures only with great difficulty and clumsiness." When she was not doing Tommy impressions, Haig was seen swimming behind a skate and pushing herself off the wall with her flipper as the skate did with its wing. When a loggerhead turtle slept on the bottom of the tank, Haig lay flat on the bottom beside it, and when it rose up to breathe, she rose up with it. After a while the stir-crazy Haig began to push the turtle down just before it took a breath. She and the other female dolphin in the tank, Lady Dimple, thought this was so funny that they drowned several turtles in the process.

A dolphin in the same aquarium, Daan, watched a diver whose job was to clean algae off the inside of a glass viewing port. Daan took to scrubbing the window with a seagull's feather, at the same time making a noise like the diver's air-demand valve and emitting a stream of bubbles like the exhaust air from the diver's equipment. When he couldn't get a feather, he would scrub the window with a rock, a piece of paper, a dead fish, or an unfortunate sea slug. He monopolized the job, threatening and shoving divers who tried to come near.* He got the window pretty clean.

Lady Dimple had a calf, Dolly, who copied her mother's show-time act. Dolly also performed one of the most famous examples of

---

*They obeyed. They saw what happened to the sea slug.

imitation in the short annals of accepted incidents. Dolly was interested in visitors and often brought feathers, seaweed, or fish skins to the viewing port to show them. If people didn't pay attention, she'd go get another object, such as her favorite rock. One day a visitor who was smoking blew a large cloud of smoke against the viewing port to which Dolly had her face pressed. Dolly immediately swam to her mother, nursed for a second, swam back to the port, and blew out a mouthful of milk in front of her side of the window, producing a cloud around her head just like the cloud of cigarette smoke. Having noticed the astonishment this created, Dolly made it a regular part of her routine.

Dolphins in captivity and in the wild love to synchronize their movements. Captive killer whales Orky and Corky worked up what Alexandra Morton calls an Esther Williams routine, in the early morning when no trainers were present. It took a while until they could back into the tank wall simultaneously, push off with their tails at the same moment, and glide in unison. A number in which they lay side by side with their tail flukes on the training platform and their right pectoral fins raised took months to perfect.

One sunset, watching wild killer whales in Discovery Passage, Canada, Morton saw two males lying on their sides in the water, head to tail. "Slowly they raised their right pectoral fins together and froze. They held their salute for fifteen seconds, then rolled to dive in perfect synchrony."

What do you have to do to get an apple around here?

The zoo in Knoxville, Tennessee, has had outstanding success in breeding red pandas. The keepers wanted to be able to perform regular vaginal swabs on the female pandas, so they could tell when the pandas were ready to mate. They didn't want this procedure to be stressful for the red pandas, so they decided to train them to accept it as routine. Over several sessions they were able to accustom the pandas to the proceedings, which involved providing them with a big bowl of apples while a keeper did a quick swab. There! All done! Have some more apples! The red pandas were very fond of apples, and this struck them as a good deal. Their cubs were on hand during the procedure, since separating mothers from cubs would be stressful. When the female cubs were mature, the keepers planned to train them to accept the vaginal swabbing procedure as the mothers had been trained, but it turned out to be unnecessary. The cubs knew all about it and cheerfully bellied up to the bowl of apples, ready for their swabs. (Whether males are bitter because they never get big bowls of apples is unknown.)

Mixed learning

A lot of things animals do are learned by a mixture of methods—a little bit of stimulus enhancement, a little bit of trial and error, maybe some imitation.*

Two groups of young red squirrels, born in a laboratory, were given hickory nuts for the first time. One group could also see a grown squirrel, a wild-born expert, eating hickory nuts. The squirrel children attacked the nuts with enthusiasm and learned how to gnaw them open. At first they gnawed randomly all over the nut, but gradually they became more purposeful and skilled. The expert squirrel had a distinctive method, beginning at a certain point on the stem end of the nut, and gnawing along the main axis until he could get at the seed on each side of the internal septum. It took this expert about 23 minutes per nut. After the first group had been allowed to observe the expert for six weeks, he was moved so the second group could watch.

Squirrels who saw the expert in the first six weeks were apt to open their hickory nuts the way he did. Squirrels who saw the expert in the second six weeks mostly did not. The expert's method was very efficient, and the young squirrels in the first group approached his 23-minute speed. The others took more than twice as long to open a nut, although they improved after six more weeks.

Thus any red squirrel who wants to eat a hickory nut seems to be able to figure out a method by trial and error. And practice helps. But being able to watch a professional is a real shortcut.

Ape this

Surely marmosets aren't the only primates who imitate. Surely apes can ape. There have been quite a few experiments in which what appeared

---

*One thing animals never do when attempting a difficult task is read the manual, but this is true of so many people.

to be imitation by primates turned out to be more like clueless stimulus enhancement. Thus chimpanzees who saw a trained animal use a T-shaped rake to pull distant food items in through the bars of their cage got the general idea of using the rake to get the food, but failed to notice the techniques that actually worked.

But there are also examples from captive-reared apes of the most blatant aping imaginable. To say that apes do not ape is impossible when you watch Chantek put on lipstick, or see Lucy sipping a gin and tonic while leafing through a magazine, or notice that Viki has stolen your key and started the car.

When Nina, a zoo gorilla, was feeling poorly, she took one of the empty burlap bags in which treats were brought to the gorilla enclosure, tore it open along the seams, and used it as a shawl. One day she brushed the leaves off a spot in the grass, laid the burlap down, smoothed out the wrinkles, straightened a dog-eared corner, and sat down on it. Then she noticed that every primate in sight, human or gorilla, was staring, and she became self-conscious. The next day her five-year-old daughter Alafia took an empty box of shredded-wheat biscuits, removed the paper sleeve the biscuits came in, tore it open along the seams, smoothed it out, and sat on it.

Anne Russon has studied imitation among orangutans at Birute Galdikas's Camp Leakey. These apes are being returned to the Indonesian rain forest after captivity. Some imitated orangutans who know the ways of the wild, not only trying the same plants for food but copying the methods used to peel and otherwise get at the edible parts. But other orangutans showed little inclination to return to the wild and hung around camp instead. They imitated such cool activities as sawing wood, washing clothes, and writing in notebooks (their notes were never legible). They borrowed canoes (but did not return them). Siswoyo, an orangutan in her twenties who was permanently hunched from spending years in a small cage, hoed the weeds along the paths exactly as a human staffer did. She piled her weeds along the path just as he did. Her work was messier, but then she had to use a stick, since no one would let her have the only hoe in camp.

Another citified rehabilitant was Supinah, who was devoted to imitation. She carried a parasol, combed her hair, and applied bug repellent. She sawed, hammered, painted, pumped water, and sharpened axes. She came impressively and disturbingly close to lighting a fire in the camp kitchen. The cooks did this in a multistep procedure involving kerosene, and one day the cook stepped out and Supinah stepped in. Having watched the cook many times, she confidently took a burning stick from the still-smoldering breakfast fire. She scooped a cup of kerosene—Russon, who had not watched as carefully as Supinah, took the clear fluid to be water and was not alarmed—and plunged the burning brand into it. By good fortune, Supinah was too fast and the kerosene smothered the flames instead of fueling them. (While Supinah was engaged in these interesting experiments, her tufty little orange infant was clinging to her, or playing nearby. That's a future firebug I'd keep an eye on.)

Meanwhile, back at the laboratory

In one experiment with captive orangutans, both human children and orangutans saw a human demonstrator get food out of a box by pulling, pushing, or rotating a rod on the box. The children imitated the demonstrator and the apes did not. They tried the orangutans again with a trained orangutan as demonstrator and they still didn't imitate.

Robert Shumaker and colleagues set up a different test situation with orangutans at the National Zoo. Instead of watching a mysteriously expert demonstrator, these apes watched another orangutan being taught to perform the task. The task, an early stage of a language acquisition study, was to label a food with an abstract symbol by pointing to the symbol with a single finger, not a usual orangutan gesture. The human teacher focused only on the orangutan being taught and ignored the other, lower-ranking orangutans who were observing. Any time the orangutan who was being taught went away for a minute, an observer immediately went over to the apparatus, touched the appropriate symbol with one finger, and then requested the reward by extending a hand or pointing their lips at the teacher.

Shumaker and colleagues suspect that one reason the orangutans imitated the task so readily is that they got to watch a teaching process, rather than simply being wowed by the inexplicable prowess of the expert.

Other forms of social learning

At a police-dog training school in South Africa, 20 litters of German shepherd puppies were bred. Some were born to mothers who had no special training, and some had mothers who were trained to sniff out sachets full of narcotics. The puppies were divided into four groups. One group left their mothers at 6 weeks, as was standard, and one group stayed with their mothers for 12 weeks. The third group stayed with their mother for 12 weeks, and between the ages of 6 and 12 weeks got to see their mother sniffing out and retrieving narcotics sachets and being praised for her work, twice a week. The fourth group also had mothers trained in this work, but since they left her at 6 weeks, they never saw her on the job.

After 6 or 12 weeks all the puppies were put in standard police-dog obedience training, but none had any work with narcotics or retrieving. At 6 months, all the pups were tested for their ability to follow directions from handlers to find and retrieve sachets of narcotics, and scored from 0 to 10. Puppies of untrained mothers did about the same whether they had been with their mothers for 6 weeks or 12 weeks. Puppies who had been with their mothers 6 weeks did the same whether their mothers were trained or untrained.

But the puppies who had stayed with a trained mother for 12 weeks and seen her at work excelled. Four of them scored 9, "the highest recorded for any untrained pups since establishment of the dog school," gloat the researchers.

Clearly the puppies had learned something from observing their mothers work. The puppies got no rewards when their mothers retrieved narcotics, so operant conditioning is unlikely to have been a factor, and their mothers weren't there when they were tested, so neither imitation nor social facilitation seem relevant.

There's nothing to do.
Let's go hang out on the pier.

Ring-tailed lemurs are prosimians—a group of animals that branched off our family tree before monkeys and apes—and they are not considered bright, although they do possess long and excellently striped tails. A group of lemurs in the Chester Zoo, in England, were found to have an unusual habit—or at least 17 out of 28 of them did. This was the custom of wetting the ends of their tails and licking the water off.

The Chester lemurs live on an island surrounded by a moat, and on one side of the island is a small wooden pier. To get a drink from its tail a lemur goes to the pier, turns, and climbs backward down a pier support until its tail tip is submerged in or floating on the water. Then the lemur climbs back up and licks water off its tail.

No one noticed when the fad began, so no lemur can be singled out as an originator. No lemurs elsewhere have been seen dipping their tails, so the researchers suspect social learning caused this practice to spread through the group. They consider it "unlikely . . . that each animal which shows the behaviour has independently discovered it for itself, as this would seem to require a lot of lemurs accidentally falling in the water in this, but not in any other, group."

This is an unnecessary, show-offy way of getting water, since the lemurs can and do lap water from the moat, lap from water dishes, lick moisture from leaves, or dip a cupped hand into the moat. It does not seem to be something baby lemurs necessarily pick up from their mothers or the other way around, since there was no correlation in mother-child pairs.

Sometimes lemurs seemed unclear on how it was done, backing up to the pier and climbing down without getting their tails wet. Baby Grace climbed down repeatedly, getting her tail wet, climbing up again, and then, instead of licking her tail, looked around "as if not knowing what to do next." Her mother, Claire, then immersed her tail and let Grace help lick the water off. Two years later, Grace was an accomplished tail-drinker.

The researchers end their paper by noting that 2 of the 15 hamadryas baboons in the Zurich Zoo dipped their tails in the moat and drank from their tails; they were watched by the other 13, who never learned to do it. The lemur researchers imply that the math (17 out of 28 lemurs versus 2 out of 15 baboons) favors the prosimians. "Perhaps it is time to re-assess our views on the evolution of primate intelligence, since a baboon is obviously no match for a cognitive lemur." (On the other hand, I must point out that the adolescent baboons at Zurich, Liba and Kalos, didn't have a pier to climb down, but had to hang by their hands. So it was harder.)

The most parsimonious explanation of how this habit spread, the researchers say, is stimulus enhancement. Lemurs noting other lemurs messing around the pier suddenly find the pier interesting, spend more time there, and invent tail-dipping. The chance to lick refreshing moat water off another lemur's tail may also promote interest. On the other hand, it might be imitation, since it's an odd, unnecessary behavior and since lemurs have been seen approximating the behavior without actually getting their tails wet. Not all lemurs do it identically, so the researchers suggest that it may be program-level imitation, or emulation.

Emulation

Emulation isn't quite imitation, but maybe it's even better. When you emulate someone, you don't copy them exactly, but you try to achieve the same result they have achieved. Such a process is also called program-level imitation.

After all, what's so great about imitation? Researchers working with dogs write that "social learning in nature does not usually involve the exact copying of the demonstrator's behaviour. Naïve observers usually want to obtain the same target as the demonstrator and develop their own method to get it." Is that so dumb? Unless you're completely without a clue as to how the demonstrator did it, you're likely to do quite well. You may do better.

The Hungarian researchers looked at how dogs learned to get past a V-shaped wire-mesh fence to get to food or a toy. Many dogs had a hard time with this, particularly if they had to start from outside the V and get something inside the V. They often barked through the fence at the desired object or tried to dig underneath when they couldn't figure out how to get to it. (Maybe dogs have more practice in trying to get out than trying to get in, suggest the researchers.)

Dogs were allowed to watch a person carry the desired food or toy to the other side of the fence and put it down. Dogs that saw a demonstration were much quicker to get around the fence, whether the demonstrator was their owner or a stranger. They didn't necessarily go around the same arm of the V as the demonstrator, which argues against simple imitation.

How to crack a safe

Austrian researchers working with captive keas (New Zealand parrots) constructed a fascinating box. It was made of plexiglass so the keas could see a treat within. To open it, birds had to undo three different gadgets: they had to pull a split pin out of a large plastic screw on the lid; unscrew the screw; and poke a bolt out of a latch. Then they had to lift the lid. Two keas were trained to open this box (or "artificial fruit," as the researchers called it). Other keas, one at a time, got to watch the trained keas open the box in an adjacent aviary numerous times, and then they got to try. The control group got to play with the box without having seen another kea open it. They all loved the box and loved trying to break into it, even when it was empty.

None of the untrained keas managed to open the box in three sessions. Those who got to watch another kea open the box attacked the locking devices, whereas the control birds wasted time nibbling on the frame. Those who had seen another kea open the box all managed to operate at least one of the locking devices, and two of them, Oldred and Green-Silver, opened all three devices (but then didn't raise the lid). Some of the control birds managed to open one device.

Green-Silver did learn from observation what kind of pulling and poking was called for. The most successful control bird, Blue, "seemed to sample from the whole repertoire of foraging and play actions to see what worked, and only by chance arrived at an efficient means of manipulating the chosen object." It took her ages to think of pulling on the split pin.

The keas who had a model to copy attacked the locks quicker, worked at them longer, and were better at opening them. The researchers ascribe this to social facilitation ("Here's a fun project for keas"), to stimulus enhancement ("Apparently the locks are the things that all the hip keas pay attention to"), and to emulation. It is possible, the researchers say, that the trained keas opened the box so fast that the observers couldn't see how. They didn't imitate in the sense of "copying . . . the response topography of the opening technique shown by the models." They didn't or couldn't think, "Okay, so after you pull out the split pin, you grab the outer end of the screw and turn it counterclockwise until it comes out," but instead thought the equivalent of, "Okay, you mess with those doohickeys up there and somehow pull out the screw."

Making up your own lesson plan

Karen Pryor and colleagues at an oceanarium in Hawaii used conditioning to teach dolphins to do many entertaining things. The idle idea came of training the performers to invent new tricks. Working with Malia, a rough-toothed porpoise, they rewarded her only when she did something she had never done before. At first Malia had no idea what she was being asked, but after a few days she got it, and session after session came up with new, unexpected, and downright bizarre activities. She did flips in the air, she glided with her tail out of the water, she skidded along the bottom of the tank. Sometimes she seemed so excited at her first session of the day that her trainers were convinced she had thought up new stuff that she couldn't wait to show them. Amazed, they repeated the experiment with another porpoise, Hou.

Hou, considered a docile and timid porpoise, spent the first few sessions being rewarded for stuff that was not that unusual, but which she hadn't done before. Pretty soon Hou was stumped. Each time she'd do all the things she'd been rewarded for before, over and over, without praise or reward. What could they possibly want? In her fifteenth and sixteenth sessions, Hou suddenly got it. She quickly began doing new things: slapping the water with her tail, somersaulting in the air, swimming in figure eights, spinning, slapping her tail upside down. Later she invented tricks ranging from sinking head downward to executing fancy jumps. By the thirty-second session "Hou's aerial behaviors became so complex that, while undoubtedly novel, the behaviors exceeded the power of the observers to discriminate and describe them." She also invented spitting at her trainer.

Before anyone says this proves how smart dolphins are, Pryor taught pigeons to do the same thing, and they originated such odd acts as lying on their backs, standing with both feet on one wing, and hovering two inches in the air.

What did I just tell you?

It is also possible to learn through being told. This almost never happens in animals due to their poor language skills, and happens in people less often than teachers like to think. The bonobo Kanzi understands a great deal of English. When he was just a little ape, he was playing with tools one day, and decided to unscrew some screws holding a plate over an electrical socket. His close friend, primatologist Sue Savage-Rumbaugh, got upset, and emphatically told him that sockets were dangerous, that something called shocks—electricity—came out of them and could hurt him. He approached the socket with caution, hair standing on end, but it didn't do anything bad to him. He threw objects at the socket, and nothing bad happened and nothing came out. He began to be skeptical.

Kanzi hid the screwdriver under a blanket. He waited until Savage-Rumbaugh was busy sorting a sheaf of pictures she was going to use to test his comprehension, sneaked the screwdriver out, and inserted it into the socket.

He got a nasty shock. His hair stood straight out from his body. "Waa! Waa!" he shouted and began throwing things at the evil socket, beckoning Savage-Rumbaugh to join the attack.

Kanzi couldn't really learn this through having it explained because he had no idea what shocks and electricity were, any more than small children do. So he learned by trial and error. He could have taken Savage-Rumbaugh's word for it, but this would have been unlikely; for this very reason many small children are not allowed to play with screwdrivers. (Also, may I point out, she lied to Kanzi about there being monsters on the roof of the building where the air-conditioning vents are, so Kanzi was not entirely wrong to be skeptical.)

Hardly any animal children have Kanzi's option of learning about the world by being told, and not every animal child can learn through imitation, but all animal children have some learning methods at their disposal. With any luck, these will help them understand the world and make their way through it. If they are born into a different world than the one in which their forebears evolved, they need not rely only on genetically fixed behaviors to carry them through, for they have ways of coming up with new solutions to the new problems that confront them.

TWO

Learning the Basics: How to Crawl, Walk, Climb, Swim, and Fly

Young California condors often think they're ready to fly when they're not. They take off from their cliffside nest, fly a short way, desperately flapping, feet dangling, and plummet into the chaparral below. It's not rare for a foolish young bird to spend a couple of days bumbling around in the chaparral (with food deliveries from its parents) before it makes its way to a clearing where it can take off again. One chick at a nest observed in 1939 spent almost a month on the ground before it could fly back up to its nest. This chick fledged so early that condor experts suggest that it hadn't really meant to fly in the first place— it merely leaned out too far from the nest cave for a little wing-flapping practice and toppled out.

A FEW MINUTES AFTER IT IS BORN, a salt desert cavy (a South American rodent) can walk, run, and sand-bathe in a coordinated fashion. Most of us took more than a year to walk and run, and still cannot sand-bathe with any degree of mastery.

One important thing that baby animals must learn is how to move from one place to another in the ways appropriate to their species, whether that includes walking, swimming, climbing, or flying. There are innate motor patterns and other clues that help some species more than others and promote some activities more than others.

The things we do that are the most clearly prescribed by our genes are the ones we hardly notice. Like many other animals, we are born knowing how to stretch and yawn and sneeze. But, as anyone who has raised a small child can tell you, we're not born knowing how to blow our noses. Sneezing may not seem like a big deal, but keeping your airways clear for breathing is vital, and it clearly gives animals a survival edge not to have to stop and learn how it's done.

We don't have to learn to see and hear, although we get better at these things as we mature. Nature and nurture interact nicely so that neurons connect up in response to light and sound impinging on our senses. We learn where the sound is coming from, however, and how to turn our heads to look at it. If young animals don't get the sensory input that the world almost invariably supplies, things can go wrong. Barn owls who have had an earplug inserted in one ear learn to localize sounds based on what they hear, and know precisely where a sound is coming from. (This is key information to a barn owl when the sound is a mouse rustling in the hay.) When the earplug is removed, the world sounds different, and they're wrong about where the noise is coming from. They learn better in a few weeks. An owlet is still growing, and its ears are getting farther apart. Since part of the way birds and mammals tell where a sound is coming from is by (unconsciously) comparing the sounds reaching the two ears, growing owls need to be able to adjust to changing sound cues.

Eureka! I have drunk it!

It's not enough to see. We also need to understand what we see. Baby chicks do not know what water looks like. They get thirsty, and when they have water in their mouths they drink gratefully, but even the thirstiest newly hatched chick will walk witlessly through puddles without suspecting that help is underfoot. But chicks automatically peck any irregularity they see, and so they peck at floating specks on water, get their beaks wet, and instantly make the connection.

Water in any form can be mysterious. Limpet, a young captive-reared otter, stepped out of her sleeping box one day and saw her first snow. Three times she huffed in suspicion, then walked out into it. Snow turned out to be wonderful, and at once Limpet shoved her head under it and raced forward, hurling a fan of snow into the air.

J. David Henry, who follows wild foxes through Prince Albert

Park, has seen several foxes learn about ice. In late November, a lake may freeze over in the night with smooth dark ice. Henry described the encounter of The Prince, a particularly noble young fox, with frozen water. The Prince came to the lake to drink and bumped his nose. He sniffed at the ice. No scent. He waved his forepaw over it and experimentally scratched at it. Hard. Carefully, one paw at a time, The Prince stepped out on the ice, which bore his weight. Solid. Gaining confidence, he walked, sniffing at the ice, and then trotted, as foxes do. A foot slipped. The Prince turned and smelled at the slippery spot. He proceeded, and several more times he slipped, stopped, turned, sniffed. "Then suddenly he burst into a galloping sprint. Legs slipping and sliding everywhere, the fox for a moment was going nowhere; it was clearly a peculiar sensation for a normally graceful animal. The fox then raced across the surface and came to a sliding, half-stumbling stop, followed by a playful twisting leap into the air." Until the snow fell, The Prince punctuated his hunting schedule with breaks for ice dancing. He loved to get up speed, galloping, and then slide across the ice crouched on all four legs or seated on his rump.

Controlling the body

Before The Prince became a fox who could gallop, sprint, slide, or even walk, he was a blind sprawling cub. He was born in a dark earth where there was little to see, and his eyes didn't open for more than a week. By the time a month had passed, he had grown remarkably, his eyes had opened, and he and his siblings were preparing to stagger out of the den, look around, and start playing. But he didn't have to learn to grasp.

Hand control was a problem for Cody, a baby orangutan. He wanted to put his fingers in his mouth and suck on them, but it was hard to get them to the right place. Keith Laidler describes eight-week-old Cody waving his hand about, jamming it in his ear, making expectant sucking motions with his mouth, and looking puzzled.

Comparing the way four different baby primates began to grasp things, Francesco Antinucci found that a baby crab-eating macaque walks (on all fours) at an early age, before it has developed hand coordination. As a result, when the little macaque wants to examine an interesting thing, it can and does walk over and explore it with its mouth, tasting, gnawing, or picking it up in its mouth.

A baby gorilla is already developing substantial hand coordination by the time it can begin to get around, and develops hand-eye coordination shortly thereafter, so it initially investigates things with its mouth but soon is about equally likely to investigate a fascinating object with mouth or hands. A baby capuchin monkey is already working on hand-eye coordination when it begins to walk, and so it investigates things mostly with its hands. In this capuchins resemble human babies, who handle things before they are mobile.

Isn't that baby walking yet?

All primates look slow compared to animals with precocial young who stand, walk, and even run shortly after birth. Although humans take around a year, it's not that we're complete idiots, it's just that we've evolved a different strategy. An infant fawn or a foal can run away from predators when it is still very young, whereas we rely on our mothers or fathers to carry us away. But even the infant prodigies of precocial species have some difficulties.

Sarah McCarthy, a veterinary equine specialist, who's seen a lot of baby horses born, says that foals know how to stand up shortly after birth. "Some foals are stronger or more coordinated than others, but the program seems to be there. Most normal foals will stand up within a few attempts, and then improve dramatically with each success such that by the time a foal has gotten up by itself even two or three times, it is generally very good at it. Within a few hours they can pop up on their feet instantly.

"In contrast, lying back down again seems not to be automatic at all. Some foals figure it out very quickly and it is not much of an issue. But others struggle with the concept. They stare at the distant ground with consternation, start to fold their legs, decide it is just too far, chicken out, try again, decide they did not really want to lie down anyway, go have a drink, try it again, give up and fall asleep on their feet, etc. They may hit the ground in quite a heap too. It can take several days for some foals to master this maneuver with any grace at all.

"Length of leg does seem to be an issue, and really leggy foals seem to have more trouble than stouter ones." The foals that figure it out generally get down by folding all four legs at once. The foals who struggle try this, which can be difficult, and also try folding the rear legs first, that is, they sit down. Or they try kneeling, folding the front legs first, which often results in "an absolute face-plant." Adult horses fold their front legs first to lie down, but foals are usually so disproportionately long-legged that this doesn't work for them.

Why is standing up easier than lying down? Standing up is important, because young hoofed animals must stand to nurse. Presumably gravity will always bring a young animal down once it gets tired enough, so standing up is the part where innate hints are needed.

The truth about great apes like us is that when we're born we sort of know how to walk. One well-known infant reflex shows this. If you hold a baby a few days old so that its feet gently touch a surface, the baby will often take steps with alternate legs. It is not possible to build on this and have your child walking by the end of the week—its head is too heavy, its muscles are too weak, its balance is wholly inadequate. We're talking about a kid that can't even hold its head up yet, so have patience. Babies walk when they've matured enough to walk, and even if they've spent much of their infancy strapped to a cradleboard, they walk at the same age.

That's just gas

If you're going to make others wait on you hand and foot for years, you'd better be charming. Human babies and chimpanzee babies accordingly smile or make play-faces much earlier than baby rhesus macaques, which are mobile at a few months old. Unlike a fox, born in a dark den but walking in weeks, primates are born with open eyes, and we can look into our parents' faces.

Toddling down to monkey town

Infant chimpanzees have one huge skill which all the primates except us have: they cling like a burr. Mother monkeys and apes can use all four limbs to get around, and don't have to use their arms to support their babies. Newborn chimpanzees need some help holding on, and their mothers get into the habit of affixing the baby more firmly to them before they move off. Very young chimpanzees cling to, and are held against, their mother's abdomen. When they are as young as eight weeks they may ride on her back. For quite some time they can't change position on their mother's back, and if their mother boosts them onto her back facing backward, backward they stay, looking over their shoulders to see where they're being carried.

As baby chimpanzees get older, their mothers encourage them to walk (on all fours, first) by putting the babies down and backing a short distance away. Then the mother looks at the infant with friendly grunts and facial expressions until the baby lurches toward her. A chimpanzee mother has also been seen encouraging an older infant to walk on two legs by taking its hands and walking backward, encouraging the infant to stagger after her. Eight-month-old baby chimpanzees begin tottering away from their mothers, since they are now able to turn around and look for her. The first time the wild chimpanzee baby Prof deliberately let go of his mother Passion was when he was dangling from a small branch in front of her. After letting go of her, Prof hung silently for a moment and then grabbed his mother again.

Of Lucy, a chimpanzee raised in a human home from two days old, it is reported, "At two months her eyes would focus and follow a bright object, and at three months she could walk about on all fours. At five months she was trying to climb out of her crib to go to people, and at six months she was pretty mobile on all four limbs, though she did not walk on two legs till she was about eight months old."

At three months old she was a cuddly infant who loved to kiss and be kissed by all and sundry. By three years old she did some of the things very mischievous human toddlers may do, such as grabbing the end of the toilet paper roll and dashing through the house with it. But she also did things toddlers seldom do, such as racing along the mantelpiece hurling ornaments to the ground or climbing a standing lamp and leaping from the lamp to a fixture.

The infant orangutan Cody was slower than Lucy, and didn't try to crawl until he was eight months old. His foster father, Keith Laidler, anxious to pour knowledge into the little ape, kept trying to get the baby to sit up. "At 12 weeks, whenever I sat him upright he would collapse forward, his hands at his sides, making no attempt to save himself from the inevitable bump on the floor. Time after time I tried it, and time after time Cody would end up with his nose against the boards, staring in shocked silence at the floor"

Many of Cody's developmental milestones came between those of chimpanzees and gorillas on the one hand, and humans on the other.

He crept at 26 weeks, later than the average chimpanzee or gorilla (20 weeks), but much earlier than human babies. Like some human babies, he later came up with a wacky mode of locomotion for which it seems unlikely there are innate motor patterns: scooting. "He could not yet manage a straight one-two-three-four movement of limbs, but instead pivoted his bottom over and around his legs, these remaining immobile until his *derrière* had reached *terra firma*. Once he had bottom-landed, his legs would move again to a forward position, and he would once more lift his rear in a little arc over his legs. His later efforts were also a bit of a hit and miss affair in that he would attempt to raise all four limbs together (no one had told him about gravity) and, with a little cry of indignant surprise, fall flat on his face." Crawling was not far off. "Within two weeks he had discovered the most efficacious form of locomotion on four limbs, and from then on my life was hell."

Learning to use your trunk

Elephant calves stand shortly after birth and soon walk. But a calf must also learn to operate its trunk, a limb with nothing like the evolutionary history of a leg, but with 50,000 muscles just waiting to be told what to do. When calves don't pay attention, they step on their trunks and trip. To drink water, the baby kneels and uses its mouth, rather than taking water in its trunk and squirting it into its mouth, as adults do. A calf spends many diligent hours trying to control the thing, flopping it about, twirling it, sucking on it, attempting to pick up ever tinier objects.

Learning to climb

Three orphaned bear cubs, Squirty, Curls, and The Boy, raised by Benjamin Kilham from the age of 7 weeks, got their first chance to climb trees at 13 weeks. "Although their climbing skills were unformed, the cubs knew immediately how to go about developing them, especially how to reach down with their feet to find a footing, and were tenacious about practicing and learning." Within three days they were adept.

Baby howler monkeys begin climbing by swarming around on their mother's body, interspersed by sticking their fingers in her mouth, ears, and eyes. Then, still holding on, they reach out and grab branches. When they're a little older they let go of her and climb on branches for 10 to 30 seconds at a time. A few months later the infants climb around, jump awkwardly in the branches, and "plop" back and forth near their mothers. They begin hanging by their tails for a minute or two and wrestling clumsily with other babies. One-year-olds "doodle," hanging by their tails for long periods, swiveling around, and touching leaves and twigs. Sometimes two monkey children wrestle, swat at each other, or hug while hanging by their tails. Once an observer saw two juveniles hanging by their tails side by side in front of an adult, who pushed them gently so that they swung like pendulums.

Captive baby rhesus monkeys without playmates practiced climbing by assigning themselves "projects." Can I leap straight up to that branch? Can I jump from the post over to the side of the cage? Can I jump back? Okay, I'm going to climb up that pole, jump over to the mesh, and climb down the mesh, and I'm going to do it really fast. Okay, this time after I climb down I'm going to hop over to the pole.

Climbing trees comes naturally to chimpanzees when they're old enough—when they're used to it. At the age of 14 months, Lucy, a chimpanzee raised in a human home, was terrified when her foster brother set her on a low branch in a mimosa tree. Later she enjoyed climbing trees, but with nothing like the ease of wild chimpanzees. In the woods, she'd climb one tree, climb down, and climb up the next one, disappointing her foster father by never swinging or even clambering from tree to tree. When, years later, as part of a reintroduction effort, she needed to climb trees in Africa to get food, she made a poor showing.

When the other chimps climbed a wide-trunked baobab tree to eat its fruits, flowers, and leaves, Lucy found the going too hard. Lucy, who had learned about 100 ASL signs in her previous life, began signing to Janis Carter, who was supervising the reintroduction effort. "More food," signed Lucy, "you go." She pointed to the tree. Her idea seemed to be that Carter should climb the tree and toss food down to Lucy. "No," signed Carter, gesturing to Lucy to climb an adjacent, slenderer tree and jump across to the baobab. Lucy led Carter to the baobab, stationed her hands on the trunk, and signed, "More food, Jan go." Carter refused (she does not say whether she could have climbed the baobab if she had wanted to, but that is not the point) and went back to camp—followed by Lucy—and found a large piece of timber, which she began to drag to the baobab. Oh, okay. Lucy helped Carter drag the timber to the tree. Then Lucy propped it against the trunk and climbed up.

What do you see when you look in the mirror, big guy?

Orangutans are, in a leisurely way, superb climbers and clamberers. They are also heavy, so branches that are superhighways to squirrels or long-tailed macaques are no use to them. Female orangutans weigh around 40 kilograms (around 90 pounds) and males weigh 80 to 100 kilograms (180 to 220 pounds). Since trees are not always obliging enough to grow large sturdy branches that reach to large sturdy branches in the next tree, orangutans have other ways of getting from tree to tree. Biruté Galdikas has described how an orangutan gets into a "pole tree" that sways under its weight and deliberately makes it sway farther and farther until the orangutan can grab the next tree over. If the ape tests a pole tree and finds that it's not strong enough, it may grab two pole trees at once, and use them together to sway in an arc to another tree.

Anne Russon describes how wild orangutans treated pole trees "as if they were down escalators, letting their weight bow the tree right to the bottom then stepping off neatly." A two-year-old apeling practiced swaying, sitting in a small tree and shifting his weight to make the tree sway in different directions for different distances.

It looks natural enough when wild orangutans do it. But the captive orangutans being released at Galdikas's rehabilitation center didn't find it so easy, never having learned to judge trees and their trajectories. They liked the idea of traveling on high. But they would try to clamber through skinny saplings that crashed beneath their weight, and they tried to sway trees that were too stiff to budge. Accustomed to swinging on steel chains, they swung on lianas that often broke. They kept on making these mistakes for months, and after years in the forest often preferred to descend to the ground and travel there, which limited them to places where there were trails.

So sophisticated are the skills needed for such a heavy animal to move around in trees that biologist Daniel Povinelli and anatomist John Cant have argued that it requires self-conception. Smaller primates (and other smaller arboreal animals) can get around in trees using a limited set of stereotyped motor patterns, they argue. But orangutans need to execute "unusually flexible locomotor patterns" and need to be more alert to the possibility of things moving, bending, or breaking beneath their weight. In a 1995 paper, Povinelli and Cant give the example of an orangutan in a tree reaching out and grabbing the ends of a bunch of small branches with one hand or foot, taking hold of a liana with another hand or foot, pausing to survey the situation, using a third hand or foot to hold another bunch of branches, letting go (with the fourth hand or foot) of the tree where it is perched, and swinging across a gap into another tree. Orangutans are notoriously slow-moving, and perhaps this is connected with their frequent need to ponder the angles.

If you weigh so much that it's that difficult to get around, Povinelli and Cant argue, an increased awareness of one's own self and one's impact on the surroundings would give a selective advantage. "Our hypothesis is that clambering in large-bodied apes who negotiate a habitat that is fragile, unstable, noncontinuous, and unpredictable as a consequence of their body weight, is underpinned by cognizance of one's actions—an ability to engage in a type of mental experimentation or simulation in which one is able to plan actions and predict their likely consequences before acting." Povinelli and Cant hypothesize that a common ancestor of all great apes, including us, was a large arboreal creature that was selected for self-conception, and who passed the capacity down to us.

I'm stuck! I'm stuck!

Scientist and animal trainer Karen Pryor writes that cats do not need to learn to climb trees, but they need to learn to climb down again. Up is easy. The cat extends its claws and goes in the direction it wants to go. When it's time to go down, the curved claws won't be any help unless the cat goes down backward, and apparently it is not obvious to young cats that this is the thing to do. If they start to go down head first, their claws slip. They stop. Often they yell for help.

Pryor argues that cats are very good at mimicking, and if they get to see their mother climbing in trees, they will copy her way of turning around and descending. But if they don't get to see this, they may be baffled when they find themselves high in the branches.

Arnold, a young starling reared by Margarete Sigl Corbo, grew up strolling around the house and did not attempt to fly. It turned out that he lacked an even more basic skill, the ability to perch. Not having been raised in a nest, he was unfamiliar with branches and other things he might have wrapped his feet around. When Corbo tossed him in the air he fluttered into a mulberry tree, aimed for a branch, and toppled off because he could not grip it. He couldn't even perch on her fingers. Corbo spent several days putting Arnold through physical therapy. She laid him on his back in her hand, held him so he couldn't squirm away, and slapped the bottoms of his feet with a dowel. He hated this. Finally he took to grabbing for the annoying dowel to ward it off. Once he had grabbed it tightly, Corbo swung Arnold upright. For a moment he stood poised, perching as a bird should, and then he lost his balance, fell forward, and hung upside down in amazement.

Corbo repeated the procedure, and the next time Arnold began to topple, she jerked the dowel down, causing him to flap his wings in alarm, which helped keep him upright. Doing this over and over, Corbo helped Arnold learn to grip and balance. As soon as he could perch, he felt safe flying, and for this he needed no lessons.

Taking wing

Learning to fly is worth it, but risky. One of the first peregrine falcons fledged in Kansas City after many decades broke her wing on her first flight out of the nest, and broke it so badly that she could never be released even after it healed. A young greylag goose flew into the wall of a house, just three feet off the ground, and instantly broke its neck. Numberless young birds have launched themselves into the reach of waiting foxes and cats.

It is routine for young birds to leave the nest before they can fly competently. They learn fast, but not always fast enough. They are in danger from cats and other predators and they are also in danger from benevolent-minded humans who pick them up and try to raise them or take them to a bird rescue center. (Most such baby birds "rescued" by humans would be more likely to survive if they were left alone. As soon as people are out of sight, the parents will return, feed the fledgling, and encourage it to get to a higher place, by flying or by other means.)

Bird-watcher Olive Thorne Miller heard a terrible squawking and outcry in the forest and found a young blue jay on the ground. She started toward him, wishing only to help, got screamed at by his parents, and was smart enough to back off. The little jaybird tried to fly but couldn't get more than two feet off the ground. Night was approaching. The young jay hurled himself at a locust tree and began climbing it. He'd fly up the trunk a few inches, cling to the bark and catch his breath, then fly up a few more inches. He got six feet up, nearly to the lowest branch, and then fell off. Undaunted, he struggled through the grass to a maple tree, which he attacked in the same manner. This time he made it to a branch, "where I saw him sitting with all the dignity of a young jay." Miller came back to check on him the next day, and he still had to fight his way up trees bit by bit, but on the third day he could fly.

Lucky, a whooping crane chick living by a lake in central Florida, had a large circle of admirers, who hoped to see his first flight. He'd flap his wings, and it is reported that whenever a breeze came up, his fond parents would flap their wings and chase Lucky until he ran, flapping his wings. No liftoff. Then one afternoon when he was nearly three months old, just after the official biologists had given up and gone home, all three cranes suddenly rose into the air and flew across the lake, leaving the remaining onlookers too awestruck even to grab a camera.

Flying late

Rarely, a bird must learn to fly as an adult. Scarlett O'Hara was a scarlet macaw that came to filmmakers David and Lyn Hancock after spending years in a small cage, doing little but acquiring a plummy English accent. Her wing muscles were atrophied, and if she had ever been able to fly, that had been long ago. She climbed, walked, or swaggered everywhere she went. The Hancocks were determined that Scarlett would fly and gave her lessons. These consisted of tossing her up in the air, for although she knew about flapping her wings, she had neither the strength nor the will to launch herself. David Hancock would hurl her into the air, and she would flap her wings and spiral gently down to the grass. Thrilled, she would waddle back to Hancock and climb his leg to be launched again. In this way she gradually built up her muscles, but she still lacked aviation experience.

One day Hancock tossed her up, and Scarlett managed to stay aloft, perhaps even caught some thermals, and ascended to 75 feet. She then made the mistake of turning downwind, got blown off course, and plummeted into a crab-apple thicket. "Not wishing to embarrass her, we went inside and looked through the window," wrote Lyn Hancock. Scarlett extricated herself from the thicket, leaving behind some tail feathers, marched back to the house, and knocked on the door with her bill, crying "Come in" to herself in a cordial voice.

Silly geese

At Konrad Lorenz's research station in the Alm Valley of Austria, many of the greylag geese in the colony there were raised by people. People don't make the best flight instructors for geese, but fortunately most of the motor patterns for flying are innate. The young geese grasp how to take off, fly, brake, and land. But they must learn to estimate distances and heights, and to judge wind conditions. They do not know that they should land against the wind, because if they land with the wind it may push them over, and "they may turn the most awful somersaults."

Since people can't fly, they can't lead the youngsters into the wind to land, but if they see the geese flying low into the wind, human foster parents can encourage landing by quickly stooping or falling flat on the ground. The young geese interpret this as landing, or a signal to land, and they follow suit. Once Lorenz mischievously encouraged four young geese to land the wrong way, with the wind. They all crashed. They weren't hurt, "but it was obvious they had lost confidence in me! For some time afterward I was unable to induce them to land by abruptly dropping to the ground." So they had to learn not only "Don't land with the wind" but also "Don't trust Daddy when he's grinning."

Parents who are actual geese discourage their children from flying too soon. Young geese let it be known that they feel like flying by spreading their wings and shaking their beaks. In the early days the parents utter warning calls which inhibit the goslings from taking off. If they do take off, the parents leap into the air, catch up, and take the lead. They choose where to land, and they choose safe spots and head into the wind.

Young geese must also learn to navigate obstacles. In *The Year of the Greylag Goose,* a photograph shows a flock of young geese returning from a practice flight. The geese, who are flying along the river, must soar upward to get over the trees along the water and then come down sharply to land in the meadow beyond. This is hard. "In gliding down, the inexperienced geese often lose their balance and veer from one side to the other," wrote Lorenz. "Their struggle to keep on an even keel is accompanied by pitiful distress cries."

Fly like an eagle, but not that eagle

Lady, a golden eagle raised by Kent and Ed Durden, also had the innate motor patterns for flight, but not air smarts and tree smarts. When she flew free she took a while to figure out that she couldn't perch on small twigs but needed to select stonger branches.

Lady had to get flying tips from humans, but Ed Durden was a pilot who knew about all the problems novice flyer Lady would face, except for the one about the right size of branch. He directed her to air currents she could catch and encouraged her to land into the wind. When summoning Lady to return to a lure, he made sure she had room to turn into the wind, but at first she didn't always see the need. She made several "highspeed tailwind landings before she learned . . . she tried desperately but to no avail to put on the brakes, always ending up in a jumble of feathers and looking rather sheepish."

Another mistake that human pilots and eagles should both avoid is flying up a narrow canyon that rises faster than the plane or bird can gain altitude, where there is no room to turn around. In this situation a plane will crash, and an eagle will be forced to land "and continue on foot to the crest of the hill."

Learning to swim

European river otters crawl on their bellies at four weeks. By seven weeks they walk and run. At three months they begin to swim. Cubs vary in their attitude toward this vital step. Philip Wayre mentions a cub who wanted to swim before his mother thought he was ready, so she kept dragging him out of the water by the scruff of his neck. But American river otter mothers have been reported grabbing reluctant cubs by the scruff to drag them in. Another European otter brought her 12-week-old cubs to the edge of a pool. The mother went in and began to swim. The cubs ran up and down the bank, squeaking. Two went in and swam clumsily, "bobbing along like corks." Young otters paddle with all four feet and don't know how to use their tails, which simply stick straight out. The third cub ran around squeaking and wouldn't go in.

Otters need to dry off after they swim, and one otter rehabber warns that this may be a learned skill. Otter orphans in human care can catch pneumonia when they get chilled. Veterinarian Frank Blaisdell advises either not giving otter pups enough water to swim in or limiting their time in the water and then encouraging them to dry off thoroughly on towels.

On the coast of Scotland, filmmaker Hugh Miles saw a mother otter lead her fluffy cubs to the sea. She swam, but the cubs stood at the brim squeaking. Their mother came back and encouraged them to follow. Her son went in, but her daughter squeaked, then turned and ran up the beach toward their den. The other two came out and raced after her. Catching up to her, they nuzzled her. Then all three walked into the sea together.

The buoyant little otters paddled, squeaking when their mother dived. After a while the male cub put his head underwater to watch his mother dive. The female tried, got a noseful of water, and paddled to shore, sputtering. The next day the cubs tried to dive, "but despite much thrashing of front and back legs only succeeded in submerging their head and shoulders."

Fury, a captive-reared European otter, was late coming to water. She had no otters to show her the joys and possibilities of the medium. When she was nearly nine months old she began going into a small pond, but for safety's sake she stayed in the shallows, swimming with her front paws and keeping her hind paws on the bottom. After days of caution, she dared to take her feet off the bottom and soon was plunging and porpoising.

Dolphins, also gorgeous swimmers, swim from the very moment of birth. Yet they still have things to learn. "Newborn calves tend to bring the head out of the water when surfacing, and it takes a month or two before they consistently roll smoothly at the surface to breathe," write marine biologists Hal Whitehead and Janet Mann.

Exploring

Once baby animals can swim, walk, or fly, and generally locomote, they may go exploring. They want to know what's out there. Indeed, adults want to know, too. Rangers in the Kruger National Park in South Africa built a causeway across the Letaba River. The day after it was finished, they went out to admire their handiwork and found it covered in elephant dung. By night the elephants had come to check it out. They didn't want to cross the river particularly, and once they had crossed the causeway they turned and recrossed it. They just wanted to know what was going on. The fact that the rangers had posted a large No Entry sign was of no concern to them, because elephants are authorized personnel wherever they go.

Many young animals go exploring while they are still with their parent or parents, and presumably may learn about areas they might disperse to if they someday strike out on their own. Young red foxes, red squirrels, and Merriam's kangaroo rats are among species known to make such excursions.

Young red-necked wallabies and eastern grey kangaroos of course go everywhere with their mothers when they are in the stage known as "pouch young." When they are a little older, they are called "young at foot." It is reported that near the end of the time when mother and child associate intensely, both red-necked wallaby and grey kangaroo mothers have been seen leading their young on "exceptionally long 'tours' around their home ranges with no obvious function . . . as if giving their young a final familiarisation with the area."

I was just playing

Play is hard to define, and researchers have wrestled with the task. The definitions show the strain. "Play is all motor activity performed post-natally that appears to be purposeless, in which motor patterns from other contexts may often be used in modified forms and altered temporal sequencing. If the activity is directed toward another living being it is called social play."

There are three differences between play and practice, Robert Fagen proposes. Animals playing come up with variations, whereas animals practicing show less variation as they master the actions. A playing lamb runs, and then it sees if it can run, jump, and kick at the same time. (Not something your average sheep needs to do.) Two kittens play-fighting mix scary sideways jumps with panicky dashes for safety. And animals keep playing after they become expert.

The study of play is recent. "Incredibly, we still do not know if most mammals do or do not play," wrote Marc Bekoff and John Byers in 1998. Play doesn't come without costs. Sometimes young animals get hurt while playing, or are snatched up by predators while distracted. Play must have value that outweighs the risk. Among the benefits of play that have been mentioned are exercise, forming neural connections, practice with objects, practice in communicating with playmates, and getting information about yourself and your playmates (who's faster? who's stronger? who's mean as a snake?). Animals who are the only children of relatively solitary species like pandas must usually fall back on their mothers as playmates.

Exuberance, exaggeration, and general witless leaping around are common in play. Susan Wilson and Devra Kleiman write that certain actions they call locomotor-rotational movements are characteristic of play. Head shaking, body twisting, running, and jumping all count when they are performed in an exaggerated way. They jump too high, they run unnecessarily fast. They found similar kinds of movement in harbor seals, pygmy hippos, pandas, and three species of South America rodents, the degu, the salt desert cavy, and the choz-choz. All had stereotyped ways of soliciting play from another animal, using postures, calls, or touches. A harbor seal pup might solicit play by leaning its chin on another pup, and a degu might gurgle softly.

Playing alone, playing with the world

A yearling polar bear cub, waiting while his mother hunted a little distance away, began racing around on the pack ice, sliding and taking headfirst dives into pools of melt water on top of the ice, making huge splashes. As he was flying through the air, a seal popped up in the water in front of him, probably wondering what all the racket was. The young bear grabbed the seal, pulled it onto the ice, and killed it. Instead of eating it, he played with it. He ran around carrying it. He tossed it into the air. He threw it into pools and dived in after it. When his mother looked over and saw him playing with his food, she galloped up and began to eat it herself.

An Anna's hummingbird jumped in the stream of water coming from a hose and floated down it. That was so much fun that the hummingbird did it over and over. Common eiders floated over a rapids and dashed back to repeat the experience. Adelie penguins floated on small ice floes in a rushing tide, and swam back to do it again.

Charmingly, honeybees make "play flights." According to James L. Gould and Carol Grant Gould, when bees are two to three weeks old, almost ready to make the transition between working in the hive and going out to forage, they go "for brief looping flights near the hive entrance" on sunny days. "The play flights, which range successively farther and farther from the entrance, may enable bees to learn to recognize the hive entrance, to land gracefully, and to become calibrated to the direction and rate of movement of the sun, their prime navigational landmark. If this is the goal of the behavior, though, it is not completely effective: attrition among foragers on their first day of serious food-gathering is nearly 50 percent." Keep in mind that even if the behavior has the effect of information gathering, and gives bees a selective advantage thereby, that does not mean that the bees are not playing. (If the alternative is that the bees are assigning themselves homework, that may be pushing the concept of the busy bee too far.)

On an uninhabited island in the Indian Ocean, visiting conservationists spent the night in a tent. They slept poorly, because the island is a nesting site for shearwaters, and toward dawn the baby shearwaters discovered that if they flew up to the top ridge of the tent, they could slide down the canvas sides, making a great scraping, screechy racket.

Object play

Playing with objects may sometimes be a precursor to tool use, something we humans consider extremely classy. Or the object could be a stand-in allowing a young animal to practice tooth use and claw use. Young hawks and eaglets love to carry objects high into the air, drop them, and then catch them in their feet, and entertain themselves this way for hours at a time.

Some object play seems to stem simply from a spirit of exploration and is hard to relate to the animal's adult life. A young garden warbler dropped a stone, which happened to fall into a glass and make a ringing tone, and then all the little warblers had to try. Aerial photographs taken by pilots looking for fish show sperm whales playing with a log. One whale grabbed the log in its jaws and the other whales chased it, spouting like madmen. Hal Whitehead, while complaining that if play is not carefully defined, it becomes a wastebasket term that can be applied to any inexplicable behavior, has also seen whales play. "Off Newfoundland we saw a solitary humpback rolling a plastic bucket. The whale, a male, finally turned upside down, so that its belly was grazing the surface, and then tried to urinate into the bucket that was floating alongside. There is no conceivable function for toilet training in whales, so we can reasonably assume that the whale was playing."

The parrots are in the trash again

Trial-and-error learning in play is illustrated by the adventures of young keas at Arthur's Pass National Park in New Zealand. Keas are zany parrots found in alpine regions. They have a hard time finding enough to eat and are apt to die of starvation. While they will eat fruit, nectar, and insects, they also eat carrion when they can get it. Also car parts. "In this rigorous and unforgiving environment the kea has evolved a level of intelligence and flexibility that rivals that of the most sophisticated monkeys," write biologists Judy Diamond and Alan Bond in their book *Kea, Bird of Paradox*. "They show more elaborate and extensive play behavior than any other bird." Diamond and Bond have spent many seasons observing keas and have found that the best place to bird-watch is the garbage dump, where keas have been coming for 40 years.

They like to roll on their backs and wave their feet in the air. When keas play together, they do such unbirdly things as throw rocks at each other, fight upside down in trees while hanging by their feet, and drag each other around by the throat.

In addition to playing with each other, keas play with things. "Fledglings spend hours at the study site mauling and destroying inedible objects, in what is essentially continuous object play." Batteries, string, foam rubber all must be examined. Some of it turns out to be food. Diamond and Bond describe a young kea trying to rip open a bag of chips by trial and error. Adult keas know that the way to do this is to grab the bag with your beak, hold it down with one foot, and jerk your head and neck back. Young keas have the general idea, but often do it in the wrong order. First they pull. Nothing happens so they stop and kick the bag. No good. They stand on the bag with both feet and have another pull. And so forth.

If you can't eat it, you can at least play with it. And an excellent way to play with things is to make an all-out, crazed attempt to destroy them. Keas are world-famous for their interest in taking cars apart and their willingness to be filmed while doing so. Keas land on cars and sample the wiper blades. "Can I eat this? No. Well, can I rip it off and toss it away? Yes! How about this gasket around the window: edible? No. Shreddable? Yes! How about this antenna: edible? Not really. Does it snap off? Yes!"

Diamond and Bond tell the sad story of a jeep parked in the mountains while its owners went on a five-day hiking trip. Keas spotted the jeep, landed on its soft roof, tore the roof to shreds, and tossed the shreds to the winds. Then they shredded the seat cushions. They topped this off by clambering under the dashboard and ripping out the wiring. It is not reported whether keas also sat in the driver's seat and shouted "Vroom vroom!" but I wouldn't put it past them.

Diamond and Bond report that tourists find this hilarious if it is not their jeep. Residents grimly "wire their garbage cans down and anchor them with concrete blocks, cover their television antennas with polyvinyl chloride piping, and close off chimneys and other attractive openings with chicken wire. Unattended backcountry sheds are often sheathed in steel siding, and the wiring and motors on ski lifts are protected with steel conduit and heavy metal casings."

Bubble stuff

Handless, dolphins still play with objects. They grab things in their jaws and they manipulate objects with water and air pressure, blowing and sucking water and air. They create toys out of thin air, by blowing and otherwise manipulating bubbles. Shilo, a captive bottlenose dolphin, once blew a series of bubble rings as three other dolphins looked on. One fetched a piece of fish and spat it into a rising ring. The turbulence spun the fish around wildly, and the fascinated dolphins did it again.

Vater, a freshwater dolphin in the Duisberg Zoological Gardens, learned to make bubble rings. His technique was to rise to the surface, take a mouthful of air, swim to the bottom of the pool, and wait for the water to still. He'd crack open his mouth to let air out while turning his head, creating a ring of bubbles. Then Vater would swim through the ring. If he could, he'd quickly turn and try to swim through the other way before the bubbles reached the surface.

Vater and his companion, Baby, played with objects, often trying to see how many they could carry at once. A photograph shows Vater with one plastic ring looped around his lower jaw, one around his upper jaw, a rugby ball gripped in his mouth, and a third ring tucked around his upper jaw after the rugby ball. They also tucked balls under their flippers and tickled each other with brushes held in their jaws.

Tabo and Golia, dolphin half-brothers at an aquarium in Italy, had learned to make bubbles with their blowholes as very young calves, and could make chains and rings this way. When they were 9 or 10 months old they invented a way of making bubble rings with their tails that hasn't been seen in other dolphins. They discovered that if you arch your body downward and slap the surface of the water—hard—with your tail, a curtain of bubbles will follow your tail down into the water, and then, if you arch your body the other way, swishing your tail through the water, you can create a large bubble ring. Tabo and Golia made bubble rings, followed them, pushed them, bit them, swam through them, and chased them to the surface.

Learning to create and manipulate bubbles is more than a charming pastime, since many grown dolphins (and orcas, and humpback whales) use bubbles to confuse and confine fish, so object play shades into tool use.

Bubbles and grabbing things in their mouths aside, dolphins have at least one other option. Shilo was once seen playing with a rock on the bottom of her tank. She swam above the rock, turned upside down, sucked it onto her blowhole, righted herself, and swam around the tank with the rock balanced on her head.

Captive octopuses played with a floating pill bottle, chasing it around the tank by jetting water at it, or grabbing it and then pushing it away, for a quarter hour at a time, as if playing with a ball.

Social play

Baby animals playing together are an engaging sight. We are enchanted as the black kitten pretends to disembowel the orange kitten with mighty kicks of its hind legs, the lop-eared puppy gnaws on the ear of the stump-tailed puppy, or the lambs race around the field leaping over imaginary obstacles. The news that the 16-month-old orangutan Chantek loved to play peekaboo is beguiling. Obviously it's fun, but that's not all play achieves.

Psychologist Samuel Wasser writes of tiger play that it is "organized to maximize social interaction in a given time period, enhance precision and behavioral flexibility of participants, facilitate assessment of one's own competitive abilities as well as those of particular play partners, and enhance ability to assess others during nonplay interactions as an adult."

Tiger cubs have two principal kinds of games they play together: "I'm going to get you, my enemy," and "I'm going to get you, my dinner." Siberian tiger cubs studied in the Milwaukee County Zoo from 9 to 13 months old did not lose their taste for these games, but they got sneakier, increasing their use of rocks and stumps as cover when stalking family members. They learned that other tigers do not always feel like playing. A tiger who notes another tiger sneaking up has three choices: stay put and be pounced on; run away and be chased; or stop what you're doing and stare at the tiger stalking you, a hint that you do not wish to be bothered. Using signals like these, tigers play reciprocally, taking turns being predator and prey.

They learn to bite softly. When a tiger bites another tiger too hard, and the bitten tiger reacts with outrage, the tiger that committed the offense at once begins licking the spot it was previously biting. "Although the licks appear to appease the recipients, biting often commences again."

When social animals like adult wolves play together, they may also learn what kind of condition the other wolves are in. While this may not be the motivation for play, we may imagine that a wolf who learns that another wolf is stronger and faster might put off its attempts at a coup, and live longer as a result.

As Marc Bekoff puts it, play is the ideal time to learn social ground rules. "What could be a better atmosphere in which to learn social skills than during social play, where there are few penalties for transgressions?" You learn how hard it is acceptable to bite or hit or peck, and you learn to play fair.

Of keas, Diamond and Bond write, "Young keas appear to emerge from the nest fully prepared to attack other birds indiscriminately, and only after receiving repeated drubbings and continuous harassment do they begin to learn discretion."

Colin Allen and Marc Bekoff note that human children begin to play with other kids before they have "a developed theory of mind," before they understand, for example, that not everyone knows what they know. Allen and Bekoff suggest that play is one way that baby animals—including children—may "learn to discriminate between [their] perceptions of a given situation and reality (learning, for example to differentiate a true threat from a pretend threat). From this perspective it would be perhaps more surprising if cognitively sophisticated creatures could get to this point without the experiences afforded by play."

Wild howler monkey children often initiate play by pulling each other's tails. Sometimes the big kids play too roughly, and the little monkeys try to leave, and the older ones hold them by the tails. Then the infants squeak and either the bigger monkeys let go or somebody's mother shows up and then the bigger monkeys let go.

Using tools is cheating

The young chimpanzee Lucy, raised in the Temerlin home, was delighted when they got a chow puppy, Nanuq. Lucy's aggressive play frightened Nanuq, who at first thought Lucy was genuinely hostile. Soon they learned to understand each other's signals. Lucy would chase Nanuq through the house, and then Nanuq would spring out and chase Lucy. Or Lucy would stick her arm in Nanuq's jaws, Nanuq would bite gently, and Lucy would laugh hysterically and run away. Often the games ended when Lucy escalated to unfair use of her primate powers. She'd pick up a chair and hold it over her head, advancing on Nanuq. The first few times this happened, Lucy either threw the chair at Nanuq or, like a lion tamer, backed her into a corner. The Temerlins felt that Lucy would have hurt the dog if they had not intervened, but soon this was unnecessary, because the minute Lucy picked up a chair, Nanuq refused to play.

Animals learn about individuals through play. If little coyotes cheat, the other pups won't play with them. On the other hand, the dingo Hercules was raised by humans, an only child, and didn't get to play with other puppies. Dingo puppies learn through play fighting with other puppies when to back down. Hercules had a full repertoire of aggressive behaviors but no submissive behaviors. When he was three months old, researchers released Hercules into the wild, where he could play with a litter of five wild dingoes of the same age. The wild pups were baffled by Hercules and his apparent belief that he was invincible. No matter how badly he was losing, he persisted in aggression. "After two days, Hercules displayed no submissive behaviour (essentially because he did not know how to), and became the leader of the group; the wild pups followed his movements and usually submitted passively whenever they made direct contact."

Turnabout is fair play

It's no fun if you always lose, and young animals take turns, more or less. Self-handicapping, so that a weaker playmate has a chance to triumph or at least to play a different role, is found in species as unexpected as rats. However, Marc Bekoff, who studies play in coyotes and other canids, writes, "At a meeting in Chicago in August 2000 dealing with social organization and social complexity, it was hinted to me that while my ideas about social morality are interesting, there really is no way that social carnivores could be said to be so decent—to behave (play) fairly—because it was unlikely that even nonhuman primates were this virtuous."

Baby squirrel monkeys love to play with others, and one small monkey who was the only child in his group treated his tail as a companion, emitting play peeps as he wrestled bravely with it. Adult squirrel monkeys don't usually play with kids, but if an infant has no one else to play with, they cooperate. Maxeen Biben writes that "it is quite amusing to see a fully adult male weighing over 1 kg going to outrageous lengths in assuming the subordinate role, lying supine, flailing his limbs, exposing his fully adult canines in an enormous open mouth grin, while a 300 gram youngster repeatedly leaps upon and attacks him."

Playing veterinarian

Sexual behaviors aren't used very often in the lives of many animals (who may have well-defined breeding seasons), and play is a chance to practice them. Wild pandas grow up alone with no other cubs to play with. Instead, they play with their mothers, and David Powell, conservation biologist at the National Zoo, says that the mother initiates most of the play. Playing, the young panda learns panda postures and vocalizations. As Lori Tarou, another panda researcher, points out, "In play we see what look like future reproductive postures like mounting and biting at the nape of the neck," so some of the problems captive pandas have had with mating could be related to the lack of chances to play with another panda.

Games

It is possible to recognize the outlines of many children's games in animal play, although no animal to my knowledge has been spotted playing a board game. King of the castle, keep-away, and follow the leader are easy to spot. Any time you have a hillock and two muskox calves, you have a game of king of the castle. Two young white-necked ravens played king of the castle on a mound. One stood on top and flaunted some random object, like a stick or a lump of manure, and the other would charge up and try to take it away.

Young white-winged choughs play keep-away with stones. Budgies play keep-away. An adolescent spinner dolphin in a Hawaiian oceanarium was trained to swim into a plastic lei and then stand on her tail and "hula." The dolphin, coincidentally named Lei, was happy to do this (the other dolphins found the plastic lei offensively prickly and refused to wear it). But she wouldn't give it back and swam all over the tank, flirting it under the beaks of the other dolphins and the hands of the trainers, then whisking it away. When not furnished with leis, dolphins play keep-away with seaweed.

Wild North American otters played hide and seek in several feet of fresh snow in a streamside meadow. One otter burrowed beneath the snow and tunneled about below the surface, while the other jumped around on top, jabbing its nose down into the snow, looking for the hiding otter, who would periodically stick its head up, chirp, and vanish again, while the seeker raced to the spot. My turn—the seeker would disappear into the snow itself, and when the other otter stuck its head up to chirp, it would start looking for the one that had been the seeker.

Juvenile magnificent frigate birds at Ascension Island play follow the leader over the rocky shore, swooping down low over the water. Hamadryas baboons in the Zurich Zoo sometimes play what Hans Kummer calls the "carrying dance." Teenaged male baboons place a small baboon across their forearms, rise to stand on two legs, and twirl around with the little one in their arms.*

Dolls and action figures

Wild chimpanzees play with dolls. Neglected by the big toy marketers, these young chimps use a stick or, in one case, a dead hyrax, as a doll. (Hyraxes are about the size of guinea pigs.) A young chimp may carry the doll tenderly, caress it, groom it, and curl up with it at night.

A six-year-old male wild chimpanzee, Kakama, began carrying a small log as if it were a newborn chimp. He cradled it tenderly for hours, and at one point built a small nest and put the log in the nest. Kakama's mother was pregnant, and primatologist Richard Wrangham wrote, "My intuition suggested a possibility that I was reluctant, as a professional skeptical scientist, to accept on the basis of a single observation: that I had just watched a young male chimpanzee invent and then play with a doll in possible anticipation of his mother giving birth."

Tamuli, a young bonobo at a primate center, liked to play with dolls. One day her older brother, Kanzi, killed a squirrel. Tamuli took the squirrel for a doll. She carried it everywhere, carefully positioning its head upright and wrapping its feet around her waist as if it were clinging to her as baby bonobos do. Her mother had a baby just a few months old, Neema. Tamuli followed her mother, doing everything with her doll that her mother did with Neema. Cruel humans took the unsanitary dead squirrel away, and Tamuli moped all day.

---

*When I say "teenaged," I don't necessarily mean it literally. I mean animals almost as big as grown-ups, but without adult powers and responsibilities.

At two years old, the orangutan child Chantek pretended to feed his Cookie Monster sips of milk from a glass. (On the other hand, Andy, a capuchin monkey, made horrible faces and screamed when the Muppets came on television.)

Play in adults

Grown animals still play from time to time. An unusual example of play in an adult comes from Pigface, a Nile soft-shelled turtle who arrived at the National Zoo in 1940. Six inches long, he was not really pig-faced, but he had the pointy nose and wide decurved mouth appropriate to his species. By the 1980s, Pigface was a yard long, and violently bored. Once a week he received a nutritious dead rat, and twice a week a dozen live goldfish. The goldfish were the highlight of the week, and he pursued them "with speed and agility," so zealously that until he caught them all he literally stopped only to breathe. This filled 20 exciting minutes a week.

In the 1980s keepers noticed that Pigface had begun to bite his front legs and rake his neck with his front claws, hurting himself badly. Suddenly grasping that he might be bored, they gave Pigface toys— basketballs and a floating hoop. This reduced Pigface's self-mutilation, and he spent hours nosing and biting at the basketballs, pushing them across the water. As for the hoop, he could not only nose it, bite at it, and push it, but also chew it, shake it, pull it, kick it, and swim back and forth through it. Best of all he liked the hose the keepers used when they refilled the tank, enjoying the sensation of the water streaming over his head at exactly the right angle, and playing tug-of-war with the keeper on cleaning day by grabbing the hose and swimming backward with it, pulling it into the tank. (There is no reason to suspect that Pigface was hoping to pull his keeper into the tank, steal his keys, and make a getaway.)

Pigface's frolicsome ways were of great interest to scientist Gordon Burghardt, since documented cases of play in reptiles are few. Burghardt speculates that Nile soft-shelled turtles like Pigface may be more inclined to play than many turtles, since they are particularly agile and speedy, and perhaps unusually good thermoregulators. Wild turtles may have to spend most of their time making a living and recovering from their exertions and so be less inclined to play, but

Burghardt notes that fishermen in the eastern Mediterranean complain that Nile soft-shelled turtles destroy their fishing nets—perhaps unrecognized bouts of tug-of-war. As for Pigface, he was warm, he was well fed, no one was trying to eat him, and he had been staring at those walls for over 40 years. No wonder he was ready to play.

Pigface died at over 50 years old in 1993. His childhood is unrecorded. Whether he was already a hoop-biting, hose-grabbing, hell-raising chelonian when he was a tiny thing the size of a silver dollar is unknown.

As baby animals stagger, flutter, and splash forth into the world, mastering the basic skills, bouncing and playing, they are preparing themselves to meet new challenges, and they are learning to learn.

THREE

Learning Your Species

---

*In the Antarctic, French explorer Jean Charcot assisted the scientists on his
expedition in documenting the nests of various birds. They took a specimen egg
from each nest they recorded, placing them in a basket. Charcot was delighted
with the Adelie penguin colony, and amused by the way pairs stole stones from
each other's nests to enlarge their own, when they could get away with it. The
advent of the explorers caused some disturbance, and Charcot noticed that one
couple, who had not yet laid an egg, was taking advantage of the turmoil to
rob their neighbors and improve their own nest. Charcot at once joined in,
heaping up stones to make a splendid nest for the thieves, and then taking an
egg from his basket and putting it in their nest ring. "Vice, once again, was
rewarded in this wicked world," wrote Charcot. "The two penguins looked at
the egg in astonishment, shaking and turning it over; one of them finally
decided to acknowledge ownership, and settled on it with a satisfied croak, the
other made proud little crowing noises. Later, no doubt, the neighbours will
all say, 'How much he takes after you both!'"*

ONE OF THE THINGS baby animals learn is so fundamental, so
omnipresent, and so transparent that it can be hard to believe it is a
learned thing. The baby must learn what sort of an animal it is. It's dis-
maying to think you could bungle this one.

Actually, what the baby animal learns is who—and what species— its
family members are. Whether it goes beyond this and assumes that it is
the same kind of animal is hard to guess. Of a kitten that grew up with
puppies and shares their activities people say "He thinks he's a dog," but

does he really? Elizabeth Marshall Thomas describes a rabbit raised with dogs who joined one of them in chasing squirrels, which they never caught. "Together [they] would slowly stalk a squirrel, together they would rush it unsuccessfully, and together they would gaze up at it as it escaped into a tree." Did the rabbit think he was a dog, or did he simply enjoy sharing the preoccupations of his friend the dog?

An orangutan named Cody was raised by humans after his mother rejected him. One day his human foster father carried him up to a cage holding an orangutan who sat placidly. Since birth Cody had never seen another orangutan, and the sight filled him with alarm. Hair on end, he scrambled behind his father, clinging so hard he left marks. The orangutan who gave him such a fright was his mother. Cody did not know he was an orangutan, but that does not necessarily mean he thought he was human. It was probably a matter he never considered.

Your mother's name here

In the process called imprinting, a baby learns to recognize its parent or parents and perhaps its siblings. This mechanism allows a young robin to learn to associate with robins, and not ducks, or cats, or kindly wildlife rehabilitators. It happens to a greater and lesser extent and at different times in different species. Beyond imprinting, social animals learn to recognize the members of their troops or flock; territorial animals learn to know their neighbors; and animals that are vulnerable to harm from others—and that's all animals, including skunks and King Kong—learn to spot their enemies. An otter rehabber warns that you shouldn't let baby otters get friendly with your dogs and cats, because once released they need to beware of them. Later in life, the imprinting undergone in babyhood is often key to figuring out who would make a nice mate. And if that goes well, animals may have a chance to imprint on their own babies and learn who they are.

Imprinting has been well studied in certain species, notably waterfowl like ducks and geese. As a result it has been assumed by some that imprinting is a bird thing, but it is important in many mammals too. Do all mammals have some form of imprinting? Do *we*? One dictionary of animal behavior notes, "Imprinting tends to occur in those species in which attachment to parents, to the family group, and to members of the opposite sex are an important aspect of their [social organization]." That's definitely us. Then it adds, "Imprinting seems to have evolved in those species in which there is a danger of such attachments being misplaced." That's not really us. Stories of children being raised by wolves are not well supported. Nor do apes snatch our children and raise them like Tarzan. We do not paddle about on lakes or graze in big herds on savannahs where our young are liable to mix with other species in a confusing way. Perhaps the primates we descended from didn't have such a clear field, however.

An animal like a turtle, which hatches from an egg long after its mother has left, doesn't have a chance to imprint on her, but a turtle, salmon, or eel may imprint on the taste of the water where it is born, so it can return there as an adult.

Filial imprinting

Imprinting takes place during certain times of an animal's life, which aren't the same for every species. These are called critical periods or sensitive periods, and often they are when the animal is very young. The classic form of imprinting is filial imprinting—imprinting on a parent or parents—and is well known to all students of cartoons. It's the one where our hero is strolling along, spots an egg, and suddenly some little creature hatches out of it, looks into our hero's eyes, and shrieks "Mama!" Usually when a creature finally smashes its way out of the egg, the individual beaming proudly really *is* a parent and things go without a snag.

It's good to be able to tell your parents from other adults, since other adults might easily kill and eat you, and almost certainly don't want to feed you and pay for your piano lessons. Why not just be born knowing what your parents should look like? Like a foundling in a Victorian novel, why not have the mental equivalent of a locket with a portrait of your parents? One answer might be that as a species evolves, what animals of that species look like (or sound like, or smell like*) may change. If we still mentally wore lockets with pictures of tree shrews (the ancestors of primates) in them, there'd be some sad scenes in maternity wards.

---

*Female mice whose parents wore Parma Violet perfume (thanks to the efforts of refined researchers) grew up to prefer male mice who wore that same aftershave.

So the flexibility of imprinting is an obvious winner in the selection sweepstakes. Imagine two species, one of which learns who its parents are by imprinting after birth and the other of which is born knowing what its parents should look like and will accept no substitutes. The first species can evolve in ways that change its looks without difficulty. The same old script—be born, look around, that's your mother, squawk "Mother, I'm *hungry*!"—will still work. But the second species won't survive if its appearance changes, because the babies of the first generation won't respond to them. The program for recognizing parents would have to evolve in lockstep with the changing appearance of the species, an unlikely thing.

Pioneering ethologist Konrad Lorenz found that greylag goslings "accept the first living being whom they meet as their mother, and run confidently after him." This is one of the classic tales of imprinting. "The gosling possesses innate information that, if translated into words, would read as follows: 'Whoever responds to your lost piping is your mother; take careful note of her appearance.'" But he was puzzled by the fact that mallard ducklings did not do this, and if the first living being they met was Lorenz, they would run away peeping desperately and hide in a corner. Yet they'd accept a big white Pekin barnyard duck as readily as a small brown mallard. Eventually Lorenz discovered that what little mallards require from a mother is constant tender quacking. He took new ducklings from an incubator, sat down in the grass with them, and quacked, saying, so he reported, "Quahg, gegegegeg, Quahg, gegegegeg." This struck just the right note, and the adoring ducklings accepted him as their parent. Unfortunately they became hysterical if he stood up, because they didn't recognize him that way, or if he stopped quacking for more than a moment. "The ducklings, in contrast to the greylag goslings, were most demanding and tiring charges, for, imagine a two-hour walk with such children, all the time squatting low and quacking without interruption! In the interests of science I submitted myself literally for hours on end to this ordeal," wrote Lorenz in *King Solomon's Ring*. He described his embarrassment on a day when a group of passers-by spotted him waddling and quacking in a field of grass so tall that they could not see any ducklings. However, he had achieved his goal in showing that "anything that emits the right quack note will be considered as mother, whether it is a fat white Pekin duck or a still fatter man."

Flying lapdogs

The spectacled flying fox, a charming Australian bat, has in recent years been dreadfully afflicted by paralysis ticks. One result has been many orphaned flying foxes. Australian animal lovers rallied to adopt (and vaccinate) the babies, who cling to their mother or foster parent for the first four weeks of life. In the wild the babies, once they are too heavy to ride on their mother as she flies, are left in crèches with other baby bats. People who adopt orphans often hang them on the shower rod at this age. When they are several months old, the babies are taken to large open pens, where they can be with other bats, eat bat food, join nearby bat colonies, and leave when they feel ready. Biologist William Laurance described the heart-rending scene in an early year of bat rescue efforts, when the foster parents bravely delivered their little charges to the pens. "Though it was a wonderful scheme, it was a sad day indeed when all the moms and dads had to say goodbye to their babies as they were left, en masse, in the pens. Most of the waifs howled as they watched their tearful parents depart. Many were petrified of other flying foxes . . . having no idea what those other frantically screaming creatures were. Some of the babies desperately grasped little stuffed animals, the final legacy of their surrogate parents."

I'd like to thank my family,
once I figure out who they are

Many animals need to recognize their siblings. Tadpoles of the spade-foot toad sometimes turn carnivorous and eat other tadpoles. The bloodthirsty tadpole first nips the other tadpole, and if it tastes unrelated proceeds to eat it. If it is a sister or brother the tadpole eats no further. (Unless it is really really hungry.)

Incest taboos exist in many species, with animals avoiding treating their brothers and sisters as potential mates. An animal that spends a long childhood playing with its siblings should have little difficulty in recognizing them, but laboratory experiments show that some animals also seem to recognize siblings they have never met. They like to hang out with these siblings, but not to mate with them. Apparently they do it by smell. Studies of house mice showed that a mouse can smell the difference between a mouse that has similar genes coding for the major histocompatibility complex (MHC) and a mouse that has different MHC genes. A mouse in the mood for romance likes to meet another mouse whose MHC genes are different.

Many people assumed that a mouse compared its own smell with that of the other mouse. But studies in which mice were switched into unrelated litters showed that the mice grew up to treat their foster siblings as they would have treated blood siblings—as less-than-ideal mate material—and were unattracted by mice with MHC genes like those of their foster siblings. They were tragically attracted to mice with MHC genes like their own, apparently not getting the "treat her the way you would treat your own sister" or "tell him you think of him as a brother" messages they would have gotten had interfering scientists not switched them in the cradle.

Rather than compare a potential mate's smell to its own, the mice showed what's called familial imprinting, learning how their family members smell.

Hamsters, on the other hand, do compare their own smell to those of other hamsters, as well as noting the smell of their siblings. Researchers Jill Mateo and Robert Johnston cross-fostered hamsters, one at a time, into unrelated hamster families. When sexually mature, the female hamsters were interested in mating with mysterious strangers and not with either their foster brothers or real brothers from whom they had been separated at birth.

Learning your community

Some sheepdogs herd sheep and others live with sheep and guard them against predators. Merrily Weisbord and Kim Kachanoff describe Flintis, an Anatolian shepherd in Namibia, whose job was to chase cheetahs, jackals, hyenas, caracals, civets, and baboons away from the sheep. As soon as he had these dastards on the run he went back to the flock. He slept outside with the sheep and spent all day with them, treating them as his friends and charges and not, like a herding dog, as prey only one step removed from dinner. He was raised with the flock from six weeks old, playing with the lambs. "Flintis had imprinted on the sheep. He thought he was one of them. They were, for all intents and purposes, his family." But when Flintis was introduced to a female dog in heat, he did not hold out for a sheep. Having spent his first six weeks with dogs, he knew he was a dog. He happened to be a dog whose community consisted of sheep.

Sexual imprinting

The extent of sexual imprinting—forming an image of a desirable mate—varies from species to species and also within species. British researchers cross-fostered lambs and kids, so that the lambs were being raised by nanny goats and the kids were being raised by ewes, but allowed the lambs to play with other lambs and the kids to play with other kids as they grew up. The lambs and kids grew up playing and grooming after the style of their foster mother, but acting like their own species in their styles of vocalizing, feeding, climbing, and fighting. When offered photographs of sheep and goat faces in a choice test, the young sheep raised by goats preferred to look at goat portraits, and the young goats raised by sheep preferred to look at sheep portraits. (Sheep are surprisingly attentive to portraits of sheep, can distinguish the faces in them, and recognize head shots of sheep they haven't seen for a year.) Once they were adult, the males wanted to hang out with females of their foster mother's species. So the billy goats raised by sheep were attracted to ewes, and the rams raised by goats were attracted to nanny goats. The preferences of the cross-fostered females were weaker and faded with time. At first they preferred to hang out with animals of the same species as their foster mother, but eventually the ewes were only interested in rams and the nanny goats were only interested in billy goats. The researchers, writing in *Nature*, argue that the finding that males are more potently influenced by their mothers "indirectly supports Freud's concept of the Oedipus complex and suggests that males may also be less able than females to adapt to altered social priorities."

Wildlife rehabilitators who work with baby animals must beware of letting chicks, kits, cubs, or pups imprint on them. They also worry about two other kinds of learning, habituation and tameness. In the process of habituation, the animal merely gets used to something that might otherwise scare it. Perhaps a captive squirrel gets used to the idea that someone is going to walk past its cage carrying food dishes and stops panicking every time it happens. Or it gets used to hearing cars drive up. This might be a problem if it greets cars after it's released.

A tame animal may go further and rush over to the wire to plead for its friend the rehabber to give it a snack. If it's a bear, this is going to be a big problem after it's released. It will probably be awkward even if it's a squirrel.

An imprinted animal is likely to want to set up housekeeping with, or at least mate with, a human. This is a severe problem, as we shall see. Much as humans may be alarmed by wild animals asking for handouts, they tend to be even more alarmed by wild animals asking them out.

Cross-fostering and confused cockatoos

When imprinting works right, which is almost always, it's invisible. It's easier to see how it works when there's a mix-up, as in cross-fostering, when one kind of animal raises another. Many animals will adopt babies of their own species who aren't related to them. And because the ability to recognize the signs of babyhood—the traits of youth and cuteness—is an ancient and generalized ability, many animals will adopt babies of other species under the right conditions.

Cross-fostering is common in captivity. It's not hard to get a mother dog or cat to accept a baby skunk or squirrel or piglet. Sometimes you can hardly keep them away from the strange baby they want to cuddle. In the wild, cross-fostering is rare but not unknown. If it happened often, natural selection would favor the evolution of mechanisms that would prevent it from happening. In one interesting place and time, it happened repeatedly.

In dry western Australia, Ian Rowley and Graeme Chapman were studying cockatoos. This arid region is blessed or cursed, depending on whether you are a birder or a farmer—let us say, equipped—with six different species of cockatoo. Rowley and Chapman were studying the most glamorous one, *Cacatua leadbeateri,* also called Major Mitchell's cockatoo after the first European to paint its picture. They're called Majors for short. Majors are huge white-and-pink parrots with tall white crests striped with red and yellow, which they raise dramatically at the slightest excuse.

There Rowley and Chapman were, following and counting Majors, putting wing-tags on them, tallying what they ate, and recording their flock movements, when they noticed an oddity. The various cockatoos form flocks with their own species, and while three or four species may feed in the same area, when they fly off, each flies off with its own kind. A flock of Majors, a flock of corellas, a flock of red-tailed black cockatoos. But in the flocks of Majors, occasionally there was a bird of the wrong species. Each time the odd bird was a common sort of cockatoo, the galah (pronounced guh-LAH). These are smaller, powdery gray-and-pink cockatoos, with small crests.

These few oddball galahs not only flew around with Majors, but they flew slowly like Majors—flap-flap-glide, flap-flap-glide—instead of flying strongly and swiftly as galahs do. Instead of saying "Chet!" like galahs, they said "Creek-ery-cree!" like Majors.

That was curious. Then other oddities were noticed. They found several cases where a male Major seemed to be keeping company with a female galah—sometimes the Major was raising fledglings. Other times Rowley and Chapman spotted a trio—male and female Major and female galah—together at the nest tree, day after day. Several times they spotted a pair of Majors with a family of chicks in which one looked an awful lot like a galah.

At first they thought they were seeing hybrids, birds that were half Major and half galah. Only after several years of research did they piece together the story of the birds they call M-galahs.

Attracted by new wheat farms in former scrubland, galahs had moved in to nest in areas where they hadn't nested before, but which had contained Majors all along. The two species had a conflict: both nest in holes in trees, which are in short supply. Sometimes a pair of galahs would begin using a nest hole, laying an egg or two, but not sitting on the eggs until the full clutch was laid. A pair of Majors might come along a few days later and choose the same nest hole as their dream home. Eventually the couples would clash.

Rowley and Chapman saw one such tussle. "The two birds fell locked together to the ground where they continued to scuffle until the [galah] together with its mate eventually flew off." So the Majors had driven off the galahs. Yet, a week later, they found a nest lining of fresh green leaves in that same hole—and lining nests with green leaves is a habit of galahs, not Majors. The galahs were sneaking in and decorating.

Ultimately the Majors won. "Possession may be nine points of the law, but when the contender is 20 percent bigger and has a bill like wire cutters and muscles to match, the battle is apt to be one-sided," writes Rowley. The Majors would toss out an egg they didn't think was theirs, but since galah and Major eggs look alike, mistakes were made.

Baby galahs and baby Majors beg their parents for food in the same way, except that baby galahs are noisier. The squeaky nestling gets the grease, and so a young galah in a nest of Majors, though smaller, was in no danger of being underfed. Thus an M-galah was created, genetically 100 percent galah, but culturally a Major.

Young galahs leave the nest at seven weeks, while young Majors wait till eight weeks. At a mixed nest, there would be a week during which the young M-galah would have left the nest and would be hanging around in the tree, shrieking for food and attention, while the young Majors were still tucked in the nest. The little M-galah would utter classic galah begging calls, "being-fed" calls, and "distant contact" calls, calls it was born knowing. Passing adult galahs would spot the lone fledgling issuing demands and fly down in a helpful spirit. But the little bird would turn away, never having seen galahs, and not liking the look of them.

By the time a young M-galah joined a larger flock, it had learned a lot about how to act. Instead of giving the galah calls they gave as fledglings, M-galahs had learned the triple call of the Major: Creekery-cree! Instead of flying fast, they adopted the majestic flap and glide of Majors, who avoid long flights.

Sometimes the M-galahs would slip up. They'd get carried away with their flying ability and speed ahead of the flock. Then they'd notice that they were alone out front, quickly circle back, join the ranks, and resume the flap-flap-glide. Or, when startled, they'd let out a galah alarm call.

Rowley and Chapman write: "Through a combination of imprinting and subsequent learning the innate patterns of [galah] behaviour were largely eclipsed so that M-galahs were accepted by Majors and allowed to feed, fly, and roost in their flocks, whereas normally the two species remained apart."

But that was as far as things went, usually. "To some extent such individuals were shunned by both species and appeared to find it very difficult to overcome the barrier of Individual Distance even though both species are very much Contact animals."

Parrots want to love and be loved, to preen and be preened. In captivity, parrots generally follow the philosophy that if you can't be with the one you love, oh well, love the one you're with. So if an M-galah had been locked up with a Major, they might have paired off. But when choice was free, the Majors didn't like the M-galahs. Maybe they looked or sounded wrong. They did say "Creek-ery-cree," but their voices were too squeaky.

Whether the M-galahs could have been accepted by galahs isn't clear, because the M-galahs weren't interested in galahs. They didn't even like each other particularly: AB and NO were two male M-galahs born the same year, raised in different Major families. Though they were often in the same flock of Majors, they weren't friendly. For these birds it seems to be inevitable that nobody they like ever likes them back, and yet they don't like anybody that likes them.

What about those pairs and trios that included M-galahs? Rowley and Chapman unraveled the stories of several of these matches. Majors mate for life in the sense that pairs usually stay together as long as they both shall live. But life in the wild is hazardous, so many cockatoos are widowed and marry again. Several M-galahs hung around widowed Majors, following them night and day but never helping with the kids. If the widow or widower found a new mate, the M-galah might persist with its attentions, creating a trio.

Maybe the love of an M-galah is sometimes returned. Rarely, hybrid cockatoos, half galah and half Major, have been found. Rowley and Chapman suspect that such pairings might occur in the wild if the galah partner is an M-galah who manages to woo the Major into over-looking certain personal quirks. Perhaps a Major who was raised with an M-galah sibling would be less close-minded about a mate with a squeaky voice and a stubby crest who's always in a hurry.

It's not hard to figure out why the M-galahs liked to be with Majors and tried to form romantic attachments with Majors. As for why they chose to fly like Majors, that seems inevitable if they wanted to be in the haven of the flock. If they flew too fast they'd be alone. But why did they stop uttering the galah "Chet" call and say "Creek-ery-cree" like the Majors? Was it only to make themselves understood, or was it also a matter of how they identified themselves?

Many people have had the sensation of not fitting into a group we wanted to belong to, even felt that we naturally did belong to. But the experience of the M-galahs goes beyond this, for galahs and Majors are different species.

To compare the cockatoo case with a human case more directly, if I were to break into a house that some French people were moving into, and kicked them out and stole their baby and raised it as my own,* so that the poor thing walked like me and talked like me—if, then, you suddenly came up behind this formerly French baby when it was all grown up and grabbed it, it would not instinctively shriek "Mon dieu!" before turning and striking you. Because—unlike those crazy cockatoos—we're all the same species.

More cross-fostering confusion

Researchers messing with the minds of zebra finches by cross-fostering them into the nests of Bengalese finches, or of mixed pairs with one zebra and one Bengalese finch, managed to come up with birds they called "dithers" because they couldn't make up their mind which species was more attractive to them. Apparently they were "double-imprinted." However, even finches that were firmly imprinted on a foster species could change their views based on experience. In such male birds, it is proposed, sexual imprinting has two stages. In the first, or sensory, stage before the bird is 40 days old, it learns what "the sexual object" should look like, but since it is only a child the question is academic. In the second, or verification, stage the male bird courts females and establishes a bond with one. "The initial preference is either consolidated if the female is the same species as the rearing parents, or it is modified if it is a different species." I call this the First Girlfriend Effect.

The black robin and the Chatham Island tit

Researchers in New Zealand were desperate to save the black robin. At one point there were no more than five of these friendly, bouncy, little birds, who are solid black from head to foot. They survived on one tiny island, on which the vegetation was in terrible shape. Researchers moved the birds to another, slightly less tiny, island on which the vegetation was in slightly better shape. To increase the number of chicks hatched, the researchers began stealing the first set of eggs the black

*I would never do this.

robins laid and putting them in the nests of other birds to raise. Meanwhile the black robins would lay another clutch of eggs. In this way the researchers could double (and sometimes triple, if they did it again) the number of chicks produced in a season.

First they tried putting black robin eggs in the nests of warblers, but the nestlings didn't survive. Then they tried putting the black robin eggs in the nests of closely related Chatham Island tits, a species which, though rare, wasn't as rare as the black robin. That worked better, and the tits raised clutch after clutch of black robins. The black robins raised by robins stayed with their parents longer, for about six weeks after fledging, whereas tit foster parents only kept the kids around for four weeks, but this didn't cause problems.

But imprinting turned out to be a problem. At first the project members thought they'd gotten away with it, when Mertie, a male black robin raised by tits, paired off with the black robin Ngaio.

But robins who'd been raised by tits were giving tit calls and singing tit songs. Alarmed researchers saw one of the black robins, Marion, encouraging the advances of a male tit, accepting gifts of insects from him. Worse, she built a nest with him, and laid an egg in it (which the researchers promptly switched with a plastic egg).

Marion wasn't the only one. The black robins raised by robins were mating and breeding, but those raised by tits were not. The exception was Mertie, but Mertie had been moved from the island where he was raised to Mangere Island, from which all the tits had been moved. In other words, Mertie may have mated with a robin, when he would have preferred to mate with a tit, only because there were no tits to be found. They began moving other malimprinted black robins to Mangere. They moved Marion, for example, and she soon took up with Mertie (Ngaio having vanished). After a while Mertie, who had previously sung a tit song, began singing a robin song. Gonzo, another male raised by tits, compromised and spent his days singing "two bars of robin calls followed by full tit song." Tiger, a male robin raised by tits, was disruptive on Mangere because he didn't respect robin territories. He was a tit, he felt, and didn't have to respect robin rules. But male black robins who saw Tiger wandering through their territories took a different view, seeing him as a wicked trespasser robin who must be resisted. They failed to understand that Tiger was merely looking for a female Chatham Island tit.

In the meantime, the researchers instituted a new plan in which they stole black robin eggs, gave them to tits to raise, and then snatched the baby robins back just before they fledged, and put them back in robin nests, on the theory that fledging is the time when imprinting takes place in this species. This meant that some robins, after laying their usual two eggs, losing them to light-fingered researchers, laying two more eggs and hatching them out successfully, would return to their nest one day and find that instead of having two chicks they had four, and not all the same ages. Fortunately the robins coped, and the babies imprinted on the robin parents, not the tit foster parents. This kept the researchers extremely busy shuttling robin eggs, tit eggs, plastic eggs, robin chicks, and tit chicks back and forth. When they weren't doing that, they were steam-cleaning nests to get rid of feather mites.

In one case, however, researchers gave a pair of tits a robin egg, which they successfully hatched and began rearing. When the researchers took it back and gave it to robins to finish rearing, the bereaved tit foster parents managed to track down their stolen darling "and were competing for the right to care for it." The researchers were forced to step in and transfer the tits to another island.

The strategy was so successful that after a few years they were able to reintroduce the Chatham Island tits to Mangere Island. By this time most of the robins who had become imprinted on tits had broken down out of sheer loneliness, accepted robins as mates, and were singing robin songs. Even Tiger took up with a black robin.

Maggie was the last of the black robins imprinted on tits. She was rather secretive, and the researchers had a hard time keeping tabs on her. They spotted her accepting insects from a courting tit, but then they saw no more of him. Later, seeing a male robin near her nest, they assumed he was her mate, but when it was time to band her chick, they were shocked to find that "the unthinkable had occurred," and the baby, Tobin, was a hybrid, half black robin, half tit.

The heart wants what?

An ill-suited dream lover is found in the case Konrad Lorenz relates of a white peacock, the sole survivor of a clutch killed by an early cold spell at the Schönbrunn Zoo in Vienna. To save the last chick, the keeper put it in the warmest place at the zoo, the reptile house. He was put in with the giant tortoises, perhaps because they could be trusted not to eat a chick. "For the rest of his life this unfortunate bird saw only in those huge reptiles the object of his desire and remained unresponsive to the charms of the prettiest peahens."

Imprinting seems to have spoiled a promising plan to protect endangered whooping cranes by duping greater sandhill cranes into raising them. By placing whooper eggs in the nests of sandhill cranes at Grays Lake National Wildlife Preserve in Idaho, the wildlife services of the United States and Canada hoped to establish a new migration pattern for whoopers. If they followed their sandhill parents from Grays Lake to their wintering grounds in New Mexico, a whole new population of whoopers could be established, increasing their numbers and reducing the chance that a single catastrophe (storm, disease, etc.) could wipe out the world's remaining wild whooping cranes. It worked to a point. The sandhill cranes were good parents to the whooper chicks, and the whooper chicks respected their parents and learned the migration route from them. But when the whoopers matured they did not breed, apparently since they had imprinted on sandhill cranes and only wished to breed with sandhill cranes. The scheme was abandoned in 1989. A few years later one of the cross-fostered whooping cranes successfully romanced a female sandhill crane. While this is nice for him, it does not lead to species survival.

On the other hand, whooping cranes who were hand-reared by humans at Patuxent Wildlife Research Center—but by humans dressed in crane suits—and then released in Florida, fell in love with whoopers and raised children.

Nice outfit, Mom. Excellent wig, Dad.

If you were raised by birds dressed as human, would you be fooled? How dumb would you have to be to be fooled? But perhaps the foster children aren't always as fooled as some might think.

Captive-reared condors were entirely too casual around people and their dwellings. Keepers had tried to keep from habituating the condor chicks to humans by serving their food via condor hand puppets, but the condors weren't really duped. "It is not clear the nestlings are truly 'deceived' by the puppets or deceived for very long. As realistic as the puppets seem, there is no way for them to be perfect mimics of adult condors in behavior," write Noel Snyder and Helen Snyder. On the other hand, the condors don't seem to imprint on beings they don't see.

Young condors who were raised by captive condors did much better once released to the wild than those who were raised by puppets. The puppet-reared condors were simply too casual about human beings, even if they didn't fall in love with them. However, in 2002 a pair of puppet-raised condors successfully hatched an egg in a cave nest. When the chick hatched, the parents were smitten with its charm, and although they had never seen it done, regurgitated food for the chick in flawless condor style.

Imprinting and international diplomacy

Another endangered species tale that may involve erroneous imprinting tells of an animal beloved of graphic designers. In the 1960s, almost the only giant pandas outside of China were An-An, who lived in the Moscow Zoo, and Chi-Chi, who lived in the London Zoo. An-An was a male and Chi-Chi was a female. In those Cold War days, it at first seemed impossible that they would ever meet. But obsessed zookeepers pleaded with bemused diplomats and finally arranged for Chi-Chi to visit Moscow.

Chi-Chi, a surly traveler, was anything but thrilled to meet An-An, and An-An was annoyed to see Chi-Chi. They were kept in separate quarters until Chi-Chi went into heat. Staff from the London Zoo flew to Moscow to be on hand for the historic date. They found Chi-Chi off her food and An-An bleating hopefully through the bars. The next day the pandas were let into the same enclosure. An-An pursued Chi-Chi, and Chi-Chi barked, snapped, and whacked him with her paw. Analyzing film of this encounter, one viewer wrote, "Her actions are those of an animal that does not understand what this other animal is attempting to do."

An-An had the idea: sex. Sex with Chi-Chi. Yes. But the idea seemed to shock and revolt Chi-Chi. Sex? Sex with An-An? Never, never, never! Well, maybe—no.

But she was in heat, and as the hours went by, Chi-Chi showed that she did grasp the idea. Sex. Sex with someone she trusted and respected, someone with whom she had a great deal in common, someone who understood the life of a lonely panda woman. Sex with her keeper.

Chi-Chi went over to one of her keepers and "presented," lifting her tail suggestively and backing into him. Far from being flattered, he was in despair. Here he was trying to fix Chi-Chi up with her dream date, the only possible guy outside of China, and here she was making perverted advances to *him*. Day after day the bamboo bears were put together, and day after day Chi-Chi's advances to her keepers became more passionate and her behavior toward An-An more irate. Finally Chi-Chi was flown home.

The zookeepers didn't give up, and in 1968, An-An flew to London to romance Chi-Chi. He made a prolonged visit, but his persistence was unrewarded. Chi-Chi and An-An are gone and neither of them left a diary, so it's hard to be certain what went wrong. Some faulted Chi-Chi for being imprinted on humans. Some faulted An-An for not getting tough with Chi-Chi. All schools of thought held that the problem was in their upbringing. Both pandas had been captured as infants and had spent most of their lives without seeing other pandas.

Chinese panda experts opined that An-An and Chi-Chi didn't connect because they hadn't learned how to act around other pandas, let alone how to act when they wanted to mate with another panda. In the wild it is likely that young pandas see adults mating, learning how to act when they are older. To remedy this, the experts explained, films of pandas mating should be shown to ignorant, innocent pandas like An-An and Chi-Chi. (This use of video as a how-to tool is distinct from its use as a purportedly motivational device, as when one zoo tried to inspire its competent but uninterested male chimpanzee by showing him videotape of mating gorillas.)

Nowadays many zoo pandas reproduce through feats of artificial insemination and the idea of showing them X-rated panda pictures is seldom mentioned, at least in the United States. There are still anecdotal accounts of pandas who learned what to do after viewing a video. But at many zoos, panda embryo transfer, rather than panda dating etiquette, is today's exciting frontier.

Was it a problem of who, or a problem of how?

In the 1970s, in a famous gesture of rapprochement, China sent two pandas to the United States. Hsing-Hsing, the male, was delayed for several weeks in Beijing so he would have a chance to see other pandas mating. Yet, despite this learning opportunity, Hsing-Hsing developed what has euphemistically been called an "orientation problem," making lustful approaches to Ling-Ling's head or side. "It took him a few years to get it right," says National Zoo behaviorist David Powell. When the pandas arrived, Powell says, "the instructions that came with them were to keep them separate. We thought that made sense, because they're solitary animals." But Powell now thinks one reason Hsing-Hsing had a hard time figuring out how to relate to Ling-Ling was that he was only around her for three days a year. Ultimately Hsing-Hsing got smart and the two pandas had five cubs through natural matings.

Really, I'm flattered

Sometimes malimprinting entirely changes an animal's definition of acceptable family members and mates, and sometimes it merely expands the category. Unlike pandas, tigers are pretty good at figuring out that other tigers are the creatures they should seek when their fancy turns to thoughts of love. But they're not infallible. Victoria, a Siberian tiger born in a zoo, was rejected by her mother and was given to Rosemary, a large dog, to raise. This is a regular practice when you need a foster mother for a great cat's kittens, and Rosemary raised her striped child successfully. But when Victoria reached adulthood she refused to have anything to do with tigers, though she was offered four potential mates. Perhaps because she was an only kitten, she apparently imprinted on her foster mother's species and found other tigers insufficiently doggy.

Innumerable lions raised by Joy and George Adamson, as well as a cheetah and a leopard, were able to work things out with their wild opposite numbers, due to a flexible attitude. Rafiki, a lioness in an orphaned litter raised by Gareth Patterson, grew up to view him as her first choice for mate material. Although she was independent, whenever she was in heat she would drop by Patterson's camp, ignore his girlfriend, moan loudly if he wasn't there, and invite him to go on little excursions into the bushes if he was there. She escalated to more graphic hints as she got older. His persistent refusal to take her meaning drove her into the arms of an actual lion, and she produced a litter of cubs. Patterson writes that George Adamson, his mentor, confided that he sometimes had to climb a tree to get away from a sex-crazed lioness.

Leopards raised and released by Arjan Singh in India found and mated with other leopards. It must be admitted that the leopard Harriet had a disturbingly open-minded attitude, and when she was in heat and the male she had been mating with was out of the area, she would turn confidingly to Singh, who was ultimately forced to sleep in a cage to "avoid [her] amorous advances. . . . Until then I had warded her off, if I had to, with a stick, but lately she had begun to seek relief from her problems by climbing on to my bed, and this was too great a nuisance to be tolerated. Not only did I lose sleep; she also kept ripping my mosquito net in trying to get at me."

Even when romance is not an issue, there can be problems with hand-reared animals. People who raise llamas have learned that a hand-raised female llama is friendly and confiding and treats you as if you were another llama. But a hand-raised male llama also treats you as if you were another llama, and he would like you to acknowledge him as a superior llama. (For, admit it, wonderful and versatile though you may be, you are a poor excuse for a wool-bearing camelid.) Llamas can be very rough with each other, so being treated like one of the gang is a bad thing. Baby llamas are among the most adorable baby animals in existence, and it's hard to keep in mind that such a big-eyed, long-lashed, fluffy thing might, if it accepts you fully, one day insist on beating you up and spitting on you to show you who's boss.

I can never be more than a rehabber to you, darling

A released animal that is merely habituated to people may be lucky enough not to get into trouble and go on to lead a normal life, but wildlife rehabilitators know that an animal that is imprinted on people may seek them at mating time, with unfortunate results. Owls, for example, readily imprint on people or, more specifically, on people's heads. If you are strolling through the woods and an owl swoops out of nowhere and comes at your face, you are unlikely to think, "This poor fellow is enchanted with my head, and who can blame him? Probably he'd like to settle down and raise a family with my head. No wonder he's following me, hooting crazily." Owls who do this tend to get clubbed to death or shot. Wildlife rehabilitator Kay McKeever writes that people usually think such an owl is rabid (even though owls don't get rabies).

Even if the owl does not try such a direct approach and merely hangs around hooting lasciviously at the head of its dreams, and swooping by as a method of displaying its own charms, trouble can ensue when the owl feels compelled to chase away possible rivals. "This might be laughable in a Robin, or even in a disorientated Saw-Whet Owl, but it is not so funny when the bird is a Great Horned Owl, with the striking power in his legs and talons and the serious intent of his ancestral behavior," writes McKeever. "Then it is truly dangerous."

To avoid such perilous imprinting, rehabilitators try to supply orphans with owl foster parents on whom they can imprint appropriately. Interestingly, McKeever writes that the ideal foster parent is a male owl of the same species who is himself imprinted on humans (so he won't attack rehabbers and will allow them to move to and fro). Even though he is imprinted on humans he will still respond to owlets. "Providing the juvenile owls make food-begging calls, the male's innate instinct to respond is a powerful drive, apparently overwhelming any visual incompatibility with the young."

If such a foster parent is unavailable, McKeever describes a method of putting the owlet in a nest box from which it can at least see an adult owl. Meanwhile the human caretaker feeds the owlet surreptitiously from the other side of the box, taking care not to be seen or heard.

The rehabilitator will have succeeded if, after having spent weeks devotedly tending to an orphan, she approaches and is greeted with a defense display of an owl crouching, spreading its wings to look scary and large, and clicking its beak in the manner that denotes "Come any closer and I'll shred your flesh!"

In the early 1970s, falconer Robert Berry had two American goshawks he hoped to breed. But they were imprinted on humans and wanted no contact with each other, the female being particularly vehement about her dislike of goshawks. (She was kind to a goshawk chick and treated it as her child until it was a few months old, whereupon she decided that she hated it too.) For the good of the species, Berry acted as go-between. He was able to convince the male goshawk to focus on Berry's leather glove as a love object. The male built a nest and courted the glove with clucking and bowing, finally (after unsatis-fying liaisons with a paper bag and Berry's shoe) mating with it more than a hundred times. Berry was able to collect semen in a syringe on 15 of these occasions. Then Berry's glove romanced the female, during which Berry was able to artificially inseminate her using the syringe. It worked, and the female laid fertile eggs.

Richard Zann writes fondly of the zebra finch Fred, whom Zann hand-raised from a nestling. Fred was imprinted on Zann or rather on his fingers and face. After running into problems in the aviary where the more bird-oriented finches lived, Fred was confined to a cage on top of a filing cabinet behind Zann's desk. The finch could not see the beloved fingers and face unless Zann turned around, but Fred learned "by trial and error" that he could flick a drop of water from his auto-matic waterer onto the back of Zann's neck, causing the graduate stu-dent to turn around, exposing his fingers and face to Fred's adoring view, whereupon Fred would launch into his courtship song and dance.

Who knows what goes on out there?

While most other examples of cross-fostering that we know about are in domestic or captive animals, there must be others in the wild we never learn about. R. M. Lockley, who spent years living on an isolated British coastal island where seabirds nested, often saw puffins and shearwaters fighting over a burrow in which one had already laid an egg. In one case, Lockley had marked a shearwater's egg (to see how long it took to hatch), which the shearwaters had already begun to brood. "The puffin, after several pitched battles, had taken over the shearwater's egg, incubated it, but failed to hatch it," Lockley wrote.

In Alaska, wildlife filmmakers found a white-tailed ptarmigan with a mixed clutch of chicks, some white-tailed ptarmigans like her, some rock ptarmigans. They speculated that she crossed paths with a rock ptarmigan family, and that when the families went their separate ways the chicks got mixed up. In any case, she treated the little rock ptarmi-gans as her own.

For four years in a row, the same long-finned pilot whale, probably an immature animal, was spotted swimming with white-sided dolphins off the coast of Massachusetts, with no other long-finned pilot whales in sight. The whale was never seen with other whales. The long, solitary companionship of the whale with the dolphins "raises questions regarding the possibility of cross-fostering and the implications of imprinting upon social development and reproduction," observers wrote.

And whose little kitten are you?

An animal that has successfully figured out who its parents are, who its friends and enemies are, and who its potential mates are should surely have no problem figuring out who its children are. But as we have seen in cases of cross-fostering, animals can make mistakes or exhibit generously broad definitions. They can also be fooled.

There are several groups of birds who do not raise their own children, but instead trick or force other birds into doing it for them, a practice known as nest parasitism. The Old World cuckoos are famous for this. The American brown-headed cowbird is so successful at this racket that it is said to threaten the survival of some other species.

Female nest parasites lurk, noticing where other birds are building their nests. When the time is right, they dart in and lay an egg. Usually they do this when the nest is unguarded in the early days of egg laying, but in some cases, they squeeze in next to the outraged occupant of the nest and attempt to lay their outlander egg before they are booted off.

So a female cuckoo may lay an egg in a reed warblers' nest, among the reed warblers' eggs, when the reed warblers have stepped out for a quick caterpillar. When the reed warblers return, they incubate the egg along with their own. The cuckoo's egg hatches sooner than the reed warblers' eggs. The baby cuckoo is bigger and squawks more loudly than the reed warbler chicks. Often the cuckoo chick pushes the other chicks out of the nest, one by one. The cuckoo grows quickly and often gets to be bigger than its foster parents, who valiantly feed their giant impostor darling. When this scam works, a cuckoo mother can get a lot of eggs raised by reed warblers and other birds, more than she could raise herself.

Many birds—typically species that have evolved in the presence of nest parasites—will kick out the foreign egg, but others do not. Cuckoo eggs often look like the eggs of their unwilling hosts, spotted if the host's eggs are spotted, brown if the host's eggs are brown, and so forth. This raises two species-recognition questions, which apply to all nest parasites. The first question is why the reed warbler can't tell the difference between its own eggs and chicks and cuckoo eggs and chicks. The second question is how the cuckoo, which is tenderly raised by reed warblers, knows to mate with another cuckoo, and not a reed warbler.

A mother's heart knows her own child

As for why birds can't recognize their own eggs, some birds can. Some species of birds abandon their nest if a cuckoo lays in it, or simply build a new nest on top of it. In the United States, studies suggest that species that have lived for a long time in the same area as cowbirds— nest parasites—are apt to be suspicious of strange eggs, and species that have not lived in the same area are apt to be too trusting.

You're no child of mine

Some birds imprint on their own eggs or babies the first time around. One researcher who quickly substituted different, paler eggs each time a couple of first-time breeding garden warblers laid an egg apparently succeeded in deluding them about what their own eggs looked like. When they laid their fourth egg among three substitutes, they viewed it as a phony and kicked it out of the nest themselves. The birds seemed to remember, after the first breeding season, what their eggs should look like, so only eggs that looked like theirs could pass muster.

Bernd Heinrich performed egg-switching experiments with ravens to see how far he could push them before they'd realize *"That* is not my egg." They would accept a chicken's egg Heinrich had painted to look like a raven's egg. They would accept a plain white chicken egg. They would, after some fussing, accept a chicken egg that had been painted red. However, they would not accept a black plastic film canister, even though it was the same weight as a raven's egg. (Consider the Laysan albatrosses who were induced to lavish parental devotion on a big yellow grapefruit: they were excused as birds with "little breeding experience.")

On another occasion, Heinrich needed to find a home for four orphaned raven chicks. He had been following the nest of a pair he knew well, Goliath and Whitefeather, who had two younger chicks, and who were nesting in a shed near his house in the woods. In the middle of the night Heinrich removed their two chicks and put in the four older orphans. At sunrise, father and mother regarded the chicks calmly, despite the fact that they were bigger and had doubled in number. But when the sun shone directly into the shed, spotlighting the nest, both parents suddenly seemed shocked. "The adults stood as if dumbstruck, intently examining the young. Both moved their heads quizzically from side to side, looking at the young in their nest first with one eye, then with the other from a different angle." Whitefeather, the mother, erected her head feathers, a sign of agitation and cried "kek-kek-kek" in alarm. This reminded the chicks to beg for food, and she pecked angrily in their direction, without touching them.

Whitefeather and Goliath then flew outside, where Whitefeather showed that she was upset by uttering alarm calls, hammering on branches, and hanging upside down by her feet. "I had never seen her do either of these behaviors before," Heinrich writes. (Should you see me hanging upside down by my feet after you have been switching my kids in the cradle, take it as a sign that I am very upset indeed.) Cleverly, Heinrich changed the subject by producing some tasty road-kill, and both parents got busy bringing meat to the nest and feeding the chicks. Heinrich then returned the original two chicks, so that Goliath and Whitefeather now had six babies of two different ages. Having gotten over her initial unhappiness, Whitefeather now cared for all six, showing no favoritism.

Master criminal seeks same

Nest parasites face a difficult situation in figuring out their species identity, and it's all their parents' fault. Having been born to families of crooks, but raised by honest citizens, they must manage to be successful crooks themselves without the tuition of master criminals. How can they find the right accomplice?

The answer is probably different for different species of nest parasites. It has only recently begun to be pieced together. Mark Hauber and colleagues have proposed that nest parasites need a password to recognize each other. This password could be a call, a display, or some other form of behavior that the species innately knows how to perform and to recognize. The password triggers learning, so that when a strange creature gives the password, the recipient learns that this is a creature of its own species.

Great party! Who is everybody?

Cowbirds don't seem to imprint on their adoptive parents. When they grow up, the females lay their eggs in suitable nests of a variety of species and do not confine themselves to the species that raised them.

When they leave the care of the suckers who raised them, they join large flocks of other teenaged cowbirds. How do they know to do this? Hauber and colleagues Stefani Russo and Paul Sherman suggest that the cowbird chatter call is the password that attracts them to the juvenile flocks. At all times of year, both males and females utter chatter calls, which don't seem to be learned. In the laboratory, nestling and fledgling cowbirds who had never heard cowbird calls showed interest in playbacks of chatter calls. The nestlings begged twice as much to chatter calls as to other cowbird calls or to the calls of other species. Fledglings approached speakers playing chatter calls more quickly than they approached speakers playing sparrow songs.

In addition, cowbirds may notice their own appearance, and perhaps look for birds who look like them. Young cowbirds are gray and streaky. When researchers gave some captive cowbirds black streaks in their plumage by coloring each pinfeather with a black marker as it emerged, the birds apparently concluded that cowbirds are supposed to have black streaks. When they were older and met both normal cowbirds and cowbirds with black streaks drawn on their plumage, they courted the black-streaked ones.

In the wild juvenile flocks, the cowbirds are among their own kind for the first time. This is the time in their life when they learn, through imprinting, who might make an attractive mate. If, in captivity, you keep young male cowbirds with canaries during this juvenile period, they will court canaries when they are older, even with lovely female cowbirds standing by.

They also learn how to act. Female cowbirds, even raised in isolation, respond to male cowbirds' song, and they indicate their opinion of it, phrase by phrase, with little wing quivers. In this way, females teach the males to perfect their songs (assuming the males are not too smitten with canaries to look at them).

So, step by step, genetic information interlocking with learned information, cowbird meets cowbird, and they continue the unscrupulous careers of parents they never met.

Other nest parasites

Great-spotted cuckoo mothers apparently return to the nests where they laid their eggs and chatter at their chicks. "That's not your real father feeding you! That's not your real mother! I'm your mother and this is what I look like! This is what I sound like! Whoops, gotta go."

This species of cuckoo has refined its criminality to a remarkable extent, at least in southern Spain, where its wicked ways were studied by Manuel Soler and colleagues. Researchers had noticed that in nests where magpies had detected and tossed out a cuckoo egg, bad things were likely to happen. The "Mafia hypothesis" suggests that if magpies do not accept a cuckoo egg laid in their nest, the cuckoos will notice and may come back and destroy the magpies' own eggs.

The experimenters tried tracking the fate of parasitized magpie nests from which the humans themselves had taken the cuckoo egg. Sure enough, more than half of the clutches (as opposed to 10 percent of the nests they left alone) were destroyed either as eggs or chicks. In one case a female cuckoo was spotted pecking the magpie eggs.

Remarkably, the point of this seems to be that it teaches the magpies a lesson. Magpies who had tossed out a cuckoo egg and then had their nest destroyed were more likely to let the cuckoo egg stay the next time they built a nest.

The good news for friends of freedom is that the Mafia tactic seems to be losing ground. In the magpie population as a whole the tendency to toss out cuckoo eggs is slowly increasing. "Just as the human Mafia might find it easier to control some victims than others, so Mafia cuckoos might be better able to control particular host individuals, say those that are less able to defend their nests, or those with fewer reproductive opportunities elsewhere," writes N. B. Davies in *Cuckoos, Cowbirds and Other Cheats*. "So we may find some individual hosts in a population that do not submit to cuckoo control, and reject cuckoo eggs, while others are forced to accept to make the best of their poor circumstances."

It's hard enough trying to raise this kid without worrying about its looks

Birds that are nest parasites don't have to look too hard to find another species of bird qualified to (unknowingly) babysit. But most animals are ill equipped to raise a baby not their own, or to be raised by a parent of another species. In Kenya, in Samburu National Wildlife Reserve, a young lioness tried desperately to be a mother to an oryx calf, and the calf did its best to be her child, but the attempt was doomed.

This mysterious pairing was seen by hundreds of Kenyans and visitors to the reserve, and was recorded by Saba and Dudu Douglas-Hamilton, sisters who are wildlife photographers and documentary filmmakers associated with Save the Elephants.

The pair was first seen together in December 2001. The oryx calf was very young and still had its umbilical cord. The lioness was two or three years old and showed no signs of having had cubs. She had formerly spent much of her time with her sister, but now she seemed to care for nothing but the calf. They were constantly together, with the lioness licking the calf and rubbing her head against it, and the calf nibbling on the lioness's ears and trying to suckle from her. "They did everything together, walked, slept, groomed, drank from the river," wrote Saba Douglas-Hamilton.

How the two of them met and how the lioness came to think of the calf as her child is unknown. A mother oryx goes away from the herd to give birth. She visits it a few times a day to feed the calf and after two or three weeks it joins the herd with her. Perhaps the lioness found the calf alone and somehow, in an atypical example of maternal imprinting, perceived it as a baby needing love and protection rather than as a quick snack. The calf was seen approaching other oryx, and rangers reported that it had suckled from one. The lioness couldn't feed it herself. When the calf approached other oryx, the lioness would watch worriedly from a distance, and if the calf moved off with the other oryx, she would follow—which would frighten the adult oryx into running away.

During the weeks they were together, the lioness didn't hunt, and grew thin. Once she was seen stalking a warthog, but broke off to race back to the calf. Another time, it is reported, she started hunting and cheetahs zoomed in and grabbed the calf—but the lioness spotted them, ran back, and drove the cheetahs away.

Two week later, the lioness and the calf were seen drinking from a river together. They were resting in some bushes, and the calf went around a bush and was seized and killed by a lion from another pride. The distressed lioness could do nothing to save the calf.

Not long afterward, the lioness was seen with another calf. After a week, the undernourished calf was taken away from her and brought to a game sanctuary to be fed and returned to its oryx mother. The lioness was then said to be following the oryx herds, apparently in search of a new child. Others said that people who wanted to benefit from the crowds of tourists who came to the reserve to see the lion lie down with the calf had provided the lioness with the second calf themselves. Whether or not she was supplied with the calf by unscrupulous persons, the fact remains that she fell for it. Try that with any ordinary lioness and the response would be less cute. By February 2003 the lioness, now named Kamuniak, had tried to adopt six oryx calves and one impala calf.

During the time that the lioness and the calf were together, many people puzzled over a way to support their relationship. If they supplied the calf with milk, could the strange pair succeed? If the lioness had been able to suckle the calf, could she have protected it long enough for it to become independent? Lion cubs are usually brought to join a pride of lions after they're about six weeks old, and are protected by the pride. It seems unlikely that the lioness's pride would have extended their affection to her strange baby.

Although the two seemed to love each other, they didn't understand each other well. "As if it were an autistic child, the lioness was unable to communicate with it, and she had no choice but to follow dotingly wherever it went," wrote Saba and Dudu Douglas-Hamilton.

A lioness in Tanzania observed by Elizabeth Marshall Thomas started to eat a dead wildebeest partly eaten by another lioness and uncovered "a full-term dead fetus." She removed it, and began tenderly licking it. "Gently and carefully, she cleaned its face by removing and swallowing the caul, then cut the umbilical cord, and licked the stump.

Having done everything necessary to deliver the fetus safely, she ate it." Thomas points out that a lioness typically eats a lion cub that is still-born, too. What if the wildebeest calf had been alive? Perhaps the train of maternal behavior that the lioness showed would have continued, and she would have treated the wildebeest as her baby.

George Schaller has described lost wildebeest calves, looking for someone to follow, who persistently followed lions. These incidents worked out as you might normally expect—badly for the wildebeest, excellently for the lions.

Soar like a goose, splash like an eagle

A somewhat more successful adoption shows the need of both parent and child to adapt to the other. Lady, a golden eagle raised by Ed and Kent Durden, laid eggs, which she brooded devotedly. The Durdens, taking the view that it was a shame that her maternal attention to the infertile eggs could lead nowhere, stole the eggs and substituted a fertile goose egg taken from one of the big white barnyard geese who lived in a flock down the hill. It was Lady's habit, when flying free, to swoop down and chase these geese, but she had never caught one.

Lady did not seem to care that her two eagle eggs had turned into a single goose egg. She continued to brood, and soon a gosling could be heard peeping within the egg. Lady chirped encouragingly, and it peeped back. The gosling began the laborious task of fighting its way out of the egg. Lady watched in fond concern and picked away bits of broken eggshell, the duty of either a loving eagle mother or a loving goose mother. Finally the gosling was out, lying damp and exhausted in the bottom of the giant eagle's nest Lady had constructed in her aviary.

Eaglets stay helpless for weeks, but goslings are up and about in no time. Within the hour the gosling was looking around, peeping. Lady chirped and tore a minute piece of flesh off a chunk of horse meat. She held it in her bill for the gosling to grab. But goslings do not grab meat from their parents' bills. They pick food—insects, seeds, and vegetation—off the ground and out of the water. The gosling, to Lady's dismay, began moving about the nest, pecking, looking for food. It was as if a day-old human baby ran over to the refrigerator and began ransacking it. Lady followed, chirping to draw attention to the meat she held. As the gosling sprinted around the nest, peeping about how hungry it was, Lady followed, offering lovely horse meat to her precocious child and being ignored. Ed Durden held the gosling up to Lady's bill, but the gosling still ignored the meat.

This went on all day, with the gosling getting nothing to eat, and Lady worrying herself into such a state that Ed Durden had to calm her by letting her feed him the horse meat. (She gently laid pieces of horse meat on his nose, and when she glanced away he palmed them.) At night mother and child snuggled up together lovingly, with the gosling tucked under Lady's feathers in a manner common to both eagles and geese.

The next day the gosling again scoured the nest in search of goose food, while Lady kept cornering him with offers of meat. The Durdens were about to take over feeding the gosling themselves when suddenly a lightbulb lit up in a little balloon over the gosling's head. "The gosling was looking up at Lady for the first time, instead of looking at the ground. Lady uttered a sound and extended her bill toward the gosling, and he responded as if his eyes had just now been opened. She held her bill still while the gosling picked the meat off the bill and swallowed it with relish. Immediately Lady tore off another piece and this time the gosling was waiting to take it." The adaptable little goose packed away the meat with pleasure.

Gosling and eagle were now perfectly happy, but after a while the Durdens began to wonder if an all-meat diet was really the best thing for a gosling, and they put a bowl of bread and milk in the aviary. Lady tasted it and thought "Why bother?" but the gosling guzzled delightedly. Lady noticed, and the next time the bread and milk was put in, she gripped an edge of the bowl in her beak, carried it over, and slapped it down in front of the gosling.

The next crisis came when the gosling, who had leapt out of the nest and was speeding around the aviary, spotted Lady's bathtub. Grown eagles like to bathe, but eaglets don't do this and might catch cold if they tried. Lady tried to head her child off, but he dodged between her legs and plunged in. He splashed and frolicked as Lady begged him to come out, leaning over the edge of the pool and calling in alarm. He ignored this, so she waded in, giving him the opportunity to dive between her legs and pop up on the other side.

As the gosling grew, the Durdens managed to get more vegetation in his diet, but even as a full-grown goose "he never turned down the chance for flesh food." Kent Durden writes that he "would tear into meat as ferociously as an eagle, and he could do remarkably well with the tools he was equipped with." Since Lady's diet was not confined to horse meat, the goose might be seen "tearing into a jackrabbit, or . . . trying to swallow a snake."

Mother and child remained fond, even after he was transferred to the flock with the other geese. Kent Durden describes a day when Lady was flying free. Seeing an eagle overhead, the geese waddled away at top speed, except for Lady's son, who called to her. She called to him. Calling, she headed for him, and calling, he stood on the grass. ("Mom! Mom! Mom! Mom!") As she swooped down upon him, he was stricken with instinctive fear, panicked, tried to run, and fell over. ("Help! A monster is attacking me!") His mother landed beside him and chirped soothingly and all was well. ("Mom! Where were you? A monster attacked me!")

In later nesting seasons the Durdens supplied Lady with more children: goslings, ducklings, some baby great horned owls (the owlets troubled Lady by staying up all night, so she had to stay awake to keep watch over them), some baby red-tailed hawks , and even a couple of eaglets. She was delighted with them all. After spending 16 years with the Durdens, Lady met a wild male eagle and eloped. She was later seen sitting on a nest on a California cliff, so perhaps her long experience of motherhood stood the species in good stead.

FOUR

How to Get Your Point Across: Being Vocal, Being Verbal, and Otherwise Communicating

---

Lucy Temerlin, a chimpanzee raised in the family of psychotherapist Maurice Temerlin, used innate chimpanzee sounds. She made food grunts when stealing yogurt from the refrigerator and said "boo" in a low voice when she saw worrying large animals like horses or cows. She used chimpanzee gestures, like an extended arm with limp hand, to turn away the wrath of a superior.

She was also taught some American Sign Language. One day she committed a crime of vandalism, sneaking into the living room just before a dinner party, pulling the leaves off a potted banana tree Temerlin had put there, tearing apart the trunk, and dumping the soil out of the pot. When Temerlin discovered this, he shouted "Goddamn you!" and raised his hand to belt her. "Instead of extending her pronated wrist, Lucy looked me directly in the eye, smiled her little girl smile, and touched her nose with her thumb, forefinger extended in the ASL sign which means, 'I'm Lucy.' I stopped in mid-gesture! I could not hit her, my eloquent chimpanzee daughter."

MOST ANIMALS ARE BORN knowing what to say. They know how to meow for their mother's attention, growl to warn another kitten away from their food, or hiss when they fear attack. When another kitten growls, or their mother purrs, they understand what that means. No one has to teach babies to cry, and no one has to teach people to want them not to cry.

A few animals can learn part of their vocal communication, but this is fairly rare. So far, the ability to learn to produce vocal communication has been found in four orders of mammals and three orders of birds. The mammalian orders are primates (featuring wonderful us), cetaceans (whales and dolphins), pinnipeds (seals, sea lions, and walruses), and bats (for example, the spear-nosed bats of the Caribbean who apparently learn the catchwords of the in-group in order to exclude the out-group from good feeding spots).

Among birds, vocal learning has been found in the psittaciformes, passerines (or passeriformes), and trochiliformes. Psittaciformes are the parrots, and we all know they can learn novel words and sounds. Passerines are a huge group of birds, sometimes called the songbirds, and they include chickadees, mynahs, and the world's biggest songbird, the raven. Trochiliformes are the hummingbirds.

Vocal learning in bats and in hummingbirds is a recent revelation, so it may be that there are other groups of birds or animals out there who learn what to say.

Why do birds sing so gay?

Why are birds in these three orders able to learn song? What do they get out of it? Maybe it's just a case of sexual selection gone wild. If birds are attracted to, mate with, and have more chicks with the best singers, and if they like novelty, then the ability to learn a little would give a bird an edge. Once the brain allows vocal learning, some species might put it to other uses, such as to establish territory or group identity.

Female great reed warblers prefer a guy who can sing anything in the songbook to those who keep singing the same old tune. Sadly, great reed warblers, both male and female, have a tendency to cheat on their mates. Female reed warblers may cheat with neighbor males if the neighbor sings more songs than their own mate, but if he sings fewer, they don't bother. Since great reed warblers expand their repertoires as they get older, knowing many songs indicates that the bird in question is a survivor, perhaps because he has a fine set of genes. Indeed, more of the fledglings fathered by the great singers survive to breed the following year.

What's this I hear?

Even species that don't learn how to make vocalizations must often learn what certain sounds mean, or at least who's making them. The first such task of the baby animal is learning the sound of its parents' voices. A baby animal hears not only its mother's heartbeat in utero, it hears her voice. Mother llamas hum to their babies, and pregnant llamas begin humming a few weeks before giving birth. Chicks in the shell hear their parents' voices.

A newborn fur seal pup will answer the call of any female, but by the time it is a few days old, it only answers its mother. Researchers studying a fur seal colony on an island in the Indian Ocean found that the mother doesn't leave the island to eat until she and her pup know each other's voice from those of all the other seals in the crowded colony. When a mother returns, she flops inland, calling. The pup calls back, and the two call until they meet. In the months to come, the mother goes to sea and forages for two or three weeks at a time—and comes back to a joyful reunion.

Brother bird

Little blue penguin chicks on Tiritiri Matangi Island, New Zealand, are raised in burrows, often with one sibling. If they were temporarily abducted and put in a fake burrow, their little hearts beat faster when they heard their brother's or sister's voice, but they remained impassive if they heard a strange penguin chick's voice. Their hearts also raced at the voices of the chicks from neighboring burrows. Music did not thrill them, nor did the calls of petrels.

Spectacled parrotlets are small, sociable birds. Although they are enthusiastically monogamous, they spend most days in flocks of up to 25 birds, and roost at night in flocks of up to 150. In the breeding season they nest in cavities, but stay in touch with their group. When all their chicks have fledged, they get together with neighbors who have kids and set up a crèche in a tree where fledglings hang out together. Within the crèches, the young birds are very close to their siblings, and these sibling groups are the main social unit that helps the young birds integrate into the life of the flock.

Suspecting strongly that birds in the parrot family are just as talented as passerines at recognizing voices, researchers did playback tests with parrotlets in aviaries and determined not only that mates recognize each other's voices, but that siblings recognize the voices of siblings, even after they have grown up. (Perhaps it pays to stay close to your siblings, researchers suggested, in case you lose your mate—you can fall back on your siblings.)

How-de-do, neighborinos

Whether you live in the city, in the country, or on a crowded coral reef, it's useful to know your neighbors. Bicolor damselfish lead lives that can easily be compared to those of some primates: they live in small, mixed-sex groups, and hold territories on the reefs. Biologists Arthur Myrberg Jr. and Robert Riggio call them "vociferous" and "highly soniferous." During the spawning season the males each maintain a residence in the colony, containing a nest where females come and lay eggs. The male tends these eggs, fanning fresh water across them and fiercely protecting them. Males on their territory perform diving displays of dips while chirping. This serves to attract females and warn other males that the territory is occupied.

Myrberg and Riggio taped the chirps of bicolor damselfish at one colony off the Florida coast, played them back, and noted the responses. Males were blasé about the chirps of their nearest neighbors, responding with a mere two-dip display. The chirps of more distant neighbors produced four to six dips, as did playbacks of their own chirps. Presumably they did not recognize their own chirps, just as most people don't recognize recordings of their voices when they first hear them (and typically reject the suggestion that the squeaky fool on the tape is their sonorous self).

To explore whether the damselfish weren't just used to the sounds of their nearest neighbors, but actually knew whose voice was whose, the scientists played the chirps of the neighbors from the wrong territories. Thus a damselfish would hear the familiar chirps of Roy Righthand coming from the territory of Leo Lefthand, or vice versa. This got them horribly upset, and they upped their dipping rate, bicolor damselfish Number Two dipping 14 times when he detected this violation of the established order of things.

Wolves also listen to their neighbors, using howling as a way to reunite with pack members and to communicate with and learn about other packs. Wolf researchers in northern Minnesota spent many evenings howling at wild wolves. For science. Many of the wolves were radio-collared, so researchers could find out not only whether they howled back, but whether they moved.

Wolves apparently treat the howls of (skilled) wolf researchers as the howls of wolves unknown to them: loud strangers. If they had a kill, or small pups, they were apt to howl back ("This is our territory—you got a problem with that?"). But if they had no food to protect and no small puppies, they were less likely to howl back. Sometimes they would silently sneak away from the howls. Lone wolves (with no territory) almost never howled back. On several occasions lone females silently headed in the direction of the strange howling, possibly hoping to meet a nice guy. ("SF seeks SM. Should enjoy long walks, big-game hunting. No smokers, humans.")

Birdsong

The basic model of birdsong learning was developed principally with white-crowned sparrows, although it applies to many species. These are bouncy little birds who like to sing from the tops of bushes. The adult males have white crowns, handsomely bordered with black stripes.

In their youth, they go through a sensitive period, when they hear and remember the sound of white-crowned sparrow songs, probably sung by their father, but also by birds on neighboring territories. They don't sing yet. When they are a little older they go through a period of subsong—quiet, inexpert song in which they seem to be rehearsing, as if under their breath, all the song types they heard when they were younger. When the males are older, setting up territories and courting females, they loudly and expertly sing just a few chosen song types, in what is called crystallized song.

Song templates

Most songbirds are not inclined to sing just anything. They have innate templates which give them the general idea of their principal song type. The template hints what a good length for a song might be, and what's a good range of pitches, and how many phrases would sound nice. "Zebra Finches must learn the details of their song-phrase since only a rough version exists without learning," writes Richard Zann. But if they are raised by a Bengalese finch foster father, they will learn his song, which will completely "mask" the zebra finch framework.

A male zebra finch will utter the classic zebra finch Distance Call no matter who raises him, but if the father who raises him isn't a zebra finch who can model the call for him, it will lack a "noise component" on the end. Females raised by Bengalese finches have essentially normal Distance Calls, if perhaps a little high-pitched.

Before you criticize zebra finches for having different male and female calls, consider the case of their near relative, the double-barred finch, whose calls are unisex. Zann writes that "strangers probably require several minutes, or possibly hours of behavioural interaction or 'interrogation' in order to determine each other's sex. Certainly, males . . . have difficulty identifying the sex of strangers in captivity and usually court everyone they encounter until they learn their sex." That's got to be awkward.

It's not that simple

Additional research has revealed all kinds of additions, exceptions, and complications to this basic outline, as is usual with real life. Sensitive periods are more flexible than was originally thought. Particularly if the birds haven't heard anything worth learning, they are able to learn new songs later than usual. Some birds, like canaries and starlings, are "open learners," who change their songs every year.

Scientists have also had to amend their assumptions about the experimental procedures they have used. It turns out that most birds learn far better (or only) if they have a live tutor, as opposed to a recording. Researchers like to use recordings because you can ensure that every bird in an experiment hears the exact same thing. But it turns out that birds are not nearly as impressed by canned music as they are by live performances.

If they can't have a live tutor, it helps if they can control their learning process. Patrice Adret trained isolated zebra finches to peck a key that turned on a recording of a zebra finch song. The young birds played the tape over and over, and fluttered in front of the loudspeaker while the song played. They learned and sang the song, whereas birds that heard the same songs on the same schedule, but didn't get to push the buttons, did not learn the songs.

Subsong and babbling

Wild zebra finches begin practicing—singing subsong—as early as four weeks old, when they are still with their family. They sing so softly that often the only way you can tell, even if you're very near, is by their posture. If a young bird could go into its room and close the door, it would.

Why so quiet? A scornful nineteenth-century auditor wrote of young song sparrows, "I actually laughed aloud at their crude, tuneless, quasi-musical efforts. They were not in good voice and, besides, had not yet fully learned the tunes that are sung in sparrowdom, and could not control their vocal chords. They made many sorry and amusing attempts to chant and trill, but their voices would break and catch in the most remarkable ways, now sliding too high up in the scale, now sliding down too low, and now veering too much to one side, so to speak. One tyro, I observed, sang the first part of a run very well, almost as well, in fact, as an adult musician could have sung it; but when he tried to finish, his voice seemed to fly all to flinders." The writer then proceeded to a swamp, where he heard adult males singing and exclaimed, "What a contrast between the crude songs of the young birds and the loud, clear, splendidly intoned and executed trills of these trained musicians!" So perhaps subsong singers fear ridicule. Or maybe the young birds aren't ready to be on an equal conversational footing with adults.

Subsong may be related to babbling, an activity that used to be considered a special human thing. But baby pygmy marmosets (tiny monkeys) babble incessantly, starting at two weeks, and add new sounds to their babble. Adult marmosets dote more when babies babble. As babies get older they babble less and get better at making the same kinds of sounds as grown marmosets.

It has recently been noticed by linguists that small human children may be better talkers than they show. In at least some cases, toddlers talking to themselves in the privacy of their cribs (but being taped) say longer, more complicated things and use grammatical forms they didn't use when talking with a parent half an hour earlier. There could be many reasons why a child might reserve this sophisticated chitchat for when it is alone, and it is interesting to wonder which might apply to subsong and other solo babble in nonhuman animals.

Only child, only bird, only living being in the world

A common method of studying birdsong has been to raise birds in isolation, often in soundproof chambers, so that their song cannot have been influenced by anything except what the experimenter chooses. Birdsong researchers West, King, and Freeberg note that although it has been known for decades that primates raised in isolation suffered from "clear and often irreversible deficits," no one seems to have worried about birds raised in isolation. "Were birds considered cognitively or socially less complex than primates and therefore less likely to suffer social deficiencies?" The focus on studying only the song of the isolated bird, and not the bird itself, may have caused researchers to miss important factors in the way a bird develops song.

When male white-crowned sparrow chicks grow up in isolation, they produce an "isolate song." This pathetic ditty consists of scratchy primitive syllables. After a certain point it is too late for them to learn better, and they will be stuck with the isolate song forever.

At the California Academy of Sciences, where scientists like the late Luis Baptista study birdsong, the bushes nearby are conveniently thronged with white-crowned sparrows. For several years a male called Weirdo held a territory there. While most of the white-crowns in the area sang the song associated with the *nuttalli* subspecies or the song associated with the *gambeli* subspecies, and a tiny minority sang *pugetensis*, Weirdo sang an isolate song, busily cranking out tasteless squeaks. How Weirdo grew up without hearing any sparrow song is unknown. But despite theories that the complexity, style, and vocal pyrotechnics of male song serve to attract females and discourage rivals, Weirdo had a desirable territory, and a charming wife, with whom he produced adorable nestlings.

The sound of a woman's voice

There are also exceptions to the notion that male birds sing and females keep their beaks shut. Female and male cardinals both sing, but they handle matters differently. Girl and boy cardinals both learn songs on their father's territory, and then the females stop, while the males learn more in the area where they set up their own territory. As a result, the female has the accent of her hometown, whereas the male sings like the other guys on his new block.

In some species where it was thought that only males sang, it's been discovered that females sing too, but not at the nest and not in breeding season, which is precisely when birdsong had been studied.

For years, white-crowned sparrow females were not believed to sing. Except once in a while. Or if you shot them full of testosterone. Biologists who studied a lowland subspecies of white-crowns found that the females sing in autumn, winter, and early spring, stopping when nesting season begins. Females of a mountain subspecies arrived singing in early summer and clammed up except for periods of late snow melt, when nest sites were hard to come by and competition was particularly savage.

Researchers characterized the female songs as shorter, softer, quavery, and lacking some terminal syllables. Sometimes the females had countersinging matches with their mates, and when they did, their songs were like male songs. Female 17, for example, had only simple syllables in her songs, but one day "she dominated and chased her mate, . . . countersang with him and subsequently sang a second theme containing a complex syllable similar to those in his song." Female 18 had a fight with her mate and, instead of her usual simple-syllabled song, sang one that matched his.

Female white-crowns accompanied singing with territorial behavior such as attacking and chasing other birds. Female 8 was widowed and sang frequently and loudly for over two weeks, until she met male BK/S, whereupon she hushed and the two got to work building a nest, suggesting that she had been singing to advertise for a mate.

So in a nest of baby white-crowned sparrows, both the female and male nestlings are listening to their father's song and committing it to memory. Although their mother can sing, the babies presumably have no inkling of this, since she is silent during the breeding season.

Learning the wrong songs

Birds learn the songs of their species, except when they don't. In a mountain meadow in the Sierra Nevada, researchers studying the white-crowned sparrow heard its familiar notes coming from a Lincoln sparrow. Fascinated, they recorded the singer. The bird sang five different themes (unique combinations of syllables), three of which contained syllables from typical white-crowned sparrow song. How did this happen? In the meadow there were 25 pairs of white-crowns and only two of Lincoln sparrows. Maybe when it was young this bird didn't hear many Lincoln sparrows, so in its quest for variety it picked up syllables from the white-crowned sparrow songs it heard on every side.

New neurons

When songbirds learn new songs on new territories, or to compete with new neighbors, they may be aided by new neurons they have grown for the purpose. It was once believed that creatures never acquired new neurons after birth, but Fernando Nottebohm and colleagues, working with canaries, have shown that males grow new neurons each spring in the part of the brain they use for learning song. Over the summer that area shrinks, but it expands again in the fall when the birds learn new songs. Moving on to chickadees, Nottebohm found that they grow new hippocampal neurons in the fall, when they need to remember where they stored the thousands of seeds they'll eat in the winter.

Dialects

If you don't recognize your neighbors as individuals, you may still recognize them as locals if your species has dialects. If it does, and you move after you grow up, you may need to learn the local expressions. Luis Baptista found a white-crown in San Francisco singing an Alaskan dialect which he presumably learned from an Alaskan bird migrating through. A month later, he had a wife, and sang San Francisco songs like his neighbors. Baptista played him recordings of Alaska songs and he acted as if he had never heard such a thing. Finally, the seventieth time Baptista played the Alaska song, the bird replied in kind.

Observing orange-winged amazon parrots in Trinidad, Fernando and Marta Nottebohm visited an area 30 miles from their usual study site and saw parrots uttering "high-pitched whistling calls . . . like nothing we had heard before." Another species? No, another dialect.

Chimpanzees do not need to learn their calls, yet they can also have dialects. Researchers compared one of the classic chimpanzee calls, the pant-hoot,* as uttered by captive male chimpanzees at U.S.A. Lion Country Safari in Florida and at the North Carolina Zoological Park. A proper pant-hoot has four sections: an introduction with "at least one long, unmodulated, low-pitched element"; the buildup, with fast, short, inhaled and exhaled sounds (going up in pitch is an option here); the climax, with at least one long scream; and the letdown, which is much like the buildup.

The Florida and North Carolina chimps showed definite differences. While the overall sequence length was the same, they emphasized different parts of the call. Moreover, the chimpanzees at Lion Country Safari often added tasteful Bronx cheers to their pant-hoots. This was due to the artistic influence of NL, a 30-year-old male who had been at Lion Country for 20 years. While chimpanzees in both colonies already made Bronx cheers when they were frustrated, only at Lion Country did they use them to adorn pant-hoots. When NL arrived as a 10-year-old, he had already mastered the cheer, and apparently it was his idea to use it in pant-hoots.

Killer whales have distinct dialects, but what's really unusual is that different pods in the same waters can have different dialects. Sometimes pods gather in larger groups with other killer whales in the population, and having your own pod-specific calls may help when it's time to break up and go home with your own folks.

Why don't you lose that accent?

What use are dialects? For one thing, in courting birds, they may signal "I'm from around here and I know the ropes." The local boy may know the terrain better, and he may be genetically adapted for local conditions, as witness the fact that he has survived.

Crows, noted for cawing, also sing. Theirs is a long and motley

---

*Jane Goodall often employs pant-hoots to add a multilingual aspect to her lectures.

song, adorned with clicks, gurgles, rattles, coos, and other sounds that please them.* Scientist Eleanor Brown, who was clearly captivated by crows, writes that their song is "not often heard in the wild due to the wariness of this persecuted species." So diverse are the elements of crow song that some ornithologists have compared it to subsong, but Brown considers it to be crystallized song of extraordinary variety. Brown suggests that crows imitate each other to come up with a group song, strengthening their bonds.

Brown raised crows in outdoor aviaries and carefully analyzed their remarks. Under "harmonic rattle," for example, she included the "rolling coo," the "rattle ow'wa," and the "short wow hoo." Under "caw" we find the "ark," the "wok," the "kek," the "haa," and more. Although the vast array of "coos" is charming, my favorite vocalization is the "harsh ahah."

Sometimes the crows duetted: RU and G would perch only a few inches apart, "uttering the same sounds and performing the same movements in almost perfect synchrony."

When the crows were a year and a half old, Brown introduced the crow J, who was the same age, by putting her in an aviary adjacent to RU and G. J changed her song elements to match those sung by RU and G. Her "coos" were much higher: she lowered them. Her "hoo" contained longer notes: she shortened them. She pronounced "hwa" as if it were "hoo-wa": she compressed it into "hwa." This was not an instant process, but at the end of two and a half months J sounded like RU and G. The more she sounded like them, the more they liked her. "Colloquially speaking, song may be the 'social grease' . . . which keeps daily life running smoothly."

Music appreciation

Clarence, a house sparrow raised from infancy by Clare Kipps, did not grow up with other sparrows. Instead, he listened as Kipps played the piano, perching on her shoulder and pinching her neck at the exciting parts. He particularly liked anything resembling a trill, and treble scales played fast.

---

*Ravens' song is also varied. It has been said that if you are in the wilderness and hear a completely inexplicable noise, you should assume it is a raven.

Clarence was born during the London blitz. Refugees living with Kipps reported that he was singing while she was out. Since she knew house sparrows do not sing, Kipps thought they had heard birds outside and mistaken them for Clarence. But while she was running water one day, she heard Clarence in the other room. "It began with twitterings; then there was a little turn, an attempt at melodic outline, a high note (far above the vocal register of a sparrow) and then—wonder of wonders!—a little trill."

Once he had perfected his art, he was delighted to sing for Kipps and for visitors. He had composed two songs, each in the key of F major. Kipps suspected that Chopin's *Berceuse* was the inspiration for his trills. The first song "began with the usual sparrow-chirpings, though less harsh in tone than those that sometimes weary us with their monotony in the early morning, and descended by a perfect fourth from tonic to dominant," Kipps wrote. "This interval was followed by a perfect fifth descending from G to C; and these two were repeated and ornamented with mordents or (sometimes) with four-note trills. Then followed a rapid triplet leading back to the tonic and repeated indefinitely."

The second, rapturous song included two eight-note trills until Clarence was five, when he elected to replace the second trill with "a harsh croaking noise that sounded as if he were clearing his throat, but of which he seemed inordinately proud."

Clarence's songs were curiously unprecedented. Looking in three bird guides, I read that the song of the house sparrow is "a long series of monotonous chirps," a mixture of "various twitters and chirps, nervous and sometimes garbled; certainly anything but musical," and "a monotonous series of nearly identical chirps." House sparrow buff D. Summers-Smith wrote that at times house sparrows string chirps together to form "a rudimentary song." Summers-Smith also wrote that a few of the captive sparrows he raised also developed something of a song, "a sustained, rambling warbling . . . built up of the adult chirrup notes but much more musical than normally associated with the house sparrow," which they sang briefly as teenagers. He also heard such songs occasionally from wild birds, which suggests that most house sparrows perform below potential. Of Clarence, Kipps wrote, "His artistry was not impeccable, but sparrows are not a musical family, and his performance at its best was an achievement for one of his species."

Hummingbirds

Hummingbird songs are high and thin, and people seldom notice them. That's okay, they weren't talking to you.

The late Luis Baptista and Karl-Ludwig Schuchmann confirmed earlier suspicions that vocal learning in birds is not a monopoly of the powerful passerine-psittacine cartel, but is also found in the hummingbirds. In particular, Anna's hummingbirds living on Guadalupe Island in Mexico sing very differently from mainland California birds.

Baptista and Schuchmann raised some baby hummingbirds where they couldn't hear adult song. A male Anna's hummingbird who was raised alone sang a monotonous song devoid of trills and vibratos, but occasionally interrupted by unorthodox bursts of buzzy chatter.

Since hummingbirds are raised by single mothers, males can't learn from their fathers. It is thought that they learn obliquely, listening to unrelated males. The song of the hummingbird raised alone was similar to the song Anna's hummingbirds sing on Guadalupe Island, suggesting that the island was colonized by hummingbirds among whom the male or males hadn't learned songs yet. They founded a dynasty that never got to hear good music and passed along their pitiful tunes, which Baptista called "baby talk."

Kathryn Rusch and Millicent Ficken followed blue-throated hummingbirds in southeastern Arizona to see if they are learners too. Sure enough, blue-throats in the Chiricahua Mountains sing parts of their delicate "whisper song" differently from blue-throats in the Huachucas, suggesting a learned component. They discovered that female blue-throats sing in the breeding season. After she has single-handedly built an adorable nest, using spider's silk, a female blue-throat goes looking for a male. She finds him perched and singing, and, if she likes him, perches next to him and sings her own whisper song. They spend a day or several days together, whispering, and then separate. She raises the twins alone.

Cetaceans

Some cetaceans have ears specialized for very high frequencies. Dolphins who live in often murky coastal waters and ascend rivers utter almost nothing but ultrasonic clicks. Hector's dolphins, for instance, are chubby little spotted dolphins who live in New Zealand's coastal waters, including harbors and river mouths. Since they are social creatures who love to gather in groups, and click like mad when they do so, it's guessed that they are using ultrasonics not just to find their way about and find food but also to communicate.

Bats

Evening bat mothers can tell their pups' calls from the calls of the other pups in the colony, and they learn this very quickly. Within a day, the pups are left in nursery crèches of up to 30 pups, while their mothers go out to forage. Mexican free-tailed bats pick their pups out by voice, in crèches that contain millions of pups. It was once assumed that no mother could locate her child in such a mob and that mothers just landed in the crèche and suckled a random pup, but it's not true. The mother flies to the general area where she last saw the kid, which only helps a little, since the pups are virtually shoulder to shoulder, up to 5,000 of them in a square meter, each cuter than the one before. She searches the area, calling. All the pups scream, "I'm hungry!" and by voice, backed up by smell, the mother picks out her own darling.

Leaf-nosed bats live in small groups with up to five females and one male. The males don't take care of the pups, but if the pups are removed from a flight cage and experimenters play tapes of a pup crying, males will fly to the mother of the pup and poke and screech until she goes over to the loudspeaker. (Mothers seem to be better than the males at telling the difference between a tape and the real thing.) The males always knew which mother was associated with which pup.

What do I get for my dues?

The female greater spear-nosed bat gives a loud screech call at foraging sites and when she is leaving on a foraging trip in the tropical rain forest. In the evening, hundreds of bats pour out of a cave, and so a distinctive call can help the group stay together when everyone around them is flapping and screeching. Female bats forage in groups of 8 to 40 regulars (each group includes one male bat), and the screeches seem to help them gather and stay in touch when they are winging through the forest. Janette Wenrick Boughman investigated whether they did this by recognizing the voices of individuals, or whether they had a group version of the screech call. It seems that group members all adopt the same screech call, to the point that "individuals are statistically indistinguishable"(as if they fly around screaming "Crips! Crips!" Or "Alpha Chi! Alpha Chi!"). Boughman transferred captive bats between groups, and found that the transferred bats change their calls to match their group.

Greater spear-nosed bats do not lead a simple life of jinking about grabbing insects on the wing, like some bats. Instead they eat a huge array of plant and animal foods scattered through the forest. Many foods occur in clusters, such as a flowering tree or a hatch of winged termites. It's useful for one group to be able to exclude everyone who's not in their group, which is easy to tell because they don't know the password. "They seem to chase them off," says Boughman. "There's lots of calling and lots of really fast flying with one hot on the tail of the other."

A young bat is fed by its mother for a few months. Then, when it's weaned, it hangs out with other teenagers in the cave and learns to find food by following older bats. After a while, young females try to join a group. Group members aren't related, and Boughman has seen sisters join different groups. It takes between one and five months for a new girl to master the screech call of her group.

Not only are greater spear-nosed bat screech calls audible to the human ear, but it's gratifying to know that it is, barely, possible to tell the group calls apart, "if your ear's well trained and you listen to a million of them."

Mimicry

Clearly a parrot who says "How are you?" is not replicating an innate template. It has long puzzled ornithologists that many birds which mimic impressively in captivity aren't known to mimic in the wild. Why should they have this remarkable ability which only manifests itself in this rare situation? Perhaps they do mimic in the wild, but not for our benefit. Listening to a four-minute tape of two wild African grey parrots vocalizing in a tree on the bank of a river in Zaire, Claude Chappuis noticed some motifs that resembled mimicry. Further analysis confirmed that the parrots (they couldn't tell if it was one or both) seemed to be mimicking the red-tailed ant thrush, Lühder's bush-shrike, the brown-throated wattle-eye, the western black-headed oriole, the bristle-nosed barbet, and a species of epauletted fruit bat.

Wild parrots are not usually so cooperative with people taping their chitchat. The authors of a paper analyzing this recording write that the fact that there were two parrots vocalizing for a long time without other birds present suggests "a song or duet of socio-sexual significance." Perhaps mimicry occurs in uncommon situations such as pair formation.

Ravens

In *Mind of the Raven*, Bernd Heinrich cites several examples of mimicry in captive ravens (one that imitates static on a portable radio and one that revs its motorcycle constantly) but also passes on an interesting incident of mimicry by a wild raven.

Biologist David Barash, studying marmots in Olympic National Park, kept hearing a countdown followed by an explosion: "Three, two one, *bccccchhh*." After hearing it several times in a row, Barash called out "Who's there?" but got no reply. Eventually he discovered that the sound was coming from a raven seated on a nearby snag. The previous week, park rangers had set off explosions for avalanche control, and the raven had heard and added this to his repertoire. Later in the season, ravens who were in the habit of perching on top of self-flushing urinals in a picnic area added the musical sound of flushing to their rhapsody.

Starlings

Starlings are intimidatingly vociferous birds who include mimicry among their accomplishments. This is obvious in captive birds like Arnold, raised by Margarete Sigl Corbo, who progressed from saying "Arnold" in every possible tone on every possible occasion to imitating Corbo whistling "Mary Had a Little Lamb" and Beethoven's Fifth to quoting small boys' "See you soon, baboon." He used this last phrase appropriately when people left the room.

Lyrebirds

The superb lyrebird of Australia is a multitalented creature who mimics brilliantly and in the case of the male, also sings, builds a mound to display on (a display which some call dancing), and grows ornate tail feathers. Young males initially learn from adult males, and later incorporate mimicry of other species.

In the 1940s, some lyrebirds were introduced from Australia into Tasmania. Soon these lyrebirds imitated a Tasmanian bird, the green rosella. Forty years later, the population still imitated whipbirds and pilotbirds, which are not found in Tasmania—their calls had been handed down from lyrebird to lyrebird.

In 1969, park ranger Neville Fenton recorded a lyrebird in the New England National Park in New South Wales, with a very odd, flutelike song in his repertoire. Asking around, Fenton was eventually told that in the 1930s a flute player living on a nearby farm had a pet lyrebird to whom he played. The bird learned tunes from him and added them to his song. Eventually the bird was released. Fenton sent the recording to a sound expert, Norman Robinson. Knowing that lyrebirds can sing two tunes at once, Robinson filtered the recording appropriately and discovered that the local lyrebirds were singing a combination of two tunes from the 1930s, "Mosquito Dance" and "The Keel Row"—modified, but recognizable.

Bird. James Bird.

The famous lyrebird James, who lived to be at least 20, was renowned for replying "Hullo, Boy" to Mrs. Wilkinson, a stately widow who greeted him every morning with "Hullo, Boy" when she saw him foraging in her garden. James was glorified in the 1940 book *The Lore of the Lyrebird* by Ambrose Pratt, which not only chronicled his feats and included a photograph of James and Mrs. Wilkinson gazing soulfully at each other, but exalted the lyrebird as "possessed of an extraordinary mentality," "one of the most sedate and respectable of all the wild creatures known to man," and with "an elementary conception of social virtue." To arrive at this conclusion, Pratt posited that the superb lyrebird is monogamous (it isn't), that father and mother care for the young for five years (the mother tends it for one year), and that if they lose their mates, the females withdraw to "some deep recess of the jungle and remain hidden until they die of loneliness or grief" (they don't). When lone males were seen, Pratt and his informants identified them as "widowers unwilling to remarry" (although he also says that "their demeanour presented . . . many aspects delicately suggestive of abnormality"—these are indeed complex birds). Pratt correctly noted that lyrebirds are territorial, although he thought that pairs owned property together, a further virtue.

James's utterance of "Hullo, Boy" and other mimicry were witnessed by many people. The plausibility of "Hullo, Boy" is also supported by the later case of a male lyrebird, this one resident in the Sir Colin Mackenzie Zoological Park, who took up replying "Hello, Chook" to the keeper who said this to him every morning. He also imitated his cagemate, a bush thick-knee.

Given that Pratt was wrong about lyrebirds raising their children together, it's interesting that decades later a male lyrebird at the Zoological Park rose to the occasion when the female he was housed with died suddenly. The chick was three and a half months old, still in need of parental care. The chick called for its mother, lost its appetite, and "developed a hunched appearance." The male let the chick roost next to him at night. He then alarmed the zookeepers by pursuing the chick with the same calls the mother had made. He began regurgitating food for the chick just as mother lyrebirds do, and after some hesitation the chick allowed itself to be fed and flourished under its father's care. The zoo documented this unheard-of behavior, which almost leads one to exclaim with Pratt that the lyrebird is extraordinary, complex, and socially virtuous. Although Pratt might have been even more impressed had father and child died of grief.

Female lyrebirds also mimic. In a footnote to Pratt's book, written by "Mr. A. G. Campbell, onetime President of the Royal Australian Ornithologists' Union," we read that "In a nest under observation in Sherbrook Forest during August, 1936, owned by a well-known pair who are used to the stare of strangers, a lusty youngster was reared. The mother, after a visit with food, would occasionally pause above the nest and sing a few delightful strains of mimicry. Thus the young Lyrebird got his first few lessons from his mother." According to current doctrine, a female lyrebird mimics when her nest is threatened, so perhaps she wasn't as used to the stare of strangers as all that.

Territorial mimicry

African robin-chats are impressive mimics. Some territorial robin-chats match their neighbors' mimicry, so if one robin-chat imitates a fish eagle, the one next door will imitate a fish eagle right back. I'll see your fish eagle and I'll raise you a francolin. Oh? Well, I'll see your francolin and raise you an oriole. Since one robin-chat was recorded imitating 36 other bird species, this could conceivably go on for quite a while.

The Beau Geste hypothesis suggests that when birds sing many types of songs they may give other birds looking for territory the impression that the place is crowded, so they should move on. Whether mimicry of other species could have such an effect is unknown.

Other uses of mimicry

A pet African black crow who ranged free found a use for mimicry. The crow had its eye on a flock of chicks, but the chicks correctly perceived the crow as a menace and went into their chicken house whenever they saw it. When the owner had scraps for the chicks, she'd call "kip-kip-kip-kip-kip," bringing them dashing to scrabble over the bits. When the owner was away, the crow learned to sneak up, call "kip-kip-kip-kip-kip," and grab the first chick that came running.

Derek Goodwin has suggested that in some birds which mimic in captivity, but which don't seem to mimic in the wild, there may be mimicry that goes unnoticed. If birds in a pair are separated, "whichever of the two is left in its home area will soon begin to utter calls or song phrases that are normally used only or mostly by its mate. This has the effect of 'calling the mate home' if it is in a position to return." But it sounds like normal bird calls unless you know the individuals.

Goodwin writes of tame European jays that associate the calls they mimic with the creatures who make them. Goodwin's tame jays barked at dogs who came into his garden and meowed at cats. He writes that some jays mobbing tawny owls imitate the hoots of the owl when they do, and cites a case in which a group of jays arrived at the usual daytime roost of an owl. Finding the owl not at home, they hooted and left for another roost. Finally, he writes that tame male jays, when they get angry with their keeper, "invariably defy him with display phrases consisting partly or wholly of words and whistles that he has often used in their hearing, usually those with which he customarily greets them."

A related use of mimicry was noticed by Terry Oatley in robin-chats. Fledglings have cryptic, camouflaged plumage, and when they leave the nest young birds perch quietly in heavy cover, occasionally uttering short, high, location calls, stirring only when their parents bring food. Robin-chats have a ratchet sound, which the adults use as an alarm call. When the little birds hear it, they fall silent and freeze. The adult robin-chats intersperse the ratchet call, which the young respond to instinctively, with imitations of the alarm calls of other birds in the area—no other calls are imitated, only alarm calls. If a mongoose, snake, or human being approaches, the parents go into a frenzy of ratcheting and mimicked alarm calls. By this means, the young birds may learn that other birds' alarm calls, and not just the ratchet, are heralds of danger.

Oatley also observed a subtle use of mimicry in chorister robin-chats. He was spying on a pair nesting near his garden, and saw them reunited after a separation of 20 minutes. The female, returning from a foraging trip, perched in the foliage and whistled softly. The male, standing sentry near the nest, peered around and cried "too-wheee," which is the call of the southern boubou shrike, not a robin-chat remark at all. The female said what a boubou shrike says to its mate in response to too-wheee: "boo-boo-boo!" They continued to exchange boubou shrike calls. Had Oatley not been focused on these birds, he would have assumed he was hearing shrikes and would have had no idea that these two used shrike calls as contact calls.

Mammalian mimicry

"Harbor seals have long been regarded as unusually quiet pinnipeds," begins an article in the *Canadian Journal of Zoology*. But adult males are not only chatty, they will mimic human speech. Two males at Boston's New England Aquarium made sounds resembling English words and phrases. One confined himself to saying hello; and the other, Hoover, had so many remarks that he was written up in *The New Yorker*.

Hoover was found as a newborn on a beach in Maine. His mother had been shot by a fisherman. Local residents Alice and George Swallow raised him for three months, naming him Hoover for the way he vacuumed up his mashed mackerel. They kept him first in the bathtub and then in a pond in the backyard, allowed him in the house, and let local kids take him for wheelbarrow rides. When Hoover made noises, George Swallow copied him and apparently Hoover copied Swallow right back. When he came home from work, Swallow would slap the side of his car and shout "Hey, stupid!" Hoover would waddle up and Swallow would say "Hello, there" to him. One day Hoover replied, approximately, "Hello, dere," but Swallow didn't think much of it. Swallow taught the seal to reply to "What's your name?" with "Hoover."

When Hoover's costly mackerel habit overwhelmed the Swallow budget, they took him to the aquarium. "I told a fellow there, 'I think he can talk,' but he gave me such a look I never mentioned it again," Swallow told *The New Yorker*. Eight or nine years later, the Swallows went to the aquarium. George Swallow shouted, "Hey, stupid!" and Hoover swam over, took Swallow's hand in his mouth as he used to do, and tried to pull him into the water.

One day when Hoover was seven, a keeper wrote in the files, "He says 'Hoover' in plain English. I have witnesses." The matter was investigated and it was decided that Hoover did indeed say his name and also said "Hello," "Hello, there," "Hello, there, how are you," "Come over here," "Get out of there," and "Hey." He also laughed. Researchers writing in the *Canadian Journal of Zoology* noted that these are things that aquarium visitors often said to the seals, and *The New Yorker* noted that these were things often said by George Swallow. All parties agreed that Hoover had a Boston accent.

It's easy to imagine that Hoover, who talked more in breeding season, learned what important grown-ups (like George Swallow) should sound like when he was an impressionable pup and repeated these things when he became an important grown-up himself.

Belugas

Beluga whales in captivity have also been known to get chatty. Logosi, a beluga who arrived at the Vancouver Aquarium as a tiny calf, said "Logosi," which is what people said to him. Logosi was flatteringly interested in conversing with people, swimming up to the windows of the beluga tank, repeatedly pressing his face against the window and making sounds, then turning and placing his ear against the glass as if to hear a response. "Logosi" was not the only thing he said, but it was the only thing anyone could understand. Logosi also did imitations of human conversation. Devoid of words, it sounded like people talking far away, or children playing.

Duets

Many bird species that form lasting pairs sing duets that may announce and cement their relationship. Wolfgang Wickler has proposed that when partners learn their duets, each is proving a commitment (or an investment in, to use the more calculating phraseology) to the other that would not be shown if duet parts didn't have to be learned. To switch to another mate, you'd have to start from scratch to work up a nice duet.

To really learn your part is also to learn your partner's part. A male white-browed robin-chat sings a four-note motif over and over, rising to a crescendo, at which point the female joins in antiphonally with loud "tsreeee"s. If she has a beakful of caterpillars, or for some other reason doesn't join in, the male sings both parts.

Of course I love you— have I ever missed a rehearsal?

Pairs of siamang gibbons sing loud and elaborate duets, which announce their territorial claims and advertise the excellence of their relationship, since their performance is so well coordinated. Observers at the Louisiana Purchase Gardens and Zoo recorded how a new couple learned to duet.

A proper siamang duet, or great call, is said to begin with two deep booms from the male. The female replies by booming once; the male booms twice more, and the female must immediately come in with accelerating high-frequency barks. After about the fifth bark, the male should utter an ascending boom, the female's barks should speed up, and the male should do a bitonal scream. At once the female must start another series of barks, and after five the male should do a scream, this time a ululating scream. The female does some fast high barks, then both of them bark and hurtle about. While hurtling, the male should do a locomotion call as well as barking.

Simple enough, you say. Well, it takes practice. The male of this pair had been caught in the wild as a mere child and had lived at the zoo for 17 years. He had had a mate, but she had died, and he had been alone for 12 years. The female was six and a half, and had lived with her parents until her arrival in Louisiana. They were put in adjacent cages and began working on their duet right away. (After a while the doors were opened so they could go back and forth.)

If one of the siamangs made a mistake in the great call sequence they'd often drop the whole matter. Over about three months the percentage of started duets that were finished rose from 24 percent to 79 percent, and in the last five bouts no mistakes were made. One of the most common mistakes was for the female to start her first set of accelerating high barks without waiting for the male's second double boom. The other common mistake was for the male to bungle his bitonal scream, giving either the ululating scream or the locomotion call instead.

Testing, testing

Guacamole, a stray dwarf macaw rescued by Martha Coyote, imitates other animals in the household and interesting phrases people say, but also finds it intriguing when people imitate him. If you mimic him, "about the third or fourth time he starts looking at you with interest." He then runs you through a few tests: can you say hello? can you bark like the poodle? can you cough? can you squawk twice? can you squawk three times? can you squawk twice low and once high? can you whistle? "He's trying to see if you can keep up," Coyote reports.

Similarly, Karen Pryor describes a wild humpback whale recorded by Bill Schevill in the Bahamas, in which the whale is calling into an underwater canyon. "The whale went 'Mroomp!' and the echo went 'Mroomp!' The whale tried it again, one note higher, until it had gone up the octave as far as it could reach. The echo answered. Having tried out higher and higher 'mroomps,' the whale essayed a few other roars and gargles, waiting for the echo each time."

Codes and signals

Many animals have calls that tell others that there's something they should know about—an opportunity or a peril. Ravens give food calls to recruit other ravens to food sources in dangerous places, because there's safety in numbers. Scientists in Austria studied the yells made by wild ravens at a boar feeding station in a game park where the ravens routinely stole food from the wolves, bears, and boars.

The rate of "haa" calls went up as soon as they could see the food, before they could get it, confirming previous suggestions that these long yells mean that the ravens have detected food they can't get. The more desirable the food, the higher the rate of "haa" yells. Ravens gave short "who" yells at their usual rate until the food was actually available in the boars' pen, when the rate went way up, particularly as they landed and advanced on the food. The rate of "who" yells went up more when there was only one bucket of food, suggesting that these yells may have something to do with claiming food.

Cats

A house cat bringing prey to her kittens has two different calls, so distinct that even humans can tell the difference, according to ethologist Paul Leyhausen. One means "I am bringing you a harmless little snack," and is used with a mouse. The other means "I am bringing you something that should be treated with caution," and is used for a rat. The mouse call brings kittens out fearlessly, and the rat call brings them out in a hesitant crouch. The rat call is actually an intensified version of the mouse call, and if the mother gets all worked up, perhaps because Leyhausen has closed the door between her and her kittens, the mother will race back and forth with the mouse in her jaws crying "Mouse! *Mouse!* MOUSE!" until she is crying "Rat!"

A cat's purr, Leyhausen speculates, originated as a signal from a suckling kitten to its mother, meaning that all is well. It has the advantage of being something a kitten can say with its mouth full. Now it is used by mothers approaching their kittens and suckling their kittens, by kittens approaching adults to play, by dominant cats reassuring other cats they approach, and in some situations by cats that are threatened or severely weakened, as by disease. This last may explain occasional mysterious purring at the veterinarian's office. "At first sight this may seem paradoxical, considering that in its original meaning the purr signals well-being. But probably in all the cases mentioned the message is basically one of appeasement and in human terms would mean: 'I am only small, helpless, inoffensive, and innocuous!'"

Alarm calls

The alarm calls given by wild vervet monkeys to warn of predators have been intensively and cleverly studied, using playback tape recorders concealed in bushes. Robert Seyfarth and Dorothy Cheney have described five different calls that vervets use to denote the dangers of large terrestrial carnivores such as leopards, eagles, snakes, baboons, and unfamiliar humans. The leopard, eagle, and snake calls have been carefully studied. Each calls for a different reaction. If you hear a leopard call, you should dash up a tree. If you hear an eagle call, you should scan the sky as you take cover in a bush. If you hear a snake call, you should stand and scan the ground for the snake. If you are tiny, clinging to your mother works for all dangers.

These calls are vital news. Only about 30 percent of baby vervets make it past the age of one year, and most of the missing were eaten by predators. Little vervets make perfectly enunciated alarm calls when they are only one month old. But they are shaky on when to give them. Adults only give eagle calls for eagles and hawks, but juveniles give them for almost any big bird—spoonbills, take cover!—and infants may call for almost anything airborne—Mommy, a dove! A leaf!

Infants who are three to seven months old may be playing away from their mother when they hear an alarm call (often a tape played back by Cheney and Seyfarth). Younger vervets usually run to their mothers, but as they get older they are more apt to take adult-type evasive actions. Initially they make mistakes, running into a bush instead of up a tree when they hear a leopard call, a bad idea, since leopards often hide in bushes. By six months, they almost always get it right.

The alarm calls of other species

Baby vervets ignore starling song, but at one or two months old they respond to starling alarm calls by looking toward the sound. By the time they are four or five months old they respond to starling terrestrial alarm calls by running up a tree. Eventually they distinguish between starling alarm calls denoting ground predators and those warning of eagles, and behave appropriately.

Zoologist Maurice Burton's family raised a family of orphaned baby rabbits to adulthood. Burton's neighbors also had a rabbit, but theirs had been caught when it was already half-grown. The alarm call of a blackbird sent the neighbors' rabbit scuttling for a dark corner, but Burton's own rabbits showed no concern, suggesting that the rabbit caught at a greater age had learned that the blackbird's alarm call was cause to worry. Burton's rabbits, perhaps because they had never noticed Burton and his family quivering with horror at a blackbird's alarm call, had not learned to understand this important signal.

A small tasty animal like a rabbit is well served by learning the alarm calls of as many other species as possible. But the rabbit has no need to produce many alarm calls. It needs to be able to thump its feet in warning to other rabbits and, apparently, it also needs to be able to scream in desperation when actually in the grip of a predator. Just why rabbits have the ability to utter a last-ditch scream has been debated— surely it's useless to cry for help if the only individuals likely to care about your well-being are also rabbits. On the other hand, what has the rabbit got to lose? Another predator might show up to see what's happening and the rabbit might get away in an ensuing confrontation. Predators are so interested in rabbit screams that hunters regularly use "rabbit calls" to attract them. So a rabbit may have a greater need to understand vocal communication than to vocalize.

I signed it with an X

A social animal might wish to identify itself to others of its group. That dolphins have "signature whistles," calls that are unique to individuals, was first suggested in 1965 by M. C. and D. K. Caldwell, but the discovery "languished" during a period when some frenzied dolphin admirers held out for the idea that dolphins might have a language fully as sophisticated as—perhaps more sophisticated than—our own, which would shortly be revealed to us. Not that anyone's bitter about Flipper and John Lilly and "floating hobbits" and the notion of ambassadors who carry a message to humanity from the divine. Not really bitter.

There is still controversy about signature whistles, with Brenda McCowan and Diana Reiss, for example, arguing that dolphins merely have individual vocal variations of shared contact calls, perhaps with regional dialectical differences. "These differing results from different methods of categorization can be resolved only by testing how dolphins themselves perceive whistles," writes Peter Tyack, whose research supports the signature whistle hypothesis. Tests with captive and wild dolphins show that they clearly recognize each other's whistles—are they recognizing a signature or merely a voice?

What good would a signature whistle do? Why would a dolphin benefit by saying "Hi, it's Bob," when he could just say "Hi" and have other dolphins know it's Bob because they recognize Bob's voice? Unlike land animals, dolphins vocalize at different water depths, which subject them to different pressures, which change their voices. So Bob the dolphin saying "Hi" when he's at the surface may sound different from Bob saying "Hi" 20 feet below the surface. But if his whistle structure says "Hi, it's Bob," then depth differences won't matter.

The observations of Janik and Slater, who recorded vocal interactions between wild dolphins in Moray Firth, Scotland, support the idea that wild dolphins—or some wild dolphins—use signature whistles. The dolphins frequently responded to a whistle with "whistle matching"—uttering a whistle of the same kind. The kind of whistles that were matched resembled the signature whistles of captive dolphins.

Baby dolphins aren't born with apparent signature whistles. Of the dolphins of Monkey Mia on the western Australian coast, Rachel Smolker writes, "They make gurgly, irregular, messy-sounding whistles at first. Gradually they refine their sound into a clean, unique whistle by the time they are about six months to a year old."

Researchers looking for signature whistles in some male dolphins at Monkey Mia in the mid-1900s weren't finding them. Babies had them, males didn't seem to. (If it were to turn out that dolphins in some areas have signature whistles and in other areas don't, that would be a fascinating cultural difference.) But they noticed that three males who swam together, Snubnose, Bibi, and Sicklefin, were uttering more of a whistle called the upcurl. "If the most common whistle a dolphin produced was its signature, then Snubnose, Bibi, and Sicklefin had all converged on the upcurl as their shared signature whistle." Records showed that these three males had formed an alliance after the period of first recording. They started doing more and more upcurls, and the upcurls sounded more alike. Perhaps they just liked it, or perhaps they were presenting a united front.

Signature thumps

There is no indication that banner-tailed kangaroo rats are capable of vocal learning, yet they too seem to have signatures. Kangaroo rats defend valuable territories consisting of mounds of dirt which the rats use as silos for seeds. These contain tunnels and galleries excavated by generations of kangaroo rats. A kangaroo rat announces its claim to one of these mansions by drumming its feet, making a sound that travels through the air and the ground. (There's also foot drumming to warn of snakes, but that's a different pattern.)

A kangaroo rat assembles foot drums into bursts called foot rolls, and assembles foot drums and foot rolls into a foot-drumming signature. Each individual's signature is consistent, and may stay the same over years, but it can also change. The reason to change your signature is to make sure that it can't be confused with those of your neighbors. Analysis of the foot-drumming signatures of kangaroo rats near Portal, Arizona, shows that signatures don't overlap with those of a given kangaroo rat's neighbors but may overlap with those of kangaroo rats who live farther away.

Kangaroo rats change territories from time to time, for reasons that are mysterious to us. When you move, you get different neighbors, and if you're a kangaroo rat, you may need to change your signature. During six years of study the kangaroo rat that changed her signature most often was a restless individual who moved four times in a year.

Primates

Compared to birds, all the primates except us seem to be vocal duds. They use innate calls and modify them surprisingly little through learning. Six infant squirrel monkeys were raised in isolated mother-child pairs. Four of the pairs were vocally normal. In the fifth pair, the infant was congenitally deaf. In the sixth pair, the infant was normal, but the mother, due to surgery on her vocal cords, could not make squirrel monkey calls. She could only groan. Researchers examined the calls of all the young monkeys as they grew. Each one made all the calls that squirrel monkeys make, and each did them perfectly.

Adult male orangutans avoid each other. Biruté Galdikas says that if they meet, there are only two possible results: one runs away, or they fight. But it rarely comes to that, because adult males utter ringing "long calls," and they use these to stay apart. These long calls may need to have no particular content, since to hear one is to understand, "There's a great big ape over there and he doesn't care who knows it."

Gorillas are famous for beating their chests, and some think that chest beating has a communicatory function. You can hear it a mile away, and one chest-beating gorilla is sometimes answered by another. Whether they are saying anything more complex than "Hey, you're not the only one with a chest" is not known.

Drumming

Primates make nonvocal sound signals at times. Male chimpanzees often drum briefly on the buttresses of forest trees, sometimes uttering pant-hoots as they do. The drumming can be heard at least a kilometer away. This clearly conveys information to other chimpanzees, but it's unclear how sophisticated the information is. In a 1991 paper, Christophe Boesch analyzed the drumming and subsequent actions of chimpanzees in the Taï Forest of Ivory Coast in the 1980s. This was a community of about 80 chimpanzees, which broke into smaller groups that foraged separately. Often they moved through the forest silently, but at other times the males pant-hooted and drummed loudly on trees with hands and feet.

Christophe Boesch and Hedwige Boesch began to suspect that the drumming conveyed not only information about where the drumming chimp was, but also what he planned to do next, "because we tended to lose contact with them just after drumming was heard. It seemed that the whole chimpanzee community abruptly and often silently changed direction following an outburst of drumming."

Boesch learned to differentiate the pant-hoots of the different males in the study area, and concluded that the chimpanzees only changed their direction of travel after the highest-ranking male, Brutus, drummed.

Boesch's analysis identified three kinds of messages sent by Brutus's drumming. By drumming first on one tree and then, within two minutes, on another tree a little distance away, Brutus identified a direction of travel. Chimpanzees hearing it would know it was Brutus and they would know what direction his group was going.

If Brutus drummed in two separate bouts on the same tree, Boesch suggests, he indicated that he was planning to rest in that place. When this happened, other chimps were also apt to stop and rest, usually for about an hour.

Lastly, Boesch argued, when Brutus drummed once on one tree and then twice in succession on another tree, or twice on one tree and then once on a second, he "proposed both a change of direction and an hour's rest." This was recorded six times. Once Brutus drummed four times on the same tree and everybody rested for two hours, making Boesch wonder if there was a connection between the many drummings and the long rest, but this was observed only once.

Excited by these findings, and now able to identify most of the adult males by voice, Boesch was preparing to gather more data, when, in February 1984, four out of ten adult males vanished, apparently killed by poachers. Within three months, Brutus "stopped sending information to the other community members."

For Brutus's messages to be useful, the other chimpanzees needed to understand them. The fact that on one occasion a group of younger chimpanzees ignored one of Brutus's messages that seemed to indicate an hour's rest, kept moving, and ended up out of hearing range of the rest of the community, a rare event at that time of year, suggests the possibility that these younger animals either hadn't learned to understand such messages or didn't realize the importance of staying in touch.

Chimpanzees need to stay in touch, Boesch writes, so they can join in defense against predators and neighboring chimpanzee groups. In the Taï Forest leopards frequently attacked chimpanzees. In areas such as the Gombe preserve, where chimpanzees have not been recorded communicating this way, there is less predation and better visibility.

Of drumming bouts, anthropologist Adam Clark Arcadi argues that "there is no way to eliminate the possibility that listeners simply deduced the drummer's current or future position, joined him, and then matched his behavior," as opposed to the drummer encoding information that was then decoded by the listening apes.

Branch dragging

Primatologist Ellen Ingmanson describes branch dragging, a behavior of wild bonobos in Zaire, which seems to signal two things: the message "Let's go" and information about the direction the bonobo proposes to take. In a resting group, one of the males will rise, break off a branch, and drag it noisily into the forest for 20 to 30 meters. He repeats this several times, and if all goes well, the others get to their feet and the group proceeds in the direction in which he was dragging the branch. Sometimes another male has a different idea and starts branch dragging in another direction. How the group decides is unclear. The directions were clear enough that Ingmanson could usually predict where the bonobos would go. Occasionally they branch-dragged when a party was in transit, perhaps to lobby for a change of direction or to goad slowpokes.

One October morning, Ingmanson started ape watching early, while the bonobos were still in their nests. Mon came down from his nest and sat, looking between a tree with another bonobo nest and a fruiting tree. Then he snapped a sapling and started branch dragging between the two trees. After ten minutes, Ika looked over the edge of the other nest. "Mon stopped branch-dragging, gave a few excited squeaks and bounces, and looked up at Ika." But Ika didn't feel like getting up, and pulled back his head. Mon resumed branch-dragging until Ika put his head out again and then slowly came down. Mon stopped branch-dragging, and the minute Ika hit the ground Mon ran to the fruit tree, slowly followed by Ika.* "Mon had succeeded in getting a friend to join him for breakfast."

Clever, clever, clever Hans

From time to time, people have tired of trying to understand animals and tried instead to make the animals do the work to communicate with us. The results have been mixed. When professors of animal behavior want to scare their students sober, they tell them the story of Clever Hans. With slides. In the early 1900s Wilhelm von Osten, who thought animals must be smarter than people said, tried to teach sim-

---

*Apparently Ika was not a morning bonobo.

ple math to a cat, who did not cooperate. He tried to teach a bear, whose attitude was very bad. He tried to teach a horse, Hans, who was patient and willing. Von Osten, with abacus and blackboard, would pose math questions to Hans, who was supposed to tap his hoof four times if the answer was 4 and so forth. Hans, a brown horse with a white star on his forehead, got pretty good at this, and at shaking his head, and pointing with his head, and at picking up one of several colored cloths. He answered questions regarding square roots correctly and did well when von Osten branched out into questions about the German language.

Von Osten announced the feats of Kluge Hans—Clever Hans—to the world and people flocked to see. Everyone was extremely impressed with the questions Hans could answer, which could be asked by anyone.

Then psychologist Oskar Pfungst performed experiments showing that Hans could only answer questions correctly if he could see someone who knew the answer, ideally von Osten. Hans determined when to stop tapping his hoof, not by noting that 4 is the square root of 16, but by noticing that on his fourth tap von Osten, who had been leaning forward slightly, looking at Hans's hoof, now straightened slightly, anticipating the end of the tapping. Hans also paid attention when von Osten raised his eyebrows or flared his nostrils minutely.

Von Osten had sincerely believed that in Hans he had created an equine mathematician and was devastated to find that he had merely created a brilliant observer of human physiology. The prevailing reaction of animal behaviorists to Hans's story is horror at the thought of being fooled that way.

Horses are good at responding to small cues, which is one reason a 60-pound child can steer a 1,000-pound horse at top speed around a course of barrels. From Hans's point of view, he simply did what he was asked.

Repeat after me: "Puh."

Hans was at least spared the demand that he speak German, but apes have been subjected to training programs intended to get them speaking English. Keith and Cathy Hayes hoped that if they raised an infant chimpanzee as they would raise an infant human, she would learn to talk just as children do. Although she was a smart and loving little ape, she learned only four words, Mama, Papa, cup, and up, and only after intensive training for six years. The chimpanzee vocal tract is not well shaped for uttering words, but Viki's comprehension was not much, either. Cathy Hayes estimated that Viki understood 50 words or phrases at three years old.

Similar heroic attempts in the 1970s to teach English to Cody, a young orangutan, were similarly frustrating. With intensive drilling, graduate student Keith Laidler taught Cody to say "kuh" for a reward. Cody enjoyed this accomplishment and would say "kuh" to himself while playing with his toys or looking out the window. Since Cody had been taught to say "kuh" for a variety of rewards, clearly "kuh" meant "gimme." Then Laidler started teaching Cody to say "puh."

Aha! Now "puh" meant gimme! Cody learned to say "puh" and refused to say "kuh." At this point Laidler decided that the sounds needed to have distinct meanings and defined "kuh" to mean "beverage, please" and "puh" to mean "food, please." With more drilling, Cody regained the use of "kuh" and could use both words appropriately. Then Laider decided that "puh" should mean "pick me up" and a new sound, "fuh," should mean "food." Okay. Cody learned these definitions in three days and used his three words appropriately, but stopped vocalizing spontaneously.

Laidler laboriously taught Cody to say "thuh," which obviously means "brush me." Cody adored being brushed.

None of Cody's words were necessary. He was perfectly capable of letting those around him know that he wanted food, drink, to be picked up, or to be brushed, using looks and gestures, but Laidler persisted in setting these maddening tasks before he would give Cody what it was obvious he wanted.

Cody tried his words on inanimate objects. When he was one, he was playing in his crib with a test table that rested across the top of the crib, banging it about and hanging from it. Somehow Cody got it wedged at the back of his crib and couldn't get it loose. He tugged and tugged without success. Finally he shouted "kuh!" Nothing happened. He shook it angrily and shouted "puh!" This was apparently what the table wanted to hear—or else the shaking dislodged it—and the table came away in Cody's hands, and he settled down to chew on it happily.

Laidler suspects that this incident reinforced Cody's hope that objects would obey his magic words. Once Cody wanted to leave his playroom and go watch TV, but he couldn't reach the handle of the door. He sat facing the door and commanded it to puh. Nothing. "There followed a puzzled silence, during which Cody's eyebrows jigged about on his forehead like two demented caterpillars. Another commanding 'puh' and another long silence. I could almost hear the wheels and cogs grinding their slow tortuous course as he tried to work out just what was amiss. His face brightened momentarily as the answer broke upon his tiny brain. Then his visage darkened and like some sorcerer conjuring unknown powers to his own fell purpose, Cody rose unsteadily to his feet, raised his fisted hand and shaking it high with tense frustration, called upon the door to open with a loud 'thuh.'"

Laidler then compounded the misunderstanding by lifting Cody up to reach the door handle.

Ask Alex what he thinks

When psychologist Irene Pepperberg set out to teach an animal to speak English, she chose a species that had the physical ability to do so, the African grey parrot. Alex is a now-famous parrot who's been taught to use words with comprehension by Pepperberg. He's been the subject of television programs, magazine stories, and journal articles exploring his linguistic prowess, not to mention Pepperberg's book, *The Alex Papers*. His working vocabulary includes numbers, colors, shapes, materials such as paper or wood, and objects of interest to parrots, such as seeds, nuts, and things it's pleasant to destroy. He has also learned words and phrases that weren't explicitly taught, such as "Come back," "Go away," "No," and "I love you."

Pepperberg's reason for teaching Alex words is not to teach him language, but to understand Alex's cognitive abilities. It is possible to question Alex and discover that he understands the class concepts of color and shape, for example.

There is no reason to think that Alex, a pet-store baby, is a particularly talented bird. The difference between him and other members of his species is the way he was trained. African greys are renowned talkers, but they seldom use words and phrases with comprehension. Often they associate a phrase with a situation, saying "Good morning" when someone comes in or muttering "Bad bird, bad bird" when yelled at, but this is limited comprehension. Geier, a talented parrot, said "Auf Wiedersehen" when people left, but despite hours of tutelage, couldn't learn to ask for water when he was thirsty.

Pepperberg's method of training Alex, the model-rival method, is based on the work of Dietmar Todt. The animal sees a model and a rival performing the task to be taught. Rewards are related to the task. This is distinct from more usual methods of teaching, in which the reward is whatever the animal likes, and is not related to the task. The usual method for teaching a parrot the word "green" might be to hold up something green, say "green, green" and give him a peanut if he says something like "green." But this doesn't work. At best you teach the bird that "green" means "give me a peanut."

Under the model-rival technique, there are two people in addition to the animal being trained. One person holds up something green, and asks the other to say "green," or to say what color it is. If they say something close to "green," they get the green object.

Meanwhile the animal is watching this procedure, perhaps thinking, "I want that green key, I bet I could do that, I can say 'green' better than that." The task that the animal is supposed to perform has been modeled, and perhaps the animal's rivalry has been aroused. This method of training works well with many animals, including people.

One of Pepperberg's twists on the Todt method is that models take turns being teacher and pupil, showing that communication is a two-way street and not just a series of tasks that one individual forces another to perform. Since parrots are rebellious types, this probably helps.

It took nine months to teach Alex to use the words "same" and "different" with comprehension. But now if you want to understand the things that Alex considers similar—can he tell the difference between two colors, two birds, two people?—you only need ask. It is now possible to ask Alex "What same?" and have him answer "color" to tell you that a piece of wood and a key are both blue. Assuming that he is in the mood for your silly questions.

Pepperberg does not consider Alex's tasks easy and natural for a parrot. She calls them "exceptional learning" and says that for that reason they require intensive training. To confirm that the model-rival method is what has worked with Alex (rather than her having happened to purchase a genius), Pepperberg tried training two young parrots with other methods. But Kyaaro and Alo did not learn words from audiotapes. They did not learn from videotapes, even videotapes featuring Alex. Even when people sat next to them and commented on the videotapes: nothing. So don't get one of those recordings to play to your parakeet unless you enjoy hearing it yourself, because the bird won't learn from it. Also, teaching yourself a foreign language from tapes? A real long shot.*

Alex finds "What color?" a boring question. He never asks for things by color, although he sometimes asks for things by shape. The first time he ever said "What color?" was one day when he was looking at his mirror image. He asked six times, and the student working with him gave him such answers as "That's gray; you're a Grey parrot." That was enough. The researchers hastily dyed some objects gray and quizzed Alex to check: yes, he said they were gray.

Alex babbles secretly to himself. "He does a lot of things in private that he doesn't do in public. You might get a lot of strange things in the evening." In these surreptitious monologues he practices things that he will not say publicly until he gets them right.

Since Alex speaks and understands English words, you or I could discuss seeds, nuts, or the color of objects with him if he were in the mood. But we couldn't chat about much else. "Alex's input is mostly limited to what I deem necessary for experiments, which primarily require object or category label," writes Pepperberg. He learns words faster when they describe things he actually wants and when he hears them in a social setting. Given how social parrots are, one wonders if he would learn more if he somehow had the chance to discuss what is really interesting to a parrot—social matters. Gossip. His environment is socially impoverished, but one can imagine that more socially fortunate parrots might wish to discuss such matters as who is cute, who flies funny, who is overrated, who has a crush on whom, and who bit whom and why.

Sign language

Meanwhile, Allen and Beatrix Gardner had had the brilliant notion of teaching chimpanzees words in a human language that didn't tax their vocal tracts: American Sign Language. They began in 1966 with

*But a nice idea. I honor you for the thought.

Washoe, a 10-month-old chimpanzee, raising her first in their home, and then in a trailer, and using only sign language in her company. Washoe acquired dozens of signs, which she principally used to request things. She combined signs in strings like "you me hide" to suggest a game of hide and seek. She made some word combinations like "open eat drink" to refer to the refrigerator. When Washoe moved to another site, the Gardners started a second project with several younger chimpanzees, and this time several native ASL speakers were among their caretakers. Eventually, in the care of Roger and Deborah Fouts, Washoe and the other signing chimps came to live together at Central Washington University. By this time Washoe knew at least 240 signs. When she adopted the 10-month-old Loulis, the researchers decided not to sign around the chimpanzees so that any signs Loulis learned would have to come from his mother or the other chimpanzees. He began learning at once, and Washoe was seen instructing him as she had been instructed. By the time humans resumed signing around the chimpanzees, Loulis had learned 55 signs.

The signing chimpanzees of Central Washington sign to each other when there are no people around, using it as a small part of their communicative repertoire. One example of this integration of sign with other modes of communication is given in a journal article: Moja was about to get some fruit juice when Tatu approached. Moja communicated her resentment of the intrusion by the classic chimpanzee technique of screaming, whereupon Tatu signed, "Smile." Moja turned away and ignored Tatu, a further and eloquent form of communication.

When the chimpanzees sign to each other, it is usually for "reassurance, social interaction, and play." They urge other chimps to hurry, and they ask to be chased. This contrasts with the pedagogical conversations humans insist on having, all "What this?" and "What color that?" When Moja was in estrus, she did a lot of signing, mostly urging a certain male chimpanzee to "hug." (At other times chimps requested hugs in a more chaste spirit.)

Inspired by the Gardners' work with Washoe, psychologist Penny Patterson wanted to see if a gorilla could learn to sign. Raised by Patterson after a year at the San Francisco Zoo, Koko is now said to have the largest vocabulary of any signing ape. Her first signed word was "food," quickly followed by "drink," "more," "out," "dog," "come-gimme," "up," "toothbrush," and "that." She understands a great deal of spoken English, is amused by rhymes, and will sign words that rhyme if spoken aloud. Like certain other signing apes, and most people, Koko dislikes being drilled on vocabulary and becomes recalcitrant if it goes on for any length of time.

Chantek

Chantek is an orangutan raised until the age of nine years old by H. Lyn Miles, who considers him her cross-fostered son. He was born in 1977 at the Yerkes Regional Primate Research Center and was loaned to Miles, an anthropology professor at the University of Tennessee, when he was nine months old. She and her colleagues taught him a modified form of sign language. Miles didn't try to teach him long lists of signs to use on quizzes, preferring to focus on signs that interested him.

The goal of Project Chantek was not to show that Chantek could or could not acquire language, but to examine his cognitive and communicative development. Miles charted Chantek's progress through the stages outlined by Piaget for human children. Chantek "was immersed in a human cultural environment and learned the rules for behavior and interaction, a process anthropologists call *enculturation*," writes Miles. They took walks, they fed the squirrels, they played with blocks, they went out for fast food, they painted, they made shadow puppets on the wall at bedtime, they talked to the nice policeman on the horse.

At first he was taught signs by molding his hands. "Food-eat" was the first thing he signed. Later he learned by imitation. When he only knew 20 to 35 words, Chantek invented the idea of signing with his feet instead of his hands.

Over seven years, it took Chantek an average of 53 days to master signs for food, 69 days for beverages, 117 days for actions (give, listen, kiss), and 565 days for pronouns. It took just 19 days to add "give" to his vocabulary, but it took 915 to add "shoe." Chantek didn't wear shoes, but he was happy to accept gifts, especially gifts of food. He learned the word "up" early and quickly, initially as a request to be picked up. But it took him over a year to learn the word "down," because he never wanted to be put down.

He combined signs into phrases, showing word order preferences. At the age of four he called contact lens solution "eye drink." "Cheese meat bread" of course means a cheeseburger. Charmingly, "noisy dog" or "noisy sky dog" means a helicopter. He invented a self-explanatory sign that means a View-master, and another sign for balloon. Chantek was fond of deception: of claiming an urgent need to use the bathroom when he merely wished to play with the taps; of claiming to have eaten his vegetables when he had actually concealed them behind the toilet; of popping an eraser in his mouth, signing "food-eat," and displaying his open and apparently empty mouth, when in fact he had wickedly hidden it in his cheek for later use.

"When Chantek saw his first video of other orangutans he called them orange dogs and he wanted no part of them," says Miles. (This is not so different from the attitude of wild orangutans toward strangers, it must be pointed out.) "He calls himself an 'orangutan person' and he now lives with other 'orangutans' but originally they were orange dogs."

When he was nine, Chantek was reclaimed by the Yerkes Center. At first Miles was able to continue some work with him, but then a decision was made to treat him as an ape, not as a child, and Miles was not allowed to see him for seven years. When they finally met again, Chantek signed, "Mommy give ice cream."

The signs Chantek remembered after his long silence were objects of human culture, Miles reports. Perhaps he could handle other concepts with normal orangutan communication, or perhaps he tried to express his apely wishes through sign and wore them out with failure.

Biruté Galdikas witnessed Gary Shapiro's endeavors to teach sign language to free-ranging orangutans being rehabilitated at Camp Leakey, her Borneo headquarters, for life in the wild. He taught 30 signs to his best pupil, Princess, although there were others she "refused to learn." These efforts were also witnessed by Galdikas's toddler child Binti Brindamour. "Naturally and effortlessly, Binti between age two and three learned more signs than Princess at age five or six learned through rigorously structured, formal lessons." While no one doubts that the child has more language aptitude than the orangutan, it is possible that this difference was accentuated by their different learning situations. In short, Princess may have been annoyed by her rigorous lesson plan, while Binti was, inadvertently, getting the benefits of a model-rival training regime.

Nim Chimpsky

Then there was the historic Nim Chimpsky, famous as a failure and a phony at signing. In the first exciting heyday of teaching chimpanzees to sign, some grandiose claims were made for the apes' abilities and the ones that they were undoubtedly about to show. In 1973 researcher Herbert Terrace got a young male chimpanzee and named him (sort of) after linguist Noam Chomsky. Nim was raised first in a family home, and then in a semi–laboratory setting. He was tutored in sign language by 60 students, most of whom had little familiarity with signing. These sessions were taped. Preparatory to writing a paper on Nim's achievements, Terrace reviewed the tapes and was shocked at what he saw. He declared that Nim repeated himself wildly and meaninglessly in hope of getting food rewards, showed no sign of syntax, rarely offered signs spontaneously, interrupted constantly, was often just imitating whatever sign his tutor had just made, and showed no signs of turn-taking, as happens in conversations. Furthermore, he charged that all signing apes did the same.

Terrace's reversal on whether apes could learn to sign meaningfully had a devastating effect on ape sign research, and quite a few apes were shipped back to the primate centers whence they had come, including Nim, not quite four years old, who was sent back to the Institute for Primate Studies, where he lived in a cage or on a monkey island with other chimpanzees.

When Nim was six (still preadolescent), some researchers at the institute tried to test Nim as he had been tested in Terrace's study, administering a drill on naming objects. "The result was a series of threats from Nim, culminating in a bite." Subsequently they videotaped some sessions with Nim to see what effect social setting had on his signs.

In conversational sessions, Nim was taken for a stroll to a grassy area, played with, and chatted with in a casual and indulgent manner. True, they did sometimes point to things and ask Nim what they were, but they tried not to be overbearing about it. Nor did they give Nim food rewards, though if he asked for someone's hat or shoes he got them. (Unlike Chantek, Nim was interested in shoes.) There was a box of toys. Nim enjoyed this, didn't bite, and displayed much more normal use of signs. During the training sample his interruption rate was 31 percent, but in conversation it was only 9 percent. He made four times as many sponta-neous contributions (using signs that others hadn't used), going from 14 percent to 60 percent.

In the same grassy setting, they tried to repeat a training-type ses-sion, complete with food rewards, but Nim was uncooperative. "The training session was terminated after eight minutes because of increas-ing irritation and volatility on the part of both participants."

In short, the better you treated Nim, the better-spoken he was. One of the complaints about Nim was that he just fired off signs ran-domly in hopes of being rewarded, but since he was taught on a reward system, this seems like a reasonable strategy. (Nim later went to an ani-mal sanctuary, where they vowed he would never be quizzed again.)

Artificial symbol systems

In the meantime, some researchers worked with apes and artificial symbol systems—magnet boards, computer screens, or computerized keyboards. Some used lexigrams, intricate geometric shapes that don't resemble the objects they symbolize, displayed on a keyboard.

Two chimpanzees, Sherman and Austin, who began training with lexigrams at the ages of two and a half and one and a half, readily used them to ask for things. When they were later asked to name those same things, they couldn't do it. They could press the lexigram for "apple" to request an apple, but if they were asked, using that lexigram, to get an apple, they didn't understand. All the symbols had to be retrained.

With time, the chimpanzees began to use their lexigrams in a new way. "We had taught Sherman and Austin key elements of communi-cation—request, naming, and comprehension. Once these were in place, other aspects of communication emerged spontaneously. The chimps began to pay close attention to each other's communications; they engaged each other before delivering their message; they gestured to emphasize or clarify messages; they took turns."

Kanzi

The same research team wanted to see how bonobos (once called pygmy chimpanzees) would do with lexigrams. An adult female bonobo, Matata was tutored for two years by Sue Savage-Rumbaugh in the meaning of 12 lexigrams. Matata was not an apt student. After two years her progress compared unfavorably to that of chimpanzees. She had learned lexigrams for banana, juice, raisin, apple, pecan, and orange, but could not generalize from the symbol for "banana" to a picture of a banana. She was patient with Savage-Rumbaugh's insistence on teaching her, just as she was patient with her exuberant and disruptive child, Kanzi, who grabbed her food, somersaulted over her head while she was trying to press a lexigram key, and slapped the keyboard to get her attention.

At this discouraging time (funding was running out), when Kanzi was nearly two and a half, Matata was sent away for a few months for breeding. Kanzi was left behind. The first day she was gone, when Kanzi was not searching for his mother, he astounded the humans by using the keyboard with unsuspected proficiency. He used all 12 lexigrams in 120 separate utterances. He combined lexigrams. He used lexigrams not just to ask for food but to comment on it. He used them to describe what he planned to do next. "Kanzi had been keeping a secret." Everything they had tried to teach Matata, Kanzi knew. He also knew things they had not even tried to teach her.* (Kanzi may be another inadvertent example of the efficacy of the model-rival method.)

"I was in a state of disbelief," Savage-Rumbaugh writes. She had to rearrange her thinking. She went back through two years of notes, trying to figure out when Kanzi had learned what no one had taught him. Kanzi was not calm enough to sit still to take a blind test and he didn't like to answer questions. That would have to come later. "From the moment the enormity of what Kanzi had done became clear to me, I knew that I would not be believed," Savage-Rumbaugh writes. She must have had the specter of Clever Hans peering over her shoulder.

Wisely, they decided not to give Kanzi the planned tutorials, but instead to interact with him through lexigrams and their own spoken English, in activities that would interest him. This turned into a sort of wilderness camp, with researchers spending their days walking and playing with Kanzi in a 50-acre forest adjacent to the laboratory. Soon Kanzi knew 256 lexigrams—in fact, he knew more, but the board would only hold 256 at a time.

---

*The other shocking discovery of the day was that he was perfectly capable of using the potty when necessary. Parents of toddlers may find this more impressive than mastery of symbols.

Kanzi uses simple rules of syntax, using word order to specify who should chase whom, for example. Conversing with lexigrams is limiting, so researchers looked at how well he understood the syntax of what people said to him. They gave him wacky requests, so he couldn't guess what was wanted. "Would you put the grapes in the swimming pool?" they asked, and he got out of the pool, found the grapes, and tossed them merrily into the pool. Still warding off Clever Hans, Savage-Rumbaugh sat behind a one-way mirror and asked Kanzi to "put the chicken in the potty." Everyone else present wore headphones so they couldn't hear Savage-Rumbaugh and inadvertently cue Kanzi, who was busily putting the chicken in the potty. His actions showed that he understood the difference between "Pour the Coke in the lemonade" and "Pour the lemonade in the Coke." "Go vacuum Liz," "Go get the noodles that are in the bedroom," "Knife the doggie," suggested Savage-Rumbaugh, and Kanzi vacuumed Liz, fetched the noodles, and knifed what I am sure was a toy doggie.

Kanzi is good at phrases like "Get the ball that's in the cereal," which might be expected to be hard, because they contain an embedded phrase ("that's in the cereal"), but terrible at even short lists like "Show me the milk and the doggie." He is apt to produce either the milk or the doggie, but not both. While Kanzi's comprehension of English and English syntax is imperfect, it casts doubt on the proposition that language is an all-or-nothing affair, possessed in its entirety by humans and not at all by any other species.

Like a young bird singing subsong or Alex babbling in the evenings, Kanzi talks to himself. He takes the keyboard and moves away from the group. "If I try to look over his shoulder to see what he is saying, he generally picks up the keyboard and moves further away."

Ai and Ayumu

Ai (pronounced eye) is a chimpanzee at the Primate Research Institute of Kyoto University in Inuyama, Japan. Tetsuro Matsuzawa and colleagues have spent two decades teaching Ai to communicate through symbols. She touches letters and numbers on a computer terminal to receive tokens she can put in a vending machine to buy snacks. She understands about 30 words in spoken Japanese. She lives with other chimpanzees, and in 2000 she gave birth to a son, Ayumu. Naturally she carried him everywhere she went, including to classes.

In February 2002, Ayumu, not quite 10 months old, suddenly decided to operate the computer himself. While Ai was off using the vending machine, Ayumu touched the start symbol on the monitor and, when the computer screen displayed the Japanese kanji symbol for brown, touched the kanji. When the computer displayed a brown square and a pink square, he touched the brown square. He had to stretch, because the brown square was above the pink square and he was still a very small ape. He had to stand on the tray under the monitor, and on his third attempt he reached the brown square, winning a 100-yen coin. This was the first time he had ever touched the computer screen.

No humans were present when Ayumu made his move—a video camera recorded the events. It's too soon to know how much Ayumu will pick up from what Ai is being taught. Like the children of immigrants who effortlessly acquire skills their parents had to toil long and hard to acquire in imperfect form, Kanzi and Ayumu make things look easy.

"The first two years of an ape's life are something of a magical time," write Sue Savage-Rumbaugh and Roger Lewin. "During this period, if exposed to brightly colored geometric symbols, apes learn to tell them apart as easily as if they were looking at different kinds of food. If exposed to human speech, they become responsive to the phonemes and the morphemes so that spoken language no longer sounds like a string of noises. If they watch television, they come to see the patterns on the screen as representations of other people and other apes in different places, rather than just flickering images."

Cat got your tongue?

Many scientists have expressed puzzlement at the fact that captive apes have the ability to learn and use elements of human and artificial languages—and yet wild apes simply don't seem to do much of the sort. Why would they have a capacity they don't use? Primatologists often seem to regard the vocal achievements of birds with restless envy and incomprehension. Birds! With tiny bird brains! How can they learn to do all these things that my apes can't do?

Christophe Boesch has suggested that chimpanzee language abilities "are present in the wild but have yet to be discovered."

Charles Snowdon and Martine Hausberger write of scientists fretting that there is "little evidence of vocal plasticity in nonhuman primates, suggesting a gap in continuity of developmental processes in the evolution from birds to humans." But when vocal learning is understood to include not just learning to make sounds, but also learning to understand sounds and learning when to use sounds, they suggest, "there is no 'gap' between birds and humans." In other words, ape and dolphin vocal learning, even though it doesn't involve much making of sounds, is sophisticated enough to form part of a spectrum between sparrow song and ridiculously complex human language.

Nonvocal communication

Perhaps we should be looking at the areas where we ourselves are not the stars—at nonvocal communication. Many animals, from dogs to captive apes, are disturbingly good at reading the emotional states of humans, for example.

In the wild, understanding the alarm calls and actions of other species increases animals' ability to gather information. The authors Anup Shah and Manoj Shah describe how, if a grazing chital deer looks up and sees a sambar deer peering into the foliage, it will stop grazing and try to spot the possible danger. Primatologist Thelma Rowell describes a patas monkey standing in tall grass peering at a possible threat: the other patas stand and, following the first one's gaze, peer too. Vocal communication—the alarm chirp—comes if there is a greater sense of threat. (If a predator is seen, the adult male of the troop will bounce up and down while making a loud racket as he retreats, while the women and children sneak silently away. This is very noble of him, and if he doesn't do a good job, the females kick him out and get a new male.)

Where do you want to meet for lunch?

In Ethiopia, hamadryas baboons sleep on cliff faces where they are safe from predators. In the morning they descend and sit around before breaking up into smaller groups. Around midday the groups meet at one or another place where there is water, and then separate again to meet at the sleeping cliffs at night. Researchers tried to figure out how the baboons managed to meet in the middle of the day. Careful observation revealed that in the morning the male baboons who lead their groups make a series of very short forays in the direction of watering places. Then they pause to see in what directions other males make forays. Having seen this, they may change the direction of their next symbolic foray or they may continue to insist on the direction they first indicated. Eventually they reach consensus and break up into their separate groups, going in separate directions for the time being, but knowing where to meet at noon. Old males, who are not the highest ranking, are the most influential in swaying other baboons to go in the direction they select. Ethologist Hans Kummer writes, "The hamadryas cannot point out this direction to one another, so at the beginning of their march they must 'walk it' for a while, even though immediately afterward they may branch off toward closer destinations in other directions."

Pointing

Pointing is a notable form of nonverbal communication that seems obvious to adults. A very small child or an inexperienced dog, seeing a person point, will look at the finger and not at where the person is pointing. Most dogs readily learn to look in the direction of the pointed finger. Small children not only learn what pointing means, but begin pointing to things themselves to make their wishes known. In fact, a small child often personally invents pointing to orient her- or himself toward an object. When the child wants to draw someone else's attention to an object, he or she will typically perform "gaze alternation," looking back and forth between the object and the person who ought to be looking at the object.

It has been suggested that this kind of pointing, "referential pointing," is the exclusive skill of the most winsome species ever to inhabit this planet: us. Primatologists leaped in, with evidence that captive apes and monkeys perform referential pointing, complete with gaze alternation. But maybe that's just from hanging around with us?

The orangutan Chantek pointed early and often, indicating things he wanted to be given, directions in which he wanted to be carried, spots where he wished to be tickled, and places he wished others to go. Kanzi, the bonobo, started pointing to indicate where he wanted to be carried when he was less than a year old, but Sue Savage-Rumbaugh notes that most primate infants can't do this, because they need their hands to cling to their mothers. Later Kanzi used pointing eloquently, pointing to a caretaker's pocket that held keys, to the lock on the door, and then in the direction of the room where he wanted to go once the door was unlocked.

In the Ikela region of Zaire, a wild bonobo spotted danger lurking concealed in the surrounding foliage, which the other members of his group had not noticed. He screamed and pointed to the two spots where menace lay hidden, using his right arm and folding back all but two fingers. He alternated his gaze between the peril and the other bonobos. The others approached him, looked where he was pointing, and beheld the horror: two ghastly groups of primatologists, feverishly making notes on his spectacular ability to communicate by pointing.

Beyond pointing, there are other gestures, more arcane in their meaning, that bonobos and chimpanzees have devised. Austin and Sherman, lexigram-using chimpanzees, used a special hand clap to indicate that they wanted to play a chasing game, and the younger Kanzi learned this from watching. He watched wistfully, because the big boys were having too much fun to play with him. He was one and a half, and it took him two days of practice to master the gesture.

An entire set of gestures used by bonobos has gotten little attention. These primates have recently become renowned for their active, cheerful, promiscuous sex lives. In the 1970s, when these apes were little known, a group of three wild-caught bonobos was observed at the Yerkes Primate Center. It was noticed that they frequently gestured, and that these gestures related to one of their big interests: sex. Analysis of the gestures that the three used before sex showed that they had an impressive array of mutually understood signals. The 21 gestures defined almost all had to do with positioning. A series of photographs in the scholarly publication reporting this shows a male bonobo making eye contact with a female and negotiating with her about positioning. She indicates interest but does not accept his initial suggestions about position. He makes an alternate suggestion and she agrees.

The researchers describe the way some of these gestures become abstracted. In a first-order gesture, to take a fairly clean example, a bonobo might push another bonobo under the chin to persuade the partner to stand bipedally. In a second-order version of the gesture the bonobo would touch the chin and then lift her hand upward. In a third-order version she would just raise her arm and flip her hand upward at the wrist.

Fingers? Who needs fingers?

Bob and Toby are two bottlenose dolphins at Walt Disney World who were being trained to interact with scuba divers via an underwater keyboard. After six months, Bob and Toby became more assertive, using the keyboard themselves (rather than simply responding to human use of the keyboard) and then looking at the humans "as if to monitor the human's response."

They also invented pointing, to the surprise of the humans. To point at a food container, the dolphin would stop swimming to hang in the water, and "align the anterior-posterior axis of his body with the object for several seconds." In other words, he'd hold his body straight and point at the object with his whole body. He'd then look back and forth between the container and the trainer. If he was lucky, the trainer might then swim up, take a herring out of the container, and hand it over. On a few occasions Toby appeared to be pointing for the benefit of the person running the underwater video camera, judging from the direction of his gaze alternation. ("Be sure you get this! Here's me, pointing. Are you getting this?")

The dolphins didn't bother to point if there were no humans around to respond. The researchers speculate that dolphins don't point among themselves because they can monitor each other's sonar attention. Pointing was a hint to sonar-blind humans.

I swear that dog is reading your mind

Dogs are good at understanding human messages, both deliberate and inadvertent. In many tests they do better than nonhuman primates at figuring out which of two boxes has food hidden in it, when a human knows the answer. They notice the orientation of the person's head, or the direction of the person's gaze, and easily grasp signals such as tapping on one box. As already noted, they catch on quickly to pointing. Trained gun dogs do better than pet dogs, and untrained gun dogs do better than untrained dogs of other breeds. Wolves do terribly.

Translation

Leo, an orphaned lion raised by a game warden's family in South Africa, loved to go for walks and loved riding in the truck. He could not discern between the words "walk" and "drive," so whenever someone suggested either, Leo would light up with excitement and gaze at his hero, the dog Wolfie. Wolfie knew the difference and would either jump in the truck or head for the gate. Whatever Wolfie did, Leo would do.

Whenever the family wanted to instruct Leo they'd tell Wolfie. If Wolfie sat, Leo sat. Wolfie, a conscientious Australian cattle dog, not only understood what people wanted him to do, he understood that Leo did *not* understand, and that it was his job to make Leo understand. Kobie Krüger, Leo's foster mother, writes, "If Leo wasn't paying attention, for instance, Wolfie would have to repeat the required action several times until Leo got the message." What a good dog.

We humans still win dramatically when it comes to vocal and verbal learning. We read and write, for heaven's sake. We sing songs with lyrics. But other animals turn out to be doing a lot more vocal learning than we knew, particularly when we examine what they understand as well as what they say. As the secret learning of Ai and of Kanzi, and the isolated learning of Alex among all grey parrots, show, we're still floundering beginners at teaching communication to animals. We're also making discoveries (hummingbird vocal learning, bat vocal learning, dolphin signatures, kangaroo rat signatures, and more) at such a rate that there must be even more exciting news in store.

FIVE

How to Make a Living

Captive-born tigers want to hunt and they have ideas on how to go about it, even if they lack practice. It is certainly unwise for mice or birds to wander through a tiger's cage. "They'll catch anything that tries to come in," says tiger expert Ron Tilson.

At the Minnesota Zoo a small flock of wild turkeys wanders the grounds. They find the zoo a charming habitat and view the humans who infest it with only mild concern. There's food for the taking in many of the enclosures, and the crowds of people keep most turkey predators away. Tilson, observing the tiger exhibit one morning, saw the turkeys strolling there just as a couple of two-year-old tigers were released from their nighttime quarters.

When one of them spotted the turkeys he knew just what to do, Tilson relates with pride. He crouched to the ground and crawled on his belly toward the clueless turkeys. The tip of his tail twitched. He placed his feet just as a wild tiger does—each hind foot set down exactly where a front foot was a moment before. When he was close, he rushed the startled turkeys, who flew up into the trees. But in the panic, one turkey didn't find a clear flight path, hit the fence of the exhibit, and bounced off. The tiger sprang into the air, also hitting the fence. Tilson was able to measure a spot where the tiger's belly had hit the fence, 10 feet off the ground. (You can visit the zoo safely: the fence is 20 feet high.) However, the turkey got away. "It just missed him!" Tilson says. One presumes the turkeys learned something that day.

A tiger with more practice might have been able to catch that turkey. In fact I can't help feeling, in species vainglory, that you or I might have caught one of those turkeys. But other hunts in the tiger enclosure have been more successful. One year a mallard duck elected to build her nest by the moat in front of the tiger exhibit and hatched a dozen ducklings. Some visitors who saw the *brood dwindling day by day were upset by this evidence that the tigers had not lost their predatory nature, and complained bitterly to zookeepers. I myself have criticisms for the duck.*

———

IT'S ALL VERY WELL to learn who you are and what to say about it, but it's even more vital for baby animals to learn how to make a living when they grow up. Some professions, such as eating grass, don't seem to require as much learning as other professions, such as sneaking up on, killing, and eating grass-eaters. On the other hand, to succeed at the job of eating grass, you must also be able to avoid those who would end your career by eating *you*.

These mean streets

When going forth to seek their fortune, one of the first things young animals must do is learn their neighborhood. Some will stay where they were born, and others will eventually go forth and look for new places, but even the ones who seek new territories need to learn what good places to live are like, and how to get around in them, the wild equivalent of how to cross the street.

Jungles are scary places for jungle animals that have never been there before. Arjan Singh raised a leopard, Prince, who was simply appalled when Singh first took him walking in the forest. There could be monsters anywhere! Only because he was with Singh and the noble dog Eelie did he dare go into the woodlands. Then the monsoon arrived. Here he'd thought he knew the worst they could throw at him and now there was water everywhere all the time. Prince was just a year old and now accustomed to making unescorted forays into the forest and even killing some of his own food, but he still returned regularly to Singh—until the monsoon, when he vanished. "I grew more and more fearful that he had died," writes Singh. When he had been gone eight days, someone "heard a faint sound across the river and saw a miserable little leopard perched in the fork of a tree. When Eelie and I went to rescue him in the boat, he cried with relief, giving pitiful *aoms*. From his ravening hunger, it was clear that he had not eaten for the whole week."

As a result of the reasonable caution a creature may feel in a strange environment, animal rehabilitators who set captive-reared animals free must often wait a long time before the animals feel bold enough to leave their cages or pens. In a release of captive-reared orangutans into an Indonesian rain forest one ape came out, looked around, went back into the transport cage, and closed the door.

Clare Kipps, who raised a crippled baby sparrow in her London apartment, eventually bought a small potted tree for him to perch in. He was terrified at the sight and dived down the collar of her shirt to hide from the tree. He wouldn't go near the ghastly thing, and she reluctantly bought a cage, which he adored.

Przewalski's horses (the only surviving wild species of horse) have been raised in captivity and released to their ancestral range. They've adapted well and a second generation has been born. The principal problem rehabilitators encountered was the horses' reluctance to leave the release area. They all wanted to stick around and fought each other for the privilege.

The released Przewalski's horses were showing "site tenacity." But in addition to learning that a particular neighborhood is home, young animals can learn that a particular *kind* of neighborhood is home. European mistle thrushes used to nest in the middle of pine forests. When many European forests were cut down, grown mistle thrushes returning to their hometowns to raise families often found themselves not in a pine forest, but in open parkland. Exhibiting site tenacity, some of the thrushes raised families there anyway. Luckily, the species thrived there. The next generation, exhibiting locality imprinting, had learned that open parkland was a fine sort of place to nest, and they increased and spread.

In the same way, curlews in Scandinavia returned to the bogs where they had always nested, only to find that the bogs had been drained and converted to fields. Insisting that no one was going to kick them out of their ancestral bog, they nested anyway and their chicks grew up to consider agricultural land their traditional nesting place.

Anemonefish are a group of fish species, and each species lives inside the sheltering tentacles of a different species of sea anemone. They lay their eggs on a nearby rock, and when the young fish hatch, they hang around the rock a little and then spend a few weeks swimming around before they settle in an anemone. But right after hatching, they learn the smell of the anemone their parents live in, the one near the rock on which they were born. When it's time to settle down, they seek an anemone of the species that smells right. If, due to the machinations of researchers, an anemonefish is born in an aquarium where there is no anemone to smell, it is not quite sure what to do when it wants to settle down and the only place available is an anemone. It can take one of these fish two days to move in to the perfect anemone, whereas a fish with the right upbringing takes only five minutes.

Learning the neighborhood may not be enough. Young animals may need to learn exactly where in the neighborhood they are safe—which street corner is the best to hang out on. In many cases when captive-reared animals are released in the wild, they lack the critical information that they should stay off the ground.

Endangered Hispaniolan parrots born and raised in captivity and released into the wild spent a lot of time foraging on the ground or in low bushes, a mistake a wild-born parrot would not make, and one which exposes the birds to great danger of predation. Indeed, researchers, trying to figure out why some released parrots they had fitted out with radio transmitters had vanished, followed the signals out of the forest and to the home of a Haitian family who had caught the birds and eaten them.

In the early days of a golden lion tamarin reintroduction project in Brazil, tamarins raised in captivity and released in the forest were obtuse about unexpected things. They were slow to leave the outdoor enclosure in which they had been allowed to adjust to the local climate, and when they did leave, they perched on top of the pen. It was not obvious to them, as it was to the scientists who were watching aghast, that for hawks and eagles this was the equivalent of a neon sign reading EATS. The scientists dismantled the pen, forcing the tamarins to move into the trees. Here it turned out that the tamarins did not grasp the principles of moving through the forest. Despite the incredible athleticism they had shown in captivity, natural vegetation struck them as untrustworthy and they became hesitant and unimaginative. They disliked the way things moved under their weight. "The animals were unable to plot a cognitive route through the forest between themselves and an incentive. Their movements were characterized by false starts, fruitless retracing of pathways to dead ends, and, finally, descent to and travel across the ground," the researchers noted. "At best, travel by the reintroductees was slow and hesitant. At worst they got disoriented and lost. Some simply sat, appearing to give up, and had to be rescued. Two perished on the ground, one taken by a feral hunting dog and one . . . likely killed by a snake."

In subsequent work with tamarins, the structures in the pens in which they were reared were made of natural materials and were changed every week to make them learn to leap from branch to branch as tamarins must. The result was better survival.

Janis Carter, trying to reintroduce captive-reared chimps into the wild, had little luck persuading two of the more civilized specimens, Lucy and Marianne, not to sleep on the ground. Merely getting up in the middle of the night, booting them out of their ground-level beds, and chasing them up trees didn't work. They'd just descend as soon as she left. Finally the desperate, sleep-deprived Carter took to creeping up to the sleeping chimps and pinching them with a pair of pliers. "I hoped to simulate the bite of an animal," Carter writes. It worked.

Captive-raised red wolves released in North Carolina didn't need to stay off the ground, but would have been wiser to stay off the street. The wolves, who weren't nearly worried enough by humans, found roads to be an extremely convenient way to bypass the annoying undergrowth in the forest, and so they were all too apt to be hit by cars or tailed down the road by incredulous motorists trying to figure out if that was really a wolf or just a weird dog.

Black-footed ferrets raised in captivity and released in a Wyoming prairie dog village adapted wonderfully in many respects, but were much more apt than wild ferrets to swank around on the surface. Wise ferrets—old ferrets—stay underground as much as possible to avoid predators.

Like the tamarins, captive-reared California condors have shown a distressingly relaxed attitude toward artificial structures. Reared in rectangular human structures full of squared-off perches and surfaces, they find these congenial when they are freed, landing on roofs, balconies, and worst of all, power poles. The trouble with landing on power poles is that it often leads to colliding with power lines. So many young condors died in collisions that rehabilitators began doing aversive conditioning before releasing condors, by putting power poles in the aviaries, wired up to give condors shocks. This lowered the rate of collision deaths.

But other congenial human structures included roofs at the Burger Barn in Cuyama, and in Pine Mountain Club, where the young birds landed 15 at a time. Once, they practiced their disemboweling skills on an air conditioner. On other occasions, they were accused of swooping down and ripping the windshield wipers off parked trucks. Eight condor hooligans broke through a screen door and into the (unoccupied) bedroom of one Pine Mountain Club resident. When he discovered them, the resident, Les Reid, is said to have ordered them, in a calm stern voice, to leave, whereupon "the raptors slunk out in single file." One of the condors released in Arizona, near the Grand Canyon, is reported to have dropped by a campground one day. A backpack was just sitting there, so this badly brought up young bird rifled through it, probably hoping to find cash or drugs. Instead, the bird found a loaded revolver. Naturally the punk took the gun and swaggered around the campground, carrying it by the trigger. Luckily, no one got shot.

Curiosity and willingness to try new things has good and bad aspects for condors. Biologists were bursting with pride when released birds fed on dead sea lions washed up on the coast, and not happy at all when the birds dropped by a ornamental backyard pond and ate all the frogs in it.

Learning fearlessness about human structures and habitations has been a good deal for a few species, notably rats, mice, and pigeons, but also for such creatures as the red fox. Some British red foxes have made a specialty of urban life. David Macdonald and his colleagues disconcerted Cambridge residents, who found them tracking radio-collared foxes in the middle of town. Macdonald writes of Unipart, whose mother gave birth to him and his siblings in a factory warehouse in Oxford. The others moved on, but Unipart made the warehouse his home, one where he could keep watch from the tops of stacks of pallets or from girders. The workers gave him food, and tossed "balls of greasy chip-paper" for him to chase. One day Unipart, prowling around a nearby auto assembly plant, got injured. He took refuge in some ventilation pipes. His meddlesome human friends caught him and took him to a veterinarian. Restored to health and his warehouse home, the last we heard of Unipart he was becoming more outdoorsy, going out in the streets and cantering behind bicycles.

What's for lunch?

Once you get your bearings, it's time to eat. As anyone who has ever argued with a small child at mealtime knows, it's not always obvious what's good to eat and what's not. Some young animals pursue the strategy of trying to eat everything they see; others are more conservative, only eating things they've eaten in the past or at least seen others eat.

Young sharp-shinned hawks are open to the charge that their eyes are bigger than their mouths. In a study in which biologists staked out sparrows, starlings, and pigeons along a shoreline that hawks follow when migrating south along Lake Michigan, young hawks were much more likely to try for pigeons, "birds that are grossly too large for suitable prey," whereas experienced adults went for smaller starlings and sparrows. The greedy young hawks had only been hunting for themselves for a month or two and hadn't learned to avoid "potentially difficult prey."

Other creatures are less alert to snack possibilities. Mississippi sandhill cranes raised in pens and released to the wild hung out with wild-born sandhill cranes, but weren't open to new menu items. Pen-reared birds joined wild birds in pecan orchards, but did not seem to grasp the concept that if you poke around in pecan orchards, you might be able to find free pecans—and you can *eat* pecans. They preferred corn, such as they had been served in the pens.

Fruitlike objects, leaflike objects

Baby howler monkeys in Costa Rica readily try to eat any fruit they see, or indeed any "frugiform object" they come across. They waste some time trying to eat woody galls and seed cases, but this doesn't do them any harm and eventually they quit trying. They are much more cautious when it comes to leaves. They may taste a leaf they have not seen an adult eat, but they won't eat it. Instead of just grabbing and gnawing, the baby follows its mother closely from tree to tree, and watches to see which leaves its mother eats. When she feeds, often hanging by her tail, the baby feeds next to her, often hanging by its own tail. As the mother pulls a branch to her mouth to eat the leaves, the baby pulls leaves off the same branch and eats them. The different attitude that baby howlers have toward leaves is probably vital because, unlike fruits, many leaves contain high concentrations of compounds that aren't good for monkeys.

Many young animals learn what others are eating—which is presumably good to eat—by sniffing the mouths of others. Benjamin Kilham, who has raised many black bear orphans, eventually discovered that the cubs liked to smell his breath and then eat whatever they found he'd been eating. He chewed on beech buds, and when the cub Yoda smelled his beechy exhalations, she excitedly began eating beech buds herself. By chewing red clover, he conveyed the news that red clover was good food.

Before he understood this, Kilham was shocked when his first pair of bear cubs examined, but did not sample, a rapidly drying puddle full of thousands of dead and dying tadpoles. How could they pass up all that fabulous protein? The answer is that he failed them—had he chewed just one moribund tadpole and let them smell his breath, they would no doubt have eaten all the rest.

Mere breath-sniffing was not enough to satisfy those urban chimpanzees being reintroduced to the wild on an island in the Gambia. Primatologist Janis Carter was working with a motley group of chimps, some of whom had spent a few years with their mothers in the wild before being captured, others who had less wild experience, and Lucy, a city girl who had been raised like a child in the home of psychotherapist Maurice Temerlin. Lucy knew more than 100 words in American Sign Language. She was 11, and more accustomed to taking cartons of raspberry yogurt out of the refrigerator than to hunting and gathering.

Carter would lead her charges through the jungle. Seeing a leaf species she knew to be favored by wild chimps, she would utter food barks, tear off young leaves, and eat them in an ostentatious manner, giving forth this-is-so-delicious food grunts. The chimps with wild experience would start eating the leaves. Those who had been indoctrinated in the dainty ways of humans were more skeptical. They would pry Carter's mouth open and look to see if she was really eating leaves. ("She is! She's really eating them! I can't believe it!") Usually that was enough to get skeptical Lucy to try the leaves. But even when Carter bravely ate them herself, it was not enough to get Lucy to try eating safari ants.

Orangutans being rehabilitated in the Indonesian rain forest learned from each other about wild foods. When Aming, a new orangutan, was brought to the camp in Sungai Wain Forest (because in his previous release area he'd been raiding orchards), he brought knowledge about certain things that were good to eat, and other orangutans learned from him. Ignorant young apes scrounged bits of what others were eating and then, if they liked it (and they usually did), tried to find some for themselves. Experts like Aming were swarmed by little apes every time they were seen eating. Shortly after his arrival, he started eating bark off a big tree next to a main trail. One of the other orangutans, Enggong, heard the bark tearing and came over and stared. Enggong dangled next to Aming for a better look. He drew closer and stared until Aming left, and then he took Aming's spot and ate bark himself.

One young orangutan, Jaja, didn't recognize most of the potential foods around her, and wasn't inclined to learn. But she was fond of a student at Camp Leakey, Femke den Haas, and den Haas showed Jaja how to eat rattan shoots. Den Haas picked a rattan shoot, tore open a sheath at the base of the shoot, and ate the tender part. Jaja picked a rattan shoot and opened it, but ate the wrong part. Den Haas did it again, and handed the edible part to Jaja to eat. Jaja picked another, and this time did everything right. She spent the rest of the day gorging on rattan shoots.

Nobody wants to be the first to try a new food. Biologist David Macdonald, studying red foxes, was delighted to discover that every evening five foxes visited the back garden of fellow biologist Hans Kruuk, seeking what they might devour. Macdonald and Kruuk encouraged these visits by scattering bread crusts, bacon rinds, and other table scraps. Perhaps the biologists felt guilty about these unnatural foods, for they were delighted when they got their hands on some dead mice, much more suitable fox food. They scattered these among the scraps and waited to see the foxes' delight. The foxes were repelled and alarmed, just as you might be. "They shrank fearfully from the mice until the bravest fox summoned up the courage to dash between them and snatch up the crusts of bread," writes Macdonald. Days passed before the foxes were willing to try the mice, combining them with bread in "a most wholesome-looking sandwich."

I'm never eating that again

As well as learning what's good to eat, animals learn what's not good to eat. The loathing that animals, including people, feel for foods that have nauseated them in the past is a primal horror. Psychologist John Garcia, who called this phenomenon learned taste aversion, discovered in the 1960s that rats learned to hate a food that was followed by nausea even if the nausea didn't occur until hours later (and even if the nausea was caused not by the food at all, but by radiation sickness, or by an injection researchers had given the rats).

This was upsetting to the then-prevalent orthodoxy which held that animals could only learn an association if the behavior (eating rat chow) was immediately followed by the reinforcer (the negative reinforcement of feeling sick). Nor did it fit that rats were much more apt to associate nausea with a food than with a visual stimulus. One diehard famously wrote that Garcia's results were about as likely as "find[ing] bird shit in a cuckoo clock."

However, not only did the results check out, they were an example of a phenomenon well known in humans. Psychologist Martin Seligman called learned taste aversion "Sauce Béarnaise Syndrome," commemorating an occasion when he went to a restaurant, selected the steak with sauce béarnaise, and got dreadfully sick later that night. Despite learning that a strain of stomach flu that was sweeping his department had undoubtedly caused his illness, Seligman now found sauce béarnaise disgusting. It was 10 years before the visceral association between the sauce and the illness wore off. Most people can rant bitterly about some random food that they once ate before they got sick, and which they plan to avoid until the day they die.

Learned taste aversion is now well accepted and studied in creatures as simply laid-out as slugs. There is obvious survival value in not taking a long time to learn what foods make you sick. The price is that you learn some false associations along the way, but it's worth it.

Despite the widespread occurrence of learned taste aversions, there are occasional exceptions. Researchers preparing golden lion tamarins for release tried to expose them to things they might meet in the wild. One day they put a large toad in the enclosure. The tamarins approached the toad cautiously and then like lightning two of them bit the toad squarely on its poisonous parotid glands. The shocked researchers snatched the toad, prizing it from the grip of a third, enthusiastic tamarin who also wished to bite the toad. Meanwhile the two that had gotten a taste began to foam at the mouth, vomit, and wail. One went into convulsions for hours. Their loved ones watched with interest. The next day, to test their assumption that the tamarins had learned a valuable Lesson About Toads, the researchers offered the toad again, in a glass jar. The tamarins were delighted to see a nice toad, and eager to get at it. For comparison, the researchers tried them with another glass jar full of their favorite grasshoppers. The tamarins, both those that had suffered toad poisoning and those that had merely witnessed toad poisoning, seemed exactly as wishful to get at the toad as the grasshoppers. The dismayed researchers suggested that maybe if the glass jar hadn't prevented them from *smelling* the toad, the tamarins might have acted more appropriately. Nevertheless, noting that tamarins tried and sometimes caught and ate snakes, but that one of the released tamarins had probably died of snakebite, they wrote, "It appears that neither the affected nor the observing animals learned much from this near-fatal encounter. We must entertain the nonadaptionist hypothesis that captive and wild tamarins learn little about reptiles and amphibians: They try to eat the small ones, and if they are unlucky they may die." Such examples of nonlearning show how valuable the ability to learn really is.

Learned taste aversion and uncertainty about what is good to eat can be exploited by humans, if they're clever enough. Ingenious wildlife biologists in San Diego wanted to protect a colony of endangered least terns. These sporty little crested birds lay their eggs in open scrapes on flat coastal land. In other words, they nest just where you might want to lay your beach towel, and that's one reason they're endangered. This colony didn't have that problem because they nested at a U.S. Marine Corps base, but they were losing a lot of eggs to ravens who nest in the coastal bluffs. Shooting ravens had not proved a successful strategy, since new ravens promptly moved in to the vacant raven territory and ate tern eggs.

The biologists took quail eggs, which look like tern eggs, made a little hole in each, and injected methiocarb, a chemical that makes birds feel extremely unwell. They sealed the hole with a bit of glue. The biologists then made fake tern nests on the bluffs near the ravens and put the nauseating eggs in them. The ravens promptly robbed the nests, but within a few days noticed that these eggs were ghastly, and stopped taking them. Learning to avoid nasty eggs at one site didn't cause ravens to stop taking eggs from other places, however.

So the biologists tried setting up an elaborate fake tern colony, complete with protective fencing festooned with electrified wires and Keep Out signs. They made 20 fake nest scrapes and put bad eggs in half of them and untreated, perfectly good quail eggs in the other half. Ravens dropped in to eat eggs. The ravens quickly learned that some eggs were good and some were bad, and learned to tell them apart by handling them for about 25 seconds, abandoning the treated ones and eating the good ones. If ravens from elsewhere came by to check out the fake colony, the resident ravens chased them away.

The fact that the ravens could learn to tell good eggs from bad was discouraging. That meant you couldn't just put bad eggs around the edges of a genuine colony with genuine tasty eggs in it.

But the following year, the biologists put out bad, treated eggs exclusively, planting them around the outskirts of genuine tern colonies—and they did this before the terns started laying eggs. The ravens speedily learned that *all* the eggs there were bad and stopped going there and taking eggs. By the time the terns began laying tasty tern eggs, the ravens had given up on foraging for eggs at the colonies. They still kept other ravens (who would not have learned about nauseating eggs) out of their territory. The result was that not a single tern egg was taken by ravens that year, and not a single raven was shot by predator control officers.

That wasn't so bad

The opposite of learned taste aversion—learning to hate the taste of things that make you feel bad—is learning to like the taste of things that make you feel good. Alcohol owes our continued support to this form of learning. So, perhaps, does zoopharmacognosy, the practice by animals of treating their own ailments with medically active plants.

Chimpanzees seek out the leaves of certain plants and swallow them whole instead of chewing them. It's hypothesized that in some cases they're doing this to scour out intestinal parasites such as tapeworms and nodule worms. Whether this is learned behavior is unclear. No one has seen a mother chimp administering leaves to her infant. No one has to teach dogs and cats to eat grass blades, a behavior which is speculated to have a therapeutic effect on occasion. But animals have been seen to eat quite a variety of apparently medicinal plants—it's hard to believe that they are such walking Merck Manuals that they have the innate ability to recognize them all.

But infant chimpanzees intently watch everything their mothers do. Two-year-olds put the same leaves in their mouths, chew, and spit them out. Not until seven years old are they seen to swallow them like adults.

In Venezuela, a troop of wedge-capped capuchin monkeys rub themselves with millipedes. Millipedes secrete powerful benzoquinones, which discourage predators from eating them. So powerful are the toxins that the monkeys who rub themselves with millipedes get enough on their fur to repel mosquitoes and parasitic bot flies, which can dangerously debilitate a monkey. The worse the bugs are, the more the capuchins anoint themselves. Sometimes the monkeys put the millipedes in their mouths for a moment to moisten them before they start rubbing, but they are never foolish enough to eat them. The benzoquinones are stronger than any mosquito repellents available for human use. One millipede may protect many capuchins, since one monkey passes it along to the next when the first one has completed the application of product. If they run out of millipede, monkeys rub against each other in an apparent attempt to get some protection.

Andy, a tame capuchin living with biologist Kathleen Gibson, is given an inch of cigar daily. Andy likes to rub himself with tobacco, orange rinds, onions, or garlic, as do many other capuchins. Gibson suspects that this self-anointing is good for Andy's skin. "Andy is particularly prone to necrotic infections," Gibson writes. "Since I began giving him tobacco on a prophylactic basis, he has had no skin infections."

It's not hard to imagine young animals learning to eat a plant that their elders often eat, and that makes them feel better, or to rub themselves with such a plant. How animals might learn to consume medicinal plants that are used only rarely is less clear. It is reported that zoologist Holly Dublin, observing elephants in Kenya, saw an elephant in an advanced state of pregnancy leave her normal range hurriedly, followed by her daughters. Several miles away she stopped at a tree and ate most of its foliage as her daughters watched. The elephant gave birth the next day. Dublin showed leaves from the tree to women in a nearby Masai village, and was told that they were sometimes used to induce labor. Did the elephant's daughters understand what she was doing, and would they remember if they felt the same need? If so, that's impressive.

Human medicine is full of examples of quack treatments and placebos. Most diseases go away, and if we happen to have taken a certain food or pill, or a certain treatment, right before we get better, we often believe that we were cured by the yogurt, or the ear-candling, or the magic spell to rid us of warts, when there was no connection. It seems possibly that animals would be just as likely to make this mistake as we are.

The junkyard bird

One culinary strategy is trying to eat everything you see. The keas of New Zealand have been dropping by the dump outside Arthur's Pass National Park every day for 40 years to see what's on the menu.

Things young keas have been seen taste-testing at the dump, according to biologists Judy Diamond and Alan Bond, include "rubber bands, string, masking tape, flashlight batteries, and foam rubber, among other unsuitable or dangerous materials. When trash was being burned, we even saw them attempt to bite the flames."

This if-it-exists-try-to-eat-it behavior is appropriate for a scavenger. Bond and Diamond theorize that keas evolved in a harsh environment that was inhabited by moas (flightless birds that came in a range of sizes), who left nourishing carcasses when they died. When the Maoris arrived in New Zealand, they wiped out the moas, and perhaps this created even harsher times for keas. When Europeans arrived, their naturalists categorized keas as feeders on fruit and nectar, but maybe this was only because they could no longer find dead moas lying around. Europeans did not confine themselves to studying wildlife, and they introduced red deer and sheep. These died at the usual rate, but in the 1960s, because the deer had become wildly destructive, vast numbers were killed, and their carcasses dumped. Bond and Diamond speculate that this may have provided short-term bonanzas for keas.

Keas also scavenge dead sheep. They enjoy landing on living sheep and going for a ride. If the sheep has an open wound, they pick at it. This is bad for the sheep, who often die of infections in the wounds, and it has led to hair-raising stories of killer keas attacking sheep and eating them alive or carrying them off in their talons. (This is slightly more likely than me leaping out of this book and carrying you off in my talons.)

One can thus speculate that keas used to snack on moa carcasses in addition to plant foods, went through a difficult period where it was even harder than usual to find anything to eat, and then started snacking on deer carcasses, sheep carcasses, and garbage.

Favorite foods

Some animals learn a good thing to look for and make it a specialty. Sea otters in Monterey Bay, California, are under close observation by researchers, who have listed 33 types of prey the species eats. No otter eats them all. Marine mammal biologist Marianne Riedman lists the whims of otters she's observed: Nosebuster ate almost nothing but turban snails, pounds and pounds of them. He'd swim into the kelp and wriggle violently, knocking snails loose "like ripe fruit shaken from a tree." Female 508 liked the wily octopus, which she was very good at locating. The Ab Queen ate crabs and sea urchins, but mostly abalone. Another otter stole squid from the bait bucket kept on the stern of a boat in the harbor. One male caught and ate seabirds, whom he seized by the feet from underwater.

By watching otters over years, Riedman and colleagues were able to confirm that sea otters learn some food preferences from their mothers. (Obviously some sea otter has to invent stealing squid out of the bucket, once buckets appear on the scene.) Thus Female 184 ate kelp crabs, mussels, and turban snails, most of which she collected in shallow water, and her daughter Female 535 ate kelp crabs, mussels, and turban snails collected in shallow water. Both had an unusual feeding technique which researchers called the "surf grass salad bar": they'd pull a clump of surf grass from the ocean floor, swim to the surface, and pick through the strands, finding and eating tiny crabs hiding there.

The refreshing beverage with tentacles

As early as the 1970s, at least one otter in Monterey Harbor had learned that soda and beer cans dotting the floor of the bay, tossed into the water by carefree litterbugs, often become starter homes for small octopuses. In a quarter of an hour, this otter was seen surfacing eight times, each time clutching a different can. Six had octopuses in them, and the otter bit open the side of the can, ate the octopus, and dived in search of another. Intrigued, researchers went down in scuba gear, looking for cans. They brought up 22 cans, 8 of which had already been ripped open, apparently by otter teeth. Of the remaining 14 cans, 7 had an octopus inside, and 2 contained onespot fringeheads, a fish which can be found lurking in tires, bottles, and shoes, if these things are lying at the bottom of a bay or slough.*

Maybe that's what other tigers eat

Wild Siberian tigers are forced to compete with increasingly efficient human beings for their principal, but dwindling prey, elk. As a result some desperate tigers have developed unusual specialties. One tiger specialized in horses. Horses are big and fierce, and one horse kicked the tiger hard enough to break two teeth and a bone in its jaw. The tiger went right on hunting horses. Another Siberian tiger specialized in eating bears. It's hard to believe that even a large tiger could get away with this for long.

A tigress in Malaya specialized in dogs, and killed and ate as many as three in one night. She ate them on the hill behind the police station, in a brushy spot littered with dog collars. She was shot by a dog-loving lieutenant colonel who enticed her using another dog as bait—after spending 10 days building a special cage for the bait dog that would keep it from being harmed by the tiger.

Becoming a specialist can be risky. Dingoes often consume a wide variety of prey, and the list of species they are known to hunt comes to 177, from magpie geese to wombats. But sometimes dingoes focus. On Australia's Barkly Plateau, there are periodic population booms of the long-haired rat. In the early 1970s there was a long boom, and several generations of dingoes grew up feasting on long-haired rats. When the rat population crashed, many dingoes seem to have been unprepared to switch, and some starved to death in the midst of abundant prey such as pipits, songlarks, and other birds nesting on the ground. Growing up in the heyday of the long-haired rat may have produced young dingoes with an insufficiently broad education.

*The onespot fringehead would make a great spokesanimal for a pro-litter campaign, should you wish to conduct one.

Dingoes who were accustomed to hunting kangaroos took a long time to learn that sheep are edible (something many people would prefer they never discover). Dingoes that grew up near sheep paddocks were skilled at hunting sheep and inept when they encountered kangaroos.

It is speculated that if dingo pups grow up hunting a wider variety of prey, they will be better able to learn to hunt new prey than dingoes that grew up as specialists. On the edge of the Simpson Desert, a population of dingoes hunted a variety of small vertebrates. When these became scarce, they switched to eating big flightless grasshoppers—and survived.

Where do you keep the food around here?

Once you know what's good to eat, you have to find it, and food will sometimes hide from you. Extensive experiments, mostly with birds, have been done to clarify the concept of the search image, a mental picture of what a creature is searching for.

In classic experiments with stick caterpillars, which look deceptively like twigs instead of like fabulous little packets of nutrition, captive-reared European jays were put in aviaries containing stick caterpillars and actual sticks. These young jays had never seen stick caterpillars before, although they had been given a session with sticks so that they wouldn't be alarmed by them. They were put in an aviary with the real and the fake sticks. Completely fooled, they had no idea there were caterpillars around. Eventually a jay would accidentally find a caterpillar, generally by stepping on it. (In the wild the caterpillar would be less apt to be in a place where it would be stepped on, and it would probably be sticking out from another stick, exactly like a little dead branch, but the experimental caterpillars were dead and so could not employ these arts.) After this some jays would start grabbing every stick they could find. Others, now that they knew that a caterpillar could look like a stick, seemed to know at once how to tell mere plant matter from mimics, and would jump around the aviary snapping up caterpillars.

The Bornean orangutans, or "brilliant botanists," studied by Biruté Galdikas and her colleagues eat more than 400 wild foods, which Galdikas summarizes as "a highly diversified mix of fruit, leaves, bark, sap, insects, shoots and stems, honey, and funguses." They seem to learn that if one tree, say a durian, is bearing fruit, then other durian trees are also likely to be fruiting. Since trees in the rain forest are scattered, rather than growing in convenient groves, orangutans need to remember where they have seen other durians. They appear to have mental maps that allow them to travel directly to the places they remember.

Young California condors who grow up in the wild can follow their parents to food once they've fledged. Then they learn to scan for other scavengers and fly over to see what they've found. In their habitat, the first creatures to arrive at a carcass are likely to be turkey vultures, which have the advantage of a fine sense of smell. Ravens and golden eagles and sometimes even bald eagles are also promising indicators that food is available.

Janis Carter taught her chimpanzee charges how to rob nests. Having gotten them used to eating eggs, she sneaked out and tied old nests in absurdly obvious spots along the trail she would lead them along in the morning, and loaded the nests with eggs. From robbing these phony nests, they learned to watch for the sight of nests, and eventually became proficient egg-hunters.

Predation skills

For many young animals, the question of how you get your paws on the food once you spot it is a serious matter. The issue is particularly acute for species that need to catch living prey. Yet even picking food up is difficult for some babies. According to Marc Hauser, the young yellow-eyed junco (a dapper species of sparrow) is "perhaps one of the most inept foragers" among birds. It can scarcely pick up a mealworm without dropping it. Eating it is harder still. Adults are sometimes seen to hop over to a juvenile and rearrange the worm in its beak, so that it's in a position in which the young bird can actually swallow it.

Wait, won't you stay for dinner?

The difficulty of picking up food that lies still is nothing compared to the difficulty of getting your beak or paws on food that runs, swims, or flies away. While teenaged great blue herons foraging in the Gaspereau

River in Nova Scotia paced along the shore or through the shallows at the same rate as adults and struck at prey items just as often, they caught prey on only 33 percent of their strikes, as compared with 62 percent for the adults. (You can tell when a heron catches something, because swallowing is so visible on their long skinny necks.)

The researchers criticized the striking technique of the juvenile herons, writing that they lacked adult "finesse." Sometimes a wildly lunging young heron would submerge its entire head, which adults never did. The young herons practiced on floating objects like sticks or clumps of algae, repeatedly striking, seizing, and then dropping the object. "Pow! Watch this! Pow!" Grown-up herons never did this.

Young herons got better fast, however. When the scientists observed 10-week-old birds, they got something to eat on only 18 percent of their strikes. A week later, in similar weather conditions, 11-week-old herons caught prey 55 percent of the time, not far behind the rate of adults.

Adult Sandwich terns fishing on the coast of Sierra Leone were more competent than teenaged terns. Zoologist Euan Dunn noted that young terns sometimes caught a fish and then dropped it, but the adults never dropped one. Young terns caught fish on fewer of their dives. Adults caught 14 fish an hour, and the kids could only manage 10 an hour. They did about as well as adults when fish were close to the surface, but less well when the fish were deeper. The deeper the fish, the harder it may be for terns to judge where it is and exactly where to dive to nab it.

After weaning, eight orphaned red bats were placed in an outdoor aviary equipped with lights to attract insects. Once they had worked on their flying skills, the little bats noticed the moths and midges and would fly into a swarm and out the other side without chasing any one moth. In the mornings their bellies would be concave until rehabilitator Barbara French gave them their morning feedings. As the days passed, the bats began darting and jinking after their prey, and within a few weeks they had bulging bellies in the mornings. One bat was self-sufficient in a week, but three of them took six weeks.

Crumbs are for babies

Dippers are elegant little passerine birds who live by streams and make most of their living wading in streams eating insects and occasionally small fish. Often dippers dive beneath the surface of the water and forage on the bottom of the stream, keeping themselves in place with their wings in what may be a significant torrent. It's hard, skilled work. Sonja Yoerg, observing young dippers in Wales, found that there is a juvenile strategy for gathering food, which involves picking a lot of tiny blackfly larvae—or "crumbs"—off rocks in shallow water. Crumbs don't try to get away, which is handy for young birds, who are clumsy. Yoerg frequently saw them pick up larger food items ineptly and drop them or be unable to keep them from struggling free.

Sometimes a juvenile's foraging netted a caddis fly larva. Caddis larvae, to protect their succulent bodies, build themselves craftsman-like tubular cases out of materials they find in the streambed, such as tiny pebbles, twigs, or pine needles. An adult dipper shucks the case off and eats the larva within, but Yoerg saw many instances of juveniles eating larvae, cases and all. "Consumption of caddis with its case may be a consequence of poor handling skills," Yoerg writes.

Juveniles also caught fish at a low rate and were as likely to drop them as to swallow them. One 19-day-old bird caught a fish by the tail and took it over to its mother, who was foraging nearby. Gripping the thrashing fish in its bill, the juvenile made begging motions. Its mother grabbed the fish and ate it herself.

The trade-off between subsisting on parental bounty and doing the work oneself has been investigated in other species, and a major factor is just how long parents will cooperate. N. B. Davies looked at the transition from begging to independent feeding in wild spotted flycatcher fledglings in an Oxford garden. At first the little birds perched on a branch and called "Tsi," which seems to mean "Feed me!" And at first the parents brought large tasty insects until the tsiing stopped. In this sated condition, the young birds looked around, pecked at things, and explored. Then it was back to Tsi! Tsi! With passing days the young flycatchers began chasing their parents with their demands, instead of calling for their parents to come to them. They began grabbing insects away from their parents and banging the bugs against a branch. (This kills a bug, so it doesn't wiggle going down, and perhaps makes it more flexible.) If they couldn't swallow it after this, the adult bird would take the insect, bash it more thoroughly, and return it to the young bird.

It became harder and harder for the young birds to wheedle food out of their parents as they grew older, but at the same time they were getting better at feeding themselves. At first the young birds tried "stand catching," in which you stand on a branch and wait for an insect to fly by in easy grabbing distance. Such insects are sadly infrequent, and by their third week young birds have more or less given this up. As they became better at flying, they attempted the skill for which they are named, fly catching, in which you spot a flying insect, fly over, and snatch it out of the air. It took the young birds about 8 to 10 days from fledging to reach a skill level at which they could catch three-quarters of the insects they went after.

As the parents became less generous to their demanding children, the young birds continued to beg for a while. They begged not only from their parents but from each other, "and sometimes briefly begged with quivering wings toward a fly as it passed them, before they flew out and caught it."

These are birds that are independent of their parents in a few weeks. It's not such a quick process for the hardworking white-winged chough.* These Australian birds laboriously sift through eucalyptus leaf litter, seeking the elusive arthropod. They live in cooperative breeding groups of 10 to 20, and all pitch in to build the nest, sit on the eggs, and raise the babies. The more choughs in a group, the better chance of survival the young ones have—a pair won't succeed in keeping their young alive without help. When they have young, group members spend "virtually all" their waking time foraging for the chicks and themselves, whereas out of the breeding season, they need spend only 80 percent of their time toiling. The babies take more than six months to achieve independence. They continue to improve their skills over a four-year adolescence. But for the first 40 days they get almost everything through incessant begging. They can't fly well yet, and so they walk along behind the foraging group, screaming for food.

Good caterpillars, bad caterpillars, and complicated caterpillars

Among the primates, squirrel monkeys, who mature quickly, need to learn foraging skills in a hurry. Once baby squirrel monkeys become mobile, they don't stick closely to adults, minimizing their chances to

---

* "Chough" is pronounced "chuff." Not my idea.

observe or imitate. Researchers observing wild squirrel monkeys in a Costa Rican national park noted that most things squirrel monkeys eat are small, and that adults, even mothers, do not share these dinky food items with babies.

When they were just a month old, baby monkeys stuck everything in their mouths—twigs, leaves, pieces of bark. But soon they became focused on things they could actually eat. Before they were two months old, baby monkeys were seen tearing apart curled leaves to find and eat caterpillars, and pulling leaves through their teeth to scrape off bugs.

The baby monkeys didn't initially have great motor skills—they were seen fumbling katydids, dropping cicadas, missing grasshoppers, and letting frogs kick free. They had to learn techniques as well as simply acquiring strength and physical fluency.

One of the more complex skills squirrel monkeys learn is that of handling caterpillars with irritating or stinging spines. When grown squirrel monkeys pick up nasty caterpillars, they hold the tufty tip of their long furry tail in their hand so that it protects their hand like an oven mitt. Then, still using their tail tuft, they rub the spines off the caterpillar before eating. Researchers Boinski and Fragaszy describe seeing a baby monkey (106 days old) eating caterpillars without using the tail technique, and then spending several minutes rubbing its hands, as if trying to wipe away the sting. The next day they saw the same little monkey eating caterpillars, this time rubbing each one with its tail first.

The one example seen of adults helping baby squirrel monkeys learn foraging skills took the form of warnings about poisonous (not merely spiny) caterpillars. One June day, for example, the troop was crossing a forest gap along a branch. An enormous black caterpillar, apparently not as delightful to eat as it looked, was lying temptingly in view on a leaf next to the branch. Adults paused to look at the caterpillar and passed on, but when four babies stopped to examine it, an adult male came back, barked a warning, and stood between the babies and the caterpillar. They gave the creature a careful stare and moved on. Adults also barked and steered babies away from opossums, snakes, and owls.

Polar bear cubs stay with their mothers for two years, and their hunting prowess gradually improves. Ian Stirling, watching wild bears in the

Arctic, found that one-year-olds spent only 4 percent of their time hunting. Two-year-olds hunted 7 percent of the time. Meanwhile, the average mother, hunting for the whole family, spent 35 to 50 percent of her time hunting. A mother might catch a seal every 4 or 5 days, but a yearling would catch one only every 22 days. The next year they might catch one every 5 or 6 days. But a one-year-old follows its mother closely, watching her, and sniffing everything she touches. When she starts to stalk prey, the cub lies down to watch and wait. Occasionally Stirling saw one get impatient and approach its mother too soon, whereupon she'd usually make a threatening charge at the cub or belt it with a forepaw. Then the cub would lie down again and wait as long as necessary.

Young dolphins observe and work to master parental technique. Dolphins at Monkey Mia in Australia round up schools of fish into a compact space and whack them with their tails, stunning a bunch of fish at once. This is apparently not something you or I could do on the first try, even if we had tails. Biologist Rachel Smolker writes, "It is not easy to smack a fast-moving fish. Once we knew what to look for, on many occasions in the shallows of Monkey Mia, we saw youngsters practicing, swiping their flukes at tiny fish, which they usually missed."

There's a special trick to eating these

When foraging techniques are observed in only one or a few areas, that may support the idea that they are learned skills (although environmental factors must first be ruled out—dolphins that live in the deep sea don't have the option of chasing fish up onto mud banks, even if they think about it all the time). Hans Fricke has observed coral reef fish, including triggerfish, for decades. At one place, Fricke found five triggerfish who each used an unusual technique for eating sea urchins: they'd bite off the spines, which allowed them to grab it and carry it to the surface, where they'd let go and take bites from the unprotected underside as the urchin slowly tumbled to the bottom. Other triggerfish eat urchins, but do so by blowing water on them in an attempt to tip them over to get to the underside. The five triggerfish with the unusual technique are thought to have learned it from each other. (Presumably one particularly ingenious triggerfish, whom I refuse to call a "triggerfish Einstein," developed the technique first.)

How to kill a gemsbok and live

Lions in the Kalahari Desert are said to use a special technique on gemsbok, antelopes with genuinely dangerous horns. F. C. Eloff reports that a gemsbok was seen going about its business with a decomposed leopard on its back: apparently the leopard pounced on the gemsbok's back and the gemsbok stabbed the leopard to death with its horns. Then, faced with the dilemma of what to do when you have a dead leopard impaled on your horns, it had no answer. Eloff also reports a hyena stabbed to death by a gemsbok, a gemsbok found with its horns stuck in a dead lion, and two gemsboks found dead with dead lions they had stabbed.

Kalahari lions still file gemsboks under Large and Tasty, but instead of biting them on the throat, they often leap onto their hindquarters and break their backs, thus staying farther from the horns. Wardens in Etosha National Park told Eloff that the lions there never broke the backs of prey, suggesting that this is a regional variation. In Etosha there are more prey species, so it may be that the Kalahari lions, who have few other options than gemsbok, have had to come up with a technique.

Eloff observed small lion prides in the Kalahari, and the pride he named S2, which consisted of a pair of young lions whose skills were not yet honed, "provided us with a lot of amusement." Extracts from the log show the bumbling duo chasing a gemsbok calf but not catching it. (Can you blame them?) Later that night they chased an aardvark. It went into its hole and they tried to dig it out. This did not work. The next day they chased a bat-eared fox into its hole and tried to dig it out. No luck. Two days later they spotted and chased an aardvark and, what do you know, it went into its hole and they were unable to dig it out. In the next two days they went after another bat-eared fox (no), a pangolin (no), and a porcupine (no). It's the kind of record that forces a lion to take another look at the Gemsbok Diet.

In South Africa, entrepreneur John Varty and cat handler Dave Salmoni introduced two seven-month-old tiger cubs, Ron and Julie, to the task of hunting prey in a fenced preserve.*As shown in a Discovery

---

* It has not escaped Varty and Salmoni's attention that tigers are not native to Africa. They don't plan to release them there, but to breed them in a semiwild captivity in which they develop their predatory skills.

Channel Quest show, Ron and Julie, although large and impressive, would never have survived without human protection. Initially, all they could think of to do with prey animals was to chase them, which is admittedly fun. When they found a dead zebra, they were thrilled and leapt on it and bit it in all the right places, but never thought of eating it.

By hanging dead antelopes from the back of a truck and driving away, Varty and Salmoni got the cubs to leap on the antelope and try to bring it down. That meant that they were teaching the tigers to chase trucks with mayhem in mind, so they switched to suspending antelopes from wires between trees.

The first prey the tigers caught was a porcupine, and if Dave hadn't pulled the quills out of Ron's face, chest, and paw, they might have worked their way in to create potentially deadly infected wounds.

They fearlessly approached Cape buffalo, who ran them off. Ron fearlessly chased a rhino, who fortunately decided to run, perhaps because he'd never seen such a creature before. Ron fearlessly attacked a herd of domestic cattle, and a cow gave him a serious kicking. But when Varty introduced a flock of tasty ostrich into their 150-acre hunting enclosure, Ron and Julie were terrified and hid for a week. Ron and Julie were given a dead ostrich to chew on, which caused them to realize that ostriches can be eaten. Shortly thereafter they killed one for themselves.

They could be seen becoming better hunters. The first warthog they chased vanished down a burrow, and they groaned in frustration. The next time they chased a warthog, Ron cut off its retreat and beat it to the burrow. The preserve provided so little cover in the form of trees and bushes that the humans had to show the tigers how to sneak along a sunken riverbed to ambush prey. The first time they killed a blesbok, Julie cleverly cut it off in a thicket, but her suffocating bite was ineptly administered. But after three years, the tigers were deemed "stalking machines"—fit to hunt for themselves.

With catlike tread upon our prey we steal

Cats have a great deal of innate technique to draw upon when hunting for their prey, as anyone knows who has seen a kitten sneak up on an unsuspecting sock, pounce on it, seize it with jaws and fearsomely clawed forefeet, and disembowel it with mighty strokes of the hind feet. Yet despite the masterful way an inexperienced kitten handles hosiery, cats and great cats do have things to learn about hunting.

Tigers, lions, leopards, and cheetahs raised by humans and then reintroduced into the wild are pretty good at adapting to the habitat, whether it be jungle, forest, or savannah, and have little problem meeting, recognizing, and mating with other cats; and the females make fine mothers. But learning to catch and kill enough prey to keep alive (and to support cubs, if any) is more difficult. They practice pouncing on their playmates, but they seem to need to learn that they can also pounce on strangers. Then they need to learn to kill edible strangers, which can be quite difficult. They need to distinguish edible strangers who can fight back and should be left alone from edible strangers who can safely be attacked. They need to refine their lying-in-wait and stalking techniques if they are to find enough prey. They also need to learn how to eat a dead animal, as "opening" a carcass turns out to be surprisingly difficult.

Mother cats of many species do not teach kittens or cubs what to do, but diligently provide perfect learning situations. Often a mother cat starts by bringing prey, safely dead, and the kittens learn what their food looks like. Then some mother cats bring prey that is living, but injured, and the kittens witness how it is killed and try to kill it themselves. Then she brings them living, uninjured prey, and they learn to chase and catch it themselves. Thus, in a preserve in northern India, the tigress Noon caught a chital fawn and gave it to her cubs, who were not quite two years old. Overjoyed, they played with it as thoughtlessly as a kitten plays with a ball of paper, but couldn't resist killing it before very long.

Paul Leyhausen, observing domestic kittens, described the situation that teaches the kitten to administer the killing bite to its prey, something that it will not necessarily learn to do, even though the motor pattern is innate. Typically, the kitten is gripping its prey in its jaws, when something makes it think there's competition, such as its mother or a fellow kitten coming close, and it bites down. Adult cats who have not already learned this in kittenhood still can, but it takes a more intense stimulus.

Other species of cats may take their cubs along on the hunt and, as they get older, permit them to try their luck on some prey. Researcher Randall Eaton has described the predatory behavior of a cheetah mother and her four cubs in a Kenyan national park. When they were old enough they went with their mother when she hunted. When she wanted them to stay put, she'd say "ughh" in a low voice, and they'd stay until she chirped.

One day when they were six and a half months old, she spotted a warthog with two very small wartpiglets, and for the first time she didn't tell them to stay behind. All five began stalking the warthogs, who were rooting in the ground with their backs to the cheetahs. When the warthogs saw them and ran, the mother dashed ahead of the cubs and chased the mother warthog. Her intent was apparently not to try to catch the adult warthog, a formidable opponent, but to separate her from the piglets. Meanwhile the cubs chased one of the piglets, running inches behind it, but never making any moves to grab it or knock it down. The other piglet raced away unmolested, and the piglet being chased by the cubs dashed down a hole.

In Eaton's analysis, stalking and chasing prey are innate behavior patterns, and cheetahs don't need significant experience to do these things—although they may need experience to figure out just which animals are prey they should stalk and chase. Seizing and killing the prey requires more experience, and are more flexible behaviors.

Cheetahs sometimes prosper

A 1980 paper describes the result when three captive-reared cheetah brothers, Jan, Rogers, and Gouws, were experimentally released in South African nature reserves. The brothers were two and a half years old. To see if they could make kills, they were put in a small enclosure with live sheep, which they managed to kill in standard cheetah fashion.

They were then released in a fenced nature reserve, wearing radio collars. There they traveled on the roads and paths and made ten kills in a month, bagging four impala, one duiker, one waterbuck, two kudu, and two giraffe calves. They were recaptured and a few months later were re-released on farm land adjoining another, larger nature reserve. Here life was even more exciting, and they got in a fight with some wild cheetahs, and Jan was injured. He couldn't get around for a week, and Rogers and Gouws stayed with him until he could hunt again. In slightly over two months they killed three giraffe calves, one kudu, one waterbuck, two impala, and an undisclosed number of chickens on the farm. After two months, they were recaptured.

The interesting thing about the prey they chose is that most of them were so large. Giraffe, even giraffe calves, are not standard cheetah prey. The brothers also made attempts on the lives of some large animals they did not succeed in killing. They attacked wildebeest 21 times and zebra 13 times, but never succeeded in getting either. One cheetah was injured first by a wildebeest and then by a zebra. Another brother was hurt attacking a buffalo.

Their impressive hunting prowess doesn't seem sustainable. Their foolhardy inclination to go for the big trophies would probably have produced fatal injuries in time. Nor was their casual attitude toward people and vehicles a good thing. Due to their strategy of going into human camps and eating chickens, they would not have survived the second experimental release if they had not been under the protection of South African game officials, who were there to "dissuade the irate owners" from spearing the cheetahs.

The bleating of the kid excites the tiger

Teenage great cats look physically readier to hunt than they are. They may be as large as adults but tiger cubs don't get their permanent teeth until they're a year old.

With the exception of the strangely public tigers of Ranthambhore Reserve in the 1980s, tigers are very hard to observe. Unlike lions, who lie around in great sleepy heaps in the sun in many African parks, or cheetahs, who sprint after prey in the open, tigers (and leopards) are stealthy hunters from concealment and live in areas that offer good cover. Human hunters who want to kill tigers, human ecotourists who want to see tigers, and biologists who want to observe the predatory behavior of tigers have all used the ancient technique of tying out livestock as bait and waiting in an elevated place to see if a tiger comes under cover of night.

On these occasions mother tigers have been seen letting their cubs practice. George Schaller repeatedly staked out buffalo and saw mothers bring their cubs to the bait. Sometimes the mother would simply lie down and let the cubs do their utmost. If the cubs were too intimidated by jabs of the buffalo's horns, or if they were simply too small or inept to knock the buffalo down, the mother might do that, and leave the rest to the cubs. Sometimes she held the buffalo down while the cubs clambered over it.

Padmini, a tiger in Ranthambhore, brought her four cubs to where a buffalo had been tied out as bait, and disabled it slightly by belting it fiercely on the hindquarters. Then she sat and looked on for half an hour while the cubs circled the buffalo and the buffalo jabbed at them. Finally Akbar got up his courage to leap on it and knock it down. His brother Hamir promptly sat on the buffalo while Akbar killed it with a throat bite. Padmini joined them in eating the buffalo while the other two cubs hung back.

Fiona and Mel Sunquist examined the kills of wild tigers in Nepal and found that the tigers had two methods of finishing off the prey they had caught. Small animals were bitten on the nape, a quick death. Larger animals were killed with a bite to the throat, which in essence suffocates the animal. Young tigers usually used the throat bite even on small prey, perhaps because it is easier and safer for them.

That certain stagger

Great cats, like house cats, have innate propensities for selecting prey. Completely naive kittens are bewitched by certain promising phenomena: rustling, squeaking, and furtive scuttling all fill them with a joyful mania to investigate. Halting gaits also cast a spell. Tippi Hedren, an actress who runs a sanctuary for great cats, describes how Needra, a lion raised from infancy by humans, whose best friends were humans, who had never witnessed an act of predation, was riveted by the sight of a small child who had come to visit, crouched, and sprang on him. Fortunately, adults were able to pull Needra off the child and he was unharmed—after all, she had never developed techniques for the next step. The sanctuary lions were also gripped with morbid fascination when one of the staff began limping due to injuries suffered in a car accident. They slunk along the fences, eyes glued to him with a new interest. On another occasion, Casey, one of Hedren's hand-raised lion cubs, now a 130- to 140-pound teenager, escaped from the family's house in Sherman Oaks, California, and began strolling downtown. To ensure that Casey would follow her home, Hedren began to limp dramatically. Enchanted, Casey turned, flattened down low, and began to creep after the apparently crippled woman. Outside the house, just as he was about to pounce, Hedren turned, flung a leash on him, and tied him to the bumper of a car.

Starter victims

Valmik Thapar, observing the cubs of a tiger called Laxmi, saw them practicing their stalking skills on peafowl, partridges, squirrels, and mice. Inexperienced great cats sometimes thrill themselves by catching something they then can't eat. The tiger cub Bhimsen, not quite a year old, caught and killed a little crocodile, took it into the bushes—and couldn't bring himself to eat it. Laxmi's cubs put a whole night into catching a civet, and when they finally succeeded, found it just too awful to eat.

Take Your Cubs to Work Day

Lions don't usually bring living prey to their cubs (although George Schaller did see one lioness bring a gazelle fawn), and lion cubs stay dependent for longer than most great cats. Two-year-old lions will probably not have hunted large prey by themselves, though they may have helped in group efforts. But cubs who are only a few months old follow and observe the adults in the pride. Schaller describes a lioness catching a zebra in a streambed in the Serengeti. The cubs who sat in a row on the bank and watched her kill it can't all have been her own, because there were 13 of them.

Older lion cubs joining adults in nighttime hunts in Etosha National Park did not simply copy their mothers' behavior, although they learned from watching adults. In these hunts, lionesses take different roles, which zoologist P. E. Stander calls "the centers and the wings." A young lioness whose mother was a center might see that a wing was needed, and stalk accordingly.

The tiger Laxmi was prowling with her two cubs, Kati Nak and Ladli, when she spotted a chital. She froze in place, paw midway through a step. The obtuse Kati Nak scampered up to nuzzle his mother, saw the chital, and tore off after it, with Ladli on his heels. The chital raced away, uttering alarm barks that ensured that every animal in earshot knew that predators were afoot. Without a sign of irritation, Laxmi lay down in the shade.

Wild tiger Noorjahan was hunting, accompanied by her three yearling cubs. Seeing a chital in a favorable location, she crouched in undergrowth. Taking her cue, the three cubs crouched too. But one, Bhimsen, couldn't stand the suspense—he couldn't *see* anything. He raised his head to peer at the chital. Noorjahan growled softly, and Bhimsen crouched back down. A moment later he stuck his head up again, Noorjahan growled, he flattened himself. But he just couldn't resist peeking and soon the chital spotted him, and another stalk was spoiled.

Sometimes when she really needed to buckle down and produce the groceries, Noorjahan would ditch her cubs so she could hunt without their "help," growling at them if they followed. While cubs can be a terrible nuisance, with time they can also become part of cooperative hunts.

Tricks of the trade

In the Ranthambhore Reserve in Rajasthan Province, India, in the 1980s, a wild tiger startled observers by taking up hunting in water. This park contains shallow lakes thick with vegetation and, in hot weather, sambar and chital deer would wade in and graze there, in addition to grazing in the meadows by the lakes. In the water they had to keep an eye out for crocodiles, but the coolness and the food made it worthwhile.

In 1983 a male tiger called Genghis adopted the lakes as his territory, and ran off all but a few tigresses. Early in the day and late in the afternoon he would stroll by the lakes without concealment, surprising and delighting the human watchers, and unnerving the deer dreadfully.

One day observers saw Genghis lurking in thickets by the lake, so settled down to await events. In the late morning he seemed to have murderous intentions toward a small group of deer walking along the shore, but a panicky peahen tipped the deer off. Hours later, as two herds of sambar grazed in the lake, Genghis suddenly rocketed out of the thicket and bounded straight into the water. Deer struggled to escape in all directions, and cameras clicked wildly as Genghis swam, swerved, and leapt on a straggler. The force of his pounce carried both deer and tiger so far underwater that only the tip of Genghis's tail can be seen in photographs.

As he surfaced with his prey, the onlookers were beside themselves with astonishment. "Never before had we seen a tiger even attempt to launch an attack in the waters of the lake; nor was it something we had ever come across in old accounts. Was this just a temporary aberration or were we seeing something really new?" wrote Valmik Thapar. "Nowhere in the literature of the past 200 years have we been able to find any other account of a tiger behaving in this way. As a strategist he is unmatched—an innovator. "

Genghis continued to hunt deer in the water and came up with several variations on the theme. The original technique was to lie unnoticed in waterside vegetation, sometimes for hours, until a group of deer grazing in the water moved close enough for Genghis to burst out and charge into the water after them. But as the summer went on, the water level in the lakes dropped, and the belt of vegetation thick enough to hide a tiger was farther from the water's edge.

Genghis, switching from lake to lake, developed the tactic of swimming into the lake, watching the sambar as he did so. As they floundered away in different directions, he'd pick out a deer that got separated from the rest, and head it off. He also discovered that if he simply sauntered to the edge of the water and stared at the sambar, they would panic and head for the sides of the lake. He'd gallop along the shore in clear view, looking sideways at the terrified sambar thrashing in the water, waiting for one to make a stupid move, and he could often cut one off as it fled the water.

Using these aquatic techniques, Genghis caught one out of five deer he went after, an unusually high success rate. (The Sunquists estimated that tigers they studied in Nepal made a kill in one of 20 hunts.)

Genghis disappeared in 1984, and the male tiger who took over his territory, Kublai, had no idea about hunting in water. But Noon, one of the tigresses who had shared the area with Genghis, took over some of his techniques. Inexpert and uncertain at first, she became more skillful and adopted the tactic of the sudden charge from cover.

Given that tigers have a variety of hunting techniques, and are capable of inventing new ones, I suspect that when tigers think—when cats of any kind think—they are musing about how to become a better mousetrap.

*I hope to teach this fearsome beast to kill and
then let it go do whatever it wants, why?*

Arjan Singh raised several leopards from cubs, letting them roam in a
nearby forest preserve in northern India. His first leopard, Prince, was
slow to master predation. He wanted very much to hunt, but it took
time. At 15 months old, he killed some prey that he then did not know
how to open and eat. He was not above eating rodents.

Finding a calf tied up in a hut, he seized it in an inexpert but
fanatic grip. Called to the scene, Singh felt that Prince would eventually succeed in suffocating the calf, but the calf's owner objected. Singh
grabbed Prince by the scruff of the neck and tried to haul him off, but
the obsessed leopard wouldn't let go. They tried to pinch Prince's nostrils shut to make him let go, and that didn't work. Finally an assistant
jammed a stick in Prince's mouth while Singh beat him with a stick of
sugarcane and Prince let go and ran out of the hut.

Despite Singh's unsportsmanlike behavior, Prince still invited him
on hunting expeditions. By 20 months old, Prince knew how to open
and eat the prey he had killed and was almost self-sufficient. Soon he
was entirely independent and stopped visiting Singh.

Two leopard sisters Singh raised, Harriet and Juliette, became independent on a similar schedule, although they were better at staying in
touch. When Harriet felt the need to move her first litter of cubs from a
den under a tree, she carried them back to Singh's house and put them in
a spare bedroom. The second-generation cubs acquired predatory skills
much earlier than their mothers had—Mameena, a cub from Harriet's
second litter, was killing for herself at 11 months old.

Mentors and role models

The story of Leo, an orphaned lion cub raised by the family of Kobus
Krüger, game ranger in South Africa, and his wife Kobie Krüger, and
recounted in Kobie's book, *The Wilderness Family*, illustrates how serious omissions in the normal learning environment can be for the
predatory prowess of a great cat. Leo knew how to crouch, stalk, and
pounce, and often practiced these skills on the Krügers. He knew how
to eat wild game he was served. But what came between these two
things was not clear to him.

He loved to walk in the game preserve, and he was fascinated by prey animals, but he had no idea what to do with them. Kobie Krüger, who showed a fainthearted unwillingness to demonstrate by leaping on an impala and sinking her teeth into its throat, was of little use to her lion pupil. The being Leo took for his mentor was not a human, but the Krügers' wise and impressive Australian cattle dog Wolfie. When they went out in the bush, Leo paid close attention. When Wolfie ran along, sniffing the ground, Leo followed, studying Wolfie's face for hints. "He strongly suspected Wolfie to be the guy who knew all the things that it was crucial for a lion to learn." If Wolfie dug frantically, Leo stood in front of him and watched thoughtfully.

Sadly, he had chosen a guy in the wrong profession. When they encountered game, Wolfie was usually well mannered enough to leave them alone, but when things got too exciting he couldn't resist chasing them—and herding them. Leo understood chasing, and he worked on herding.

When Wolfie stuck his head in a burrow and a warthog and her three piglets burst out, Wolfie took off after the warthog and the screaming piglets. Leo raced after them too. Wolfie ran on one side of the warthog family, herding them. So did Leo, looking happy. When Wolfie had had enough fun, he ran back to the Krügers for praise (they praised him "for being kind to animals and for not hurting them"). They were forced to pat Leo too, even though they were beginning to realize with sickening clarity that Leo wanted to be an Australian cattle dog when he grew up.

Kobie Krüger tried to give Leo hunting lessons. She took Leo and Wolfie out to hunt impala and succeeded in locating an unsuspecting herd. She crouched low in the long grass and told Wolfie to get down. Wolfie obeyed. Although Leo did not understand when Krüger gave instructions like this, he did what his idol Wolfie did. She ordered Wolfie to stay, and dog and lion both stayed in place while Krüger crept through the grass until she was on the other side of the herd, where the impala would get her scent. She jumped up, spooking the impala into running straight into the dog-lion ambush. She shouted to Wolfie to get the impala. Dog and lion burst out, right into the path of the panicked impala. "I hoped with all my heart that Leo would respond to some inborn instinct and grab an impala."

But no. Inborn instinct told Leo nothing. Leo stood and stared at the hysterical impala, excited but baffled. "His instincts were obviously signaling that something very important was happening and that there was something urgent that he should do." In that state of confusion, he spotted Krüger, lit up with certainty, bounded through the herd, and pounced on her.

One of Leo's favorite foods was an impala leg. But perhaps the fact that he liked eating impala legs but didn't think of leaping on an impala and killing it is no more surprising than the fact that children who like to eat chicken drumsticks seldom think of leaping on hens and killing them, even if they're really hungry.

A few days later, the threesome startled some buffalo. Krüger managed to keep Wolfie from pursuing, but Leo ran after them, herded them downriver, and came back for praise.

Lacking a suitable role model, Leo did not learn to kill. The idea of letting him go free had to be abandoned. He eventually found a home at a game park maintained for the benefit of photographers and filmmakers, where he was fed, hung out with lionesses, and behaved in a photogenic manner.

Like Leo, Penny, a leopard raised and released by Joy and George Adamson, did not seem to know what to do after a certain point in the hunt. One day when she was a little over a year old she spotted an aardwolf sitting in the sun and having a wash. With admirable form she crouched, used cover to get close to him, sprang from hiding, and gave him *such a slap*. The aardwolf took off running, and Penny gazed at the Adamsons in perplexity, "as though to ask, 'What next?'"

Even when released great cats are not able to catch quite enough prey to feed themselves, they are able to raise cubs that can. The second-generation leopard Mameena, whose mother was raised by humans, was self-supporting. Pippa, a cheetah raised by humans from cubhood and rehabilitated by Joy Adamson, experienced no difficulty in relating to other cheetahs, successfully mating with wild males and raising several litters in the wild. Her only deficit was in hunting. She wasn't very good at it, but Adamson supplied her and her cubs with a steady stream of goat carcasses for most of her adult life. When she didn't get these supplies, sometimes she managed to get enough food on her own and sometimes she didn't. Her cubs, however, who began hunting at an earlier age, became self-sufficient and raised litters of their own.

An Englishcat in India

Perhaps the only captive-reared tiger ever to be released to the wild was Tara, born in the Twycross Zoo in the UK. Arjan Singh had Tara shipped to India, where he raised her at his home, Tiger Haven, adjoining a forest preserve prowled by wild tigers. He had previously raised and released leopards there, with the aid of his yellow dog Eelie.

Tara spent her days playing and walking in the forest with Singh and Eelie. At night Singh kept her in the house or, after she was older, in a large cage.

Like cubs reared by their mothers, Tara seems to have started out as a floppy, bounding, undisciplined creature who enjoyed chasing animals and seeing them flee. If Eelie chased an animal, Tara would join in. Wild pigs did not appeal to her at first, but they appealed to Eelie greatly, so Tara chased them too. On one occasion she chased a bicyclist until he fell off his bike, and she lost interest.

Tara was rather bad at taking cover, which Singh tentatively ascribed to Eelie's influence.* When she was nine months old, Tara made her first kill, a chital fawn that had gotten tangled in some creepers when Tara put the rest of the herd to flight. Eelie felt he had just as much right to it as Tara, but Tara carried it up a tree and wouldn't share.

Tara had a feud with a troop of otters who chittered maddeningly at Eelie and her one day as they walked with Singh by the river. Later that day she abandoned her companions, stayed out all night, and was found the next day covered with tiny bite marks in odd places. As Singh reconstructed it, Tara had gone back to teach the otters a lesson, and had caught one. Not yet having permanent teeth, she sat on it. It bit her, as did the rest of the gang, until she let it go. Some months later, when Tara had her new canine teeth, she intercepted a group of otters who were crossing a field and killed one. The rest of the otters surrounded her, chittering, and she "looked from side to side, all the time grinning nervously." A possible mob scene was averted by the arrival of a human on Tara's team.

As she grew older, Tara was more and more drawn to the forest, leaving earlier in the morning and coming back later at night. She was

---

* In other words, the way to become a tiger may be to have a better role model than a dog.

just over 20 months old when she stopped coming back—as anyone might have guessed, she had met a guy. A slightly older guy, one who got his own dinner and was delighted to share with Tara. She didn't come home, didn't write, didn't call. Whenever Singh heard that livestock had been killed by tigers, he rushed to the scene to see if Tara had been there. At first it was easy to tell, since Tara's methods were inexpert. If it had been killed with a single throat bite, that wasn't Tara's work. If a tiger had leapt on the hindquarters of an animal lying down, and clawed at it wildly before killing it with a nape bite, that was Tara. Singh fretted—a nape bite is dangerous on a buffalo, Tara. It could gore you.

She had been with Singh for 17 months before she stopped coming back. "It will be obvious that nowhere do I claim to have taught Tara anything, for the simple reason that a human cannot teach an animal: his lifestyle is too different," writes Singh. But experience and, perhaps, the examples of the wild tiger, did teach Tara, and with time she became self-sufficient. She successfully raised at least three cubs.* Happily for the neighbors, she seems to have had no further interest in associating with humans.

How to fix dinner

Some food requires preparation. Mountain gorillas at Karisoke, Rwanda, who eat herbaceous plants, have no difficulty finding food. When they sit in the rain forest, they are surrounded by it, and it isn't trying to get away. Very little of it is toxic or indigestible, so learning what's good to eat is not a major task for a gorilla child. But many of the best food plants have stings, spines, hooks, or tough cases, so learning how to eat these is harder. Mountain gorillas are fond of nettles, but dislike being stung. Adult gorillas pick nettle leaves and fold the stinging top and edges into the middle of a bundle before they eat them. Gorilla children haven't figured this out yet, and get stung when they first start eating nettles. The only adult gorillas at Karisoke who

---

* It is unclear whether Tara was descended from Indian tigers, as Singh writes that he was assured. It is said that testing has shown Siberian tiger genes present in wild tigers now living in the area, a source of great displeasure for those who hope to save not only the tiger but the different subspecies of tiger.

didn't fold their nettle leaves were a female with maimed hands and a female named Picasso, who transferred into the area as an adult. Picasso grew up at a lower elevation, where few nettles grow. Investigators Richard and Jennifer Byrne suggest that in her youth Picasso didn't get to observe other gorillas folding nettles and so didn't learn to do it herself. Interestingly, her son Ineza, three years old, was not very good at processing his nettles.

The Byrnes suggest that the gorillas learn how to process difficult foods by trial and error, but that young gorillas may be showing program-level imitation. Thus a gorilla child gets the idea to fold nettles from noticing that its mother folds nettles before eating, but it doesn't precisely copy her movements. Instead it figures out for itself how to fold nettles; as a result, each gorilla has its own idiosyncratic ways of executing the same processing techniques.

Thick-billed parrots eat a lot of pine nuts, which they extract from pinecones. Learning to do this takes young parrots several months. Captive-reared birds who were given pinecones for six months before being released eventually learned how to get food out of them—but then when they were free, couldn't figure out where to find pinecones.

Squirrels? How could it be squirrels?

In the early 1980s a high school biology teacher in Israel, Ran Aisner, took students on a field trip to some plantations of Jerusalem pine. These had been planted in the last 50 years and did not constitute a particularly natural habitat. Aisner noticed piles of pinecone scales and bare pinecone shafts under some trees on the edge of the plantation and brought them to zoologist Joseph Terkel. Clearly, Terkel said, some creature had been stripping the scales off the cones to get at the pine nuts inside, and the culprit in such a case is generally a squirrel, but there are no squirrels in Israel.

Eventually, by placing traps in the branches of the pines, Aisner caught some black rats. There turned out to be a population of shy rats that ate only pine nuts, drank only water they licked from pine needles, and lived in tree nests made of pine needles.

Through surreptitious observation it was discovered that the rats selected a ripe cone, gnawed it off the branch, carried it to a better spot for a sustained bout of gnawing, and took it apart in a quick and systematic manner. This technique involves starting at the base of the cone and pulling the scales off in a manner that entails a minimum of gnawing. The scales wind around the shaft of the cone in a spiral, and the efficient thing to do is to take them off in a spiral path. After a few turns—a few rows of scales—the rat will find a pine nut under each scale. The bits fall to the forest floor, potentially alerting sharp-eyed nature buffs.

In Terkel's laboratory they began to investigate these rats and their methods. Rats of the same species who were trapped in urban sites such as warehouses were clueless about cones and not interested in obtaining clues. If they were really hungry, and there was nothing else in the cage, they would eventually chew on the cones, but their method was simply to start gnawing inward from a random point, a procedure so inefficient that they would have starved had the researchers not eventually given them rat chow. Not one developed the spiral technique.

If the city rats were housed with a rat from the piney woods who knew how to strip cones, that didn't help. After three months of bunking with an expert, they still had no inkling how it was done.

Unsurprisingly, rat pups who grew up with mothers who got their pine nuts by stripping pinecones with the spiral technique also stripped cones with the spiral technique. Of 33 rat pups raised this way in the laboratory, 31 could do it. Two could not and gnawed at random.

The helpful researchers then provided a set of clues to adult rats who could not imagine how to strip cones. They gave them pinecones that had been started. The first four rows of scales had been stripped off, so a rat that proceeded to remove scales in a spiral manner would find a pine nut under each one. Of 51 grown rats, 35 could finish the job on a started cone, using the spiral method.

Twenty of these rats then received further tuition. Having succeeded in doing this with cones that had had four rows removed, they were then given cones with three rows removed, and if they succeeded with those they got cones with two rows removed, and so forth. Finally they got intact cones. At the end of the course all of the rats could open cones that had had just one row removed, and 18 passed the final exam. They could open an intact pinecone and extract all the seeds with the spiral technique.

How this whole thing got started is unknown. There might have been an innovative female rat who came up with the technique and passed it on. (An innovative male who came up with the technique would not have had a chance to pass it on.) Or perhaps there was a female who ate some pine nuts and some other foods (since without good technique a rat cannot get enough pine nuts to survive) and who left partly eaten pinecones around, and this was sufficiently illustrative to let the pups figure it out.

More techniques

Sea otters are famous for their use of rocks to smash shellfish. In the canonical example, a sea otter floating photogenically on its back places a flattish rock on its chest, holds a clam in its paws, and smashes the clam on the rock. When an otter has found a good rock, it will carry it around in the web of loose skin under its forearm (the term "pawpit" is tempting, but not quite right). An otter may keep track of a really good rock for years. Sometimes a second rock is used to hammer the hapless shellfish.

Underwater, otters also use rocks or other objects as tools to pry or bash abalone and sea urchins loose from rocks. If they catch more than one crab, otters will wrap the ones they aren't eating in strands of seaweed to keep them from getting away while they eat the first one. In Monterey Bay, Female 532 was seen using a piece of abalone shell to scoop abalone meat out of its shell.

Pups stay with their mothers for six months or more, getting much of their food from their mothers and improving their skills. In one study the pups came up with food on 13 percent of their dives, while their mothers got food on 70 percent. Pups aren't initially sure what things are food. Marianne Riedman describes a pup named Josie industriously propelling an old car tire to the surface and giving it a serious chewing before deciding that it wasn't food.

Female 190 passed on her liking for rock oysters to her daughters Josie and Tubehead. To knock them off the rocks, Female 190 favored a glass bottle as a tool. (People are always tossing cans and bottles into the water, which is disgusting but not without utility to otters and octopuses. People seldom toss in can openers, corkscrews, hacksaws, and other really useful tools.) When they were old enough, Josie and Tubehead also used glass bottles to get their oysters.

Whose brilliant idea was childproof packaging?

For some predators, killing their prey is easy compared to actually eating it. It's too big. They don't know where to start. The skin is tough. Or maybe it's covered with weird stuff—feathers or hair or scales or something. Biologists often refer to this as "opening" carcasses.

Prem, a hunting guide who assisted Fiona and Mel Sunquist in Nepal, once tried to feed some wild orphaned tiger cubs, about six months old, by tying out a goat for them. When he came back the next day he found the goat dead but uneaten. Somehow the cubs had managed to kill it, but with their baby teeth and lack of experience, they had been unable to open it. They had, however, licked it bald.

Even if eating it isn't a problem, getting dinner home can be difficult. One wild Malayan tiger developed a labor-saving way of moving the carcasses of his prey. He dragged them into the nearest stream, and then towed them through the shallow water until he got to a good place to begin eating. In the water they were easier to pull and perhaps less likely to catch on roots and saplings. This was a great source of annoyance to the tiger hunter who was trying to follow him, as he left no tracks.

Save some for later

Niff, a red fox cub raised by David Macdonald, was already caching food when she was one month old, although she had never seen anyone else cache food and had never known shortage. Macdonald's house, not being floored with dirt for burying things, offered her insufficient scope, but she persevered, stuffing food into corners with her nose and then attempting to conceal it by pushing notebooks over it. Eventually she came up with the brilliant idea of hiding food in the bedclothes, an idea many human children have devised. It was awkward, because she shared the bed with Macdonald, and he did not like the idea of finding a dead chick under the pillow, nor did she like the idea of him stealing her dead chick.

When Niff was older, Macdonald took her for walks through the fields. Sometimes he had arranged for the path they took to be sprinkled with dead mice and other taste treats, to see what Niff would do.

After eating a few, Niff began caching mice, and given the proper setting she performed superbly. She'd carry the mice off the path, dig a shallow hole and bury the mouse in it, sweep dirt over it with her nose, and then muss up the grass stems over the spot, which prevented the site from being obvious. This behavior, clearly innate, gave Niff great satisfaction. She remembered her caches, and might go back to check on them more than a week later.

At one point, when local farmers had caught many mice raiding their chicken houses and donated them to Niff's cause, she was particularly well fed and began to get slapdash. One day she didn't even cache all the mice she found, and those she did cache she buried so carelessly that their tails stuck out. By evening, when they walked the path again, Niff was hungry and went straight for her caches. Horror awaited—magpies had stolen her mice. She ran from spot to spot, but all the mice were gone. The next day, all the mice she did not eat were buried with exacting craftsmanship. Niff had learned the importance of doing something well she innately knew how to do.

Career choices

Supporting oneself is of course vital, and animals who survive often have the help of many innate hints. But learning is still tremendously valuable—if the food available in the environment changes (if caterpillars start pretending to be sticks, if leaves evolve toxins, if you're confronted with ostriches instead of peacocks), you'll have a chance of figuring out how to eat.

SIX

How Not to Be Eaten

In the Brazilian forest, biologist Karen Strier was watching a group of muriqui monkeys, whom she had been quietly following for six months, trying to habituate them to her presence. It was hot, and the muriquis were resting, when a male from another group dropped by. Suddenly he spotted Strier and launched into a threat display, uttering frenzied calls, swinging wildly around, breaking branches and dropping them near Strier. She worried that his alarm might be infectious, and undo her patient work of months. Four females hurried over to see what was so awful. Arriving at the scene, they hesitated and huddled together, looking back and forth between the raucous male and the source of his alarm. After a few seconds the four charged at the male and threatened him. He seemed astonished, froze, and then took off, hotly pursued by the females.

A few minutes later, they returned. "The females began to embrace one another, chuckling softly as they hung suspended by their tails, wrapping their long arms and legs around each other. Two of the females disengaged themselves from the others. Still suspended by their tails, they hung side by side holding hands and chuckling. Then they extended their arms toward me, in a gesture that, among muriquis, is a way to offer a reassuring hug."

Strier didn't hug them, badly though she wanted to, because she intended to remain a passive observer, not an active participant in the muriquis' life, and because she wanted to be certain that no disease or parasites were passed from her to the monkeys or the other way around. But she knew she had been accepted as a benign presence in their lives and ruled out as a predator.

Learning the bogeyman's name

A vital thing that most baby animals must learn is who might want to eat them. It's almost certain that somebody does. Even animals at the top of the food chain, huge and armored or bristling with teeth, start as infants, relatively tiny and tender and defenseless. And even those who don't plan to eat a baby animal might harm it. On the other hand, being afraid of everybody is a lot of work, it disrupts your life, and it's not even accurate—not everyone means you harm. It's best to distinguish between enemies, friends, and noncombatants, as the muriquis did. Some of the ways animals do this are innate, some are innate but triggered by learning, and some must be learned more or less from scratch.

In early experiments on enemy recognition in birds, ornithologists Margaret Nice and Joost ter Pelkwyk raised song sparrow nestlings indoors and measured their reactions to things they might or might not view as threats. Song sparrows were an ideal species for such tests, they wrote, because they showed gradations of worry, unlike goldfinches, which were either indifferent or hysterical, with nothing in between. In the first, or "alarm," stage of concern, Nice and ter Pelkwyk write, the song sparrow says "tchunk" and may raise its crest, raise and flip its tail, flip its wings, and change location. In the second, or "fear," stage the bird says "tik" and compresses its feathers, elongates its neck, and crouches. In the third, "fright," stage the bird says "tik-tik-tik" and flees, hides, flutters desperately, and pants with open bill. Most of the stimuli presented to their sparrows elicited varying numbers of tchunk calls per minute, which they graphed on what I can only term a tchunkogram.

Young hand-raised birds did not tchunk much at cats, although one of them not only tchunked but escalated to tik-tik-tik after spending some time on a screened porch where cats came to sit on the railings. Stationary (stuffed) hawks produced tchunking and tikking, but moving models of hawks elicited the tik-tik-tik of terror. Snakes weren't scary, even when they were zipping around the room. A stationary (stuffed) barred owl produced steady tchunking and even some tiks, and was even more worrisome when it was placed on the Victrola. However, the young birds were getting used to it until they were treated to the spooky sight of the owl perched on the piano in the next room. For some reason this bloodcurdling vision made a lasting impression, to the extent that afterward they tchunked at the piano even when it didn't have an owl on it.

Nice and ter Pelkwyk concluded that song sparrows know their enemies by a combination of inborn and learned patterns. Thus owls were innately scary, but piano fear was learned.

It's a bird! It's a plane!
Oh no, it really is a bird!

Many young birds (and some mammals) respond with terror when a generic bird shape with a long tail and short head passes overhead. But if the shape passes overhead in the reverse direction, so that, although unchanged, it now appears to have a long neck and a short tail, all is calm and good cheer. In the first orientation the shape looks likes a hawk (particularly like a falcon or accipiter), a worrisome thing if you are small and tasty, but in the second orientation it looks like a goose, and few birds need to fear goose attacks.

This behavioral program has room for learning. Ellen Thaler, at the Alpenzoo in Innsbruck, compared the reaction to possible predators in chicks raised by mothers and hand-raised chicks. These were chicks of rock partridges, hazel hens, and rock ptarmigans. In their first week, all chicks were terrified by dummies of predatory birds being moved overhead; by an actual sparrowhawk that occasionally dropped by the zoo to see if lunch was being served; and by just about anything that moved quickly. They would give an alarm call, crouch, and freeze.

But as the weeks went by, the hand-raised chicks became blasé. They scarcely bothered to crouch and glanced only briefly at the predator. They often didn't give alarm calls. By six weeks old, they didn't even stop eating. They only responded when the wild sparrowhawk (a dedicated volunteer) flung itself against the bars of the aviary in attempts to grab them. They panicked, "more 'hysterically'" than the chicks raised by their mothers.

Meanwhile chicks raised by their mothers continued vigilant, even though they had experienced the same number of threats. The difference seemed to be that they saw their mother's reaction to predators, and that convinced them that they were right to worry.

Thick-billed parrots, some captive-born, some wild-born, were kept in outdoor aviaries in Arizona, awaiting release into the wild. From the aviaries they could see predatory birds, and captive-born parrots could witness the alarm of their wild-born companions. To drive the point home, goshawks sometimes went so far as to swoop down and hit the cage in attempts to find out how parrot tastes. The captive-born birds learned to respond to the thick-billed parrot alarm call. Despite this cautionary experience, many captive-born parrots were killed by goshawks and red-tailed hawks shortly after release, perhaps due to being insufficiently watchful. They knew they should be worried, but they hadn't learned how to act on their worries.

Hawk shapes are scary, and so are eyes. As a result, many creatures are decked with eyespots that may cause predators to hesitate before grabbing them. Some researchers suspect that primates and perhaps other animals may have an innate response to leopard spots. The rosette spots of leopards, jaguars, and many smaller cats are believed to be an ancient pattern. Since spotted cats have been around for so long, primates have been around them for a long time. Bonnet macaques in an urban troop that had probably never seen a leopard were appalled when they saw a spotted model leopard and alarm-called like crazy. Not only were they upset by the sight of the nicely painted model, they were upset by the sight of a model made out of leopard-printed towels.

Researchers Richard Coss and Uma Ramakrishnan presented wild troops of bonnet macaques with leopard models and measured the level of their alarm. A realistic spotted leopard upset them much more than the same model upside down, which still upset them more than a leopard model in solid brown, which upset them more than the solid model upside down. When seeing a model with anomalous qualities, such as being upside down, they looked longer at it and also looked at other members of the troop, apparently to see how they were reacting.

We blame the snake

Many animals, including many primates, including many humans, are afraid of snakes. Whether this fear is learned or innate has been hotly debated, especially in the company of persons who state that they have an instinctive fear of snakes and other persons who want to prove to them that their pet Kaa, or Nag, or Ourobouros, is not slimy at all, but pleasingly smooth and dry.

As is so often the case, the truth is in between. It seems that many creatures are very ready to fear snake-shaped things that move in a snaky way, but although this often leads to fear of snakes, it does not inevitably do so.

If a rhesus monkey raised in a laboratory sees another rhesus behave fearfully around a snake, the first monkey will fear snakes too. Eight minutes of seeing the other monkey act fearfully at the sight of a snake is enough to produce a lasting fear. The newly cautious rhesus won't go into a room where it can see a snake through a window, it won't reach across a fake snake to get food, and its fear can infect a third rhesus with the same fear.

Snakes have been around for a long time, and many of them are dangerous. If you have to learn this by being bitten or even by seeing others bitten, that's going to be expensive in terms of species survival.

Snakes are special. Researchers tried hard to instill fear of flowers in monkeys, using the same techniques that had produced fear of snakes, and they couldn't do it. One researcher did induce observational fear of "arbitrary objects such as kitchen utensils" in male rhesus monkeys. The first monkeys were shown the utensils and then frightened by blasts of air. The second group of four monkeys saw a monkey reacting with fear to the sight of the utensil, and subsequently three of them were fearful at the sight of the utensil. (Sadly, the crucial question of what sort of kitchen utensil it was is missing from my reference. If it was a potato masher, that is one thing, but anyone might fear one of the more elaborate corkscrews.)

A rhesus monkey can also learn that snakes are okay. If it sees that other monkeys are calm around snakes, it will itself be calm about them. It becomes "immunized," in the sense that it is less likely to fear snakes through witnessing another monkey's fear.

Viki, a chimpanzee raised by Cathy and Keith Hayes, had never seen a snake. One day, when Cathy Hayes was carrying Viki and a portable heater, the cord came loose and trailed behind them, and Viki flew into hysterics. "After that whenever I carelessly dangled the heater cord, she went into paroxysms of fear, and I philosophized upon the ancient enmity of primate and serpent."

But then one day as they played on their Florida lawn, a black racer snake slithered within inches of Hayes and Viki, who looked "mystified," but no worse. Hayes gave Viki a dead garter snake to play with. She didn't find it creepy, even when Hayes trailed it over Viki's toes. "But the instant her eyes fell on the heater cord, a frown appeared, her face flushed, and she burst out screaming. What was I to conclude but that Viki had an instinctive fear of electric-heater cords?"

Viki continued to find real snakes mildly interesting and enjoyed chasing them off the lawn. One morning Hayes and Viki went outside and nearly stepped on a large rattlesnake. Hayes grabbed Viki by the skirt of her adorable outfit, and hurled her away from the snake. Viki screamed, and the snake feinted at Hayes and slithered away. A week later they saw a coral snake in the garden, and both froze, Viki clutching Hayes's skirt until the snake departed peacefully. Viki never chased another snake. (Whether she continued to fear the heater cord is not recorded.)

It seems that Viki had a predisposition to fear snake-shaped things moving snakily, which was triggered by the sight of the heater cord pursuing her. Real snakes didn't trigger this predisposition until the day of the rattlesnake, when Hayes's reaction left Viki in no doubt about how awful snakes are.

Afraid of snakes? Me?

For some animals, fear is not the best reaction to snakes. Bernd Heinrich presented a garter snake to his hand-raised great horned owl, Bubo. Great horned owls eat snakes and have no use for innate snake fear. This snake tried hard to be scary. It coiled, it flattened, it lashed about, and it lunged toward Bubo, showing the pink inside of its mouth. Somewhat flustered, Bubo leapt onto the snake, who promptly sank its fangs into Bubo's foot. Bubo flew off, trailing the snake, and the snake let go and escaped into the woods. The next time Heinrich produced a snake, Bubo was more decisive, and after watching the snake do its intimidating maneuvers, leapt on it with both feet, crushed its head with his bill, and swallowed it by the head first.

Flenter, a bull terrier belonging to Kobie Krüger, also viewed snakes with hostile intent. He harassed a spitting cobra living in a flower bed in the garden of the Krügers' house in a South African wildlife reserve. Naturally, being a spitting cobra, it spat venom in his eyes. The first time this happened, one of the family rushed Flenter to the nearest garden spigot and rinsed out his eyes. The second time it happened, Flenter rushed himself to the tap and waited to have his eyes flushed out. Subsequently, if there was no one in the garden to witness his need, he'd go to the kitchen door and howl for help and when someone came out, he'd dash to the tap. (His being a bull terrier, it did not occur to him to leave the snake alone. The Krügers' Australian shepherd had no trouble with this concept.)

Mobbing

Birds sometimes assemble in mobs to scream at, taunt, threaten, and even attack predators, often going after birds of prey they find perched, uttering alarm calls, fluttering around them, perhaps swooping over their back or head and pecking at them. Mixed flocks of species, attracted by the sound of mobbing birds, may form if the target doesn't move on. Daphne Sheldrick describes how Red Head, a wild red-headed weaver living in her garden in Kenya, became so tame that he would enlist her help in mobbing snakes, flying to her or her husband, uttering an alarm note, then swooping back toward the snake.

One bird-watcher described finding two robins, a catbird, two chewinks (towhees), and a hummingbird hysterically mobbing something on the ground in a cottonwood grove. When the birdwatcher investigated, she found a tiny kitten, eyes barely open. Well, they were right in principle.

In experiments on mobbing, Eberhard Curio put European blackbirds in compartments from which they could see another blackbird mobbing some object. By rigging the compartments suitably, Curio could delude the watching bird about what was being mobbed. For example, it might see a blackbird that was mobbing, and think it was mobbing a honeyeater, when in fact the mobbing blackbird was seeing and directing its mobbing at an owl. Curio designed experiments to test how easily blackbirds could be induced to mob a stuffed honeyeater, a large Australian bird that the blackbirds would not encounter in nature, as opposed to a multicolored plastic bottle about the size of a honeyeater. They learned to mob both but could never be convinced to fear the bottle as much as the honeyeater. The news that honeyeaters are enemies to be mobbed could be passed along a chain of six birds, each learning from the one before, without losing its power to inflame and outrage.*

New Zealand researchers used mobbing to teach New Zealand robins that stoats are dangerous. Robins on the mainland are appropriately scared at the sight of stoats. But Motuara Island robins know nothing of predatory animals, including stoats. These island birds did not fret when experimenters put a stuffed stoat near their nest, any more than they fretted when experimenters put a stoat-sized cardboard box near their nest. Experienced robins from the mainland were upset about the stoat but not about the box. To try to convince the naive islanders that stoats are bad, the experimenters set up fake mobbing events, involving a moving stuffed stoat with a dead robin in its mouth, playbacks of robin alarm calls, playbacks of robin distress calls, playbacks of blackbird alarm calls, a stuffed robin in mobbing position, and a stuffed blackbird in mobbing position. The blackbirds and blackbird calls didn't impress the robins, but they found the other materials very persuasive and after only one exposure conceived a lasting distrust of stoats.

The takahe, a gorgeous blue-green flightless rail, must also contend with stoats, and researchers put on cautionary stoat acts for hand-reared takahe chicks. They were subjected to attacks by a stuffed stoat on a stick and by a stoat puppet, which swooped on them from above. They beheld a tragic melodrama in which the stuffed stoat attacked a takahe dummy. The dummy uttered distress calls until the stoat "killed" it. They also beheld a drama with a message, in which the stuffed stoat attacked a dummy takahe and the takahe fought back, pecking the stoat to "death." They watched "intensely," but although the chicks seemed more aware of stoats as a bad thing, and more cautious around them, they declined to actually attack the stuffed stoat themselves. Once released, the captive-reared takahes had as good a survival rate as wild-born takahes.

Dinner parties make me uncomfortable

Once you have learned to mob a scary bird like a hawk, it is possible to learn fine distinctions about which hawks are the scariest. Researcher

* This species, *Philemon corniculatus*, is also called the noisy friarbird or the leather head. As the name "honeyeater" suggests, it is fond of nectar and flowers, but it is not above snatching nestlings from other birds' nests, so mob away.

Frances Hamerstrom wrote that it's easy to tell by looking at a hawk whether it's hungry. Falconers know that a hungry hawk looks "sharp-set" and will fly at game. Wildlife painters don't always have this information. Hamerstrom writes that experts "are often amused by paintings of hawks with feathers and attitude of the body showing repose bordering upon somnolence but with talons 'fiercely' clutching prey." If a falconer can tell when a hawk is hungry, can prospective prey do the same?

Hamerstrom took a tame red-tailed hawk out into fields around Plainfield, Wisconsin, and stationed him in plain view to see if birds would mob him. For half the tests he was sharp-set and for the other half he had just eaten. He was much less upsetting to little birds when he was well fed, being mobbed by 12, as opposed to the approximately 100 birds who mobbed him when he was sharp-set. One kingbird was so maddened by the sight of him even when he was well fed that it dive-bombed the hawk and pecked him on the head, but kingbirds are touchy at the best of times.

Just as birds may learn that one class of hawks (hungry ones) is scarier than another class (sated ones), they can learn to distinguish individuals. Bernd Heinrich writes that the crows at Cornell have learned the menace that is Kevin McGowan, an ornithologist who climbs up to nests and bands baby crows. "He is singled out among all the students and professors for attack by crows when he walks across campus."

Young magpies have a suite of innate fears, so that a hand-reared magpie rattled when she was introduced to a dog, and she was horrified when she saw her first hawk on the wing. She didn't like the look of deer or bicyclists at first, but eventually learned that they were not much of a threat. Newly fledged wild magpies rattled at the sight of a vulture on the wing, but a week later they followed their parents in remaining calm and silent at the sight of a vulture. (While a vulture would happily eat a dead magpie, living creatures are in no danger from a vulture. It is a waste of time to panic at the sight of one unless you are concerned about the beauty of your corpse.)

My pet peeve

According to Konrad Lorenz, a baby jackdaw has no idea who its enemies are and is willing to be friends with such undesirables as a fox or a cat. The one exception to this tolerance is a creature gripping anything resembling a jackdaw. "Any living being that carries a black thing, dangling or fluttering, becomes the object of a furious onslaught." Lorenz's hand-raised baby jackdaw would attack him if he picked up the other jackdaw in the nest. Lorenz was attacked by a flock who spotted him carrying a wet black bathing suit, and he saw jackdaws attack another jackdaw who made the mistake of trying to carry a raven's feather to her nest.

Birds and ground squirrels are the best-studied mobbers, but not the only ones. Three humpbacked dolphins mobbed a shark on the South African coast, repeatedly chasing it and driving it into coves until it took off for the open sea. Fish mobbing has been seen in which damselfishes, butterfly fishes, and surgeonfishes, alone or in gangs, approach a moray eel or a lizardfish with harassment on their minds.

Mobbing sometimes works wonders. Jon Rood saw a martial eagle swoop down and grab a banded mongoose out of a pack foraging in the Serengeti. The eagle, flapping low, landed in a nearby acacia. As it did, the other 13 mongooses raced for the tree. Several climbed the tree and the dominant male, the oldest mongoose, swarmed up to the branch where the eagle sat, lunged at it, and startled it into dropping its prey, which it had not yet killed. For the next seven minutes, mongooses boiled around the base of the tree and the lower part of the trunk, and then they all ran off together, including the rescued animal, who was limping slightly.

Let's go see what all the screaming is about

Mobbing provides clues to a naive young animal as to who should be regarded with concern. Another behavior that provides such clues is inspection. Wildlife observers are sometimes surprised when they notice that prey species don't always flee at the sight of predators and acts of predation, and may wander over for a hard stare.

Thomson's gazelles inspect many predators. They put in more time inspecting stalking predators like cheetahs and lions than coursing predators like wild dogs and hyenas. Clare FitzGibbon analyzed 90 cases of gazelles inspecting cheetahs. Gazelles who spot a cheetah are apt to walk over and look at it. The whole gang often comes along to mill around gawking. If the cheetah walks away, they follow, sometimes for over an hour. Up to a thousand gazelles have been seen following a cheetah. And some of them were snorting. This makes it hard for a cheetah to sneak up on its prey.

Gazelle children and teenagers were the most likely to approach and inspect cheetahs. Adults were the next most frequent inspectors. Fawns did not inspect cheetahs at all: either their mothers led them away or the fawn lay down with its head flattened to the ground, literally keeping a low profile. Inspecting is not risk-free. FitzGibbon saw one adult and two half-grown gazelles attacked and killed from groups of gazelles following cheetahs.

Cheetahs moved away farther and faster when gazelles inspected them than when they were left alone. The more gazelles, the farther away the cheetahs went. The gazelles may also be gaining information about cheetahs—what they look like, how they move, what they smell like, and so forth. "The fact that younger gazelles were more likely to approach than older animals suggests that they may be using the opportunity to learn more about the predators," FitzGibbon writes.

In the Serengeti, George Schaller saw buffalo inspecting crime scenes where buffalo had been killed by lions. They came in such numbers that they drove the lions away. Thus, he saw five male lions kill a buffalo in the late morning, and when he returned in the early evening, 100 buffalo were there, milling around and sniffing the ground while the lions stood back and waited. The habit of prey animals of staring at the sight of predators eating one of their number is sometimes called the bystander phenomenon. A buffalo bystander would surely learn that lions are to be distrusted and that the stakes are high.

Kobie Krüger describes walking in the South African bush with her dog Wolfie and Leo, a tame teenaged lion, and seeing the reaction of the waterbuck, wildebeest, kudu, and impala, who would stop and stare at this inexplicable combination. One male waterbuck seemed particularly perplexed and would often drift over to stare at the three, occasionally stamping a hoof "as if demanding an explanation." Krüger liked to throw her arms around the dog and the lion just to boggle the waterbuck further.

Even minnows seek information about pike. Pike are large predatory fish that eat minnows. When a pike is put in a tank with a school of minnows, the smaller fish at first huddle together. After a little while one or a few minnows will sally forth and examine the pike, staying away from its mouth. The interactions of minnows and pike have been intensively studied, perhaps because it is so hard to keep lions and buffalo in the laboratory. What makes it worthwhile to engage in such risky actions?

After inspecting a pike, minnows return to the safety of the school. This is real safety, since it is harder for a predator to pick a fish out of a school than it is to grab an isolated fish. Sometimes the inspecting fish returns with a skitter that may convey information to the other minnows. If inspection only carried risk and no benefit, the genes of minnows who indulged in it would all be eaten by pike. Hypotheses about possible benefits include not only getting information about the pike—whether it really is a pike, and if so, how hungry it is—but also providing information about the pike to the other fish. If the rest of the school knows how dangerous the pike is at this moment, the school can act accordingly, and each fish benefits if the school acts appropriately. Being approached may discourage the pike by letting it know it has been seen. It has also been suggested that female minnows may be favorably impressed by the boldness of males who inspect predators, and more inclined to mate with them.

Enemies? I have enemies?

Wildlife rehabilitators often need to teach animals raised in captivity how to spot a predator and what to do if they spot one. This comes more easily to some animals than others. Species that have lived for millennia without predators—on islands, for example—may be much too trusting.

Researchers Andrea Griffin, Daniel Blumstein, and Christopher Evans worked with tammar wallabies from an island population where there were no predatory mammals. They thus hadn't seen predators in 9,500 years.

The researchers set out to make wallabies fear foxes. Seeing a model fox while wallaby alarm thumps were heard did not trouble the wallabies. What did trouble them was being caught in a net. "Approaching humans consistently elicit alarm response in captive marsupials, probably because animals associate them with being caught, bagged and handled." They decided to pair the sight of the fox model with that of a human carrying a net. The model was a stuffed fox on a wheeled cart, moved by a string and pulley system. The fox model would roll out and within seconds a human with a net would fake trying to capture the wallaby. The fox rolled back and forth, the human pursued the wallaby back and forth, and then human and fox exited.

Wallabies subjected to this training were noticeably less calm about the fox model afterward, and when a cat model was produced, they didn't like that either. A stuffed goat didn't bother them. It was considered encouraging that the wallabies held their bad fox experiences against cats, suggesting that they have a generalized concept of "predator" that can be triggered.

Another group of Australian researchers trying to make quokkas and rufous bettongs (two species of small marsupials) fear foxes didn't have much luck with a stuffed fox (and a different protocol), but a series of experiences with a very sweet dog were so hideously misunderstood that afterward quokkas and bettongs viewed both dogs and foxes with the greatest alarm. The large white dog was trained to chase but not harm animals. She was careful not to crash into them, and if she cornered one, she licked it. Neither quokkas nor bettongs appreciated what a great dog this was, and furthermore, foxes reminded them of the dog.

Everywhere, foxes

Wildlife biologists reintroducing captive-raised houbara bustards (think: turkeys of the desert) in Saudi Arabia found that more than half of the young birds didn't make it through the first year, being gobbled up by predators, particularly red foxes. Most bustards perished within three weeks. One group was "trained" with a mobile fox model intended to make them fear foxes. Instead of learning "foxes are scary, avoid them" the bustards habituated, learning "our friend the mobile fox model won't hurt us, relax."

The bustards did not instinctively regard foxes as scary. The first time they saw the model fox they were a bit upset, but no more so than the first time they saw a model penguin. (Where you get a penguin model in the middle of Saudi Arabia is not explained in the articles I have read. Scientists are resourceful people.) Training with the model fox didn't reduce bustard mortality.

Since the fake fox didn't work, the researchers got Sophie, whose "delightful personality threatened to shift our allegiance from bustards to foxes." When she was old enough, bustards were treated to "encounters with a live red fox, muzzled and on a long lead." These meetings took place at dusk, in a cage with two wild bustards "to provide examples of appropriate fear responses." Tapes of bustard alarm calls were played. This haunted-house program worked well—foxes now freaked the bustards out.

Subsequent data showed that bustards who had been exposed to Sophie before release were much more apt to survive than bustards who had merely seen a fox model. The actual fox made deadly serious attempts to catch and kill the bustards, and the bustards seem to have noted the vibes. Although Sophie was muzzled, she several times managed to slip the muzzle and seize a bustard, although she never got to keep or kill it. Sophie did get to participate in the release of the bustards by chasing them out of their cages.

Dingo puppies in central Australia aspired to catch rabbits but had terrible technique. Their running was floppy and awkward. They lacked stealth. To see a rabbit was to hurtle toward the rabbit, and the result was that the rabbit, easily spotting the puppy, scooted down a burrow. Not in great terror, and not until the last moment, since "the rabbits seemed to know that the pups were only amateur hunters," according to researcher Laurie Corbett. As the season progressed, the pups got slyer and the rabbits got jumpier. The pups stalked from cover and didn't make a dash until they were close, and they now ran smoothly. By this time the merest glimpse of a dingo was enough to make the rabbits dive for a hole.

Wild black-footed ferret kits are wary when they first stick their noses above ground. But captive-reared ferrets may be all too bold and lollop around on top of the earth as if there were no such things as eagles, coyotes, and badgers. Researchers sought ways to make the incredibly endangered animals more cautious. Rather than practice on actual black-footed ferrets, they worked with related Siberian ferrets.

In initial work, they used model predators. The model terrestrial predator was a stuffed badger mounted on a remote-controlled model truck frame, later famed as RoboBadger. The ferrets would be chased by a model, which they could escape by darting down a burrow. A photograph of a Siberian ferret being chased by RoboBadger shows a ferret that is running, but does not seem terrorized. And although the ferrets acquired a dislike of the models, a dislike that the scientists tried to enhance by snapping rubber bands at them while they were being chased, they got over it within a week.*

In subsequent work, they used a dog, a soft-mouthed retriever that would never hurt a ferret. A photograph shows a very happy dog chasing a very unhappy ferret, its mouth wide with horror. This experience proved somewhat more memorable than the model predators.

What, us worry?

Moose mothers are vigilant on behalf of their children and are particularly perturbed by the sound, scent, or sight of wolves and grizzly bears. If they have any idea what those are. In the Grand Teton National Park, where both predators have been gone for 50 to 75 years, mother moose scarcely reacted to the sound of wolf howls played for them. They went on eating. Weird sound—try the grass over here. When researcher Joel Berger sneaked up and tossed either snowballs soaked with wolf urine or grizzly bear scat wrapped in biodegradable toilet paper to within a few meters of mother moose, they weren't bothered by the wolf smell. Sometimes they strolled over for a sniff. The bear smell worried them, suggesting that they may have some innate ability to recognize grizzly bears. (Or perhaps the smell reminded them of black bears.)

Moose mothers in Alaska, where wolves and grizzlies are an ongoing issue, reacted differently. Not only did wolf howls upset them, so

---

* When the *Washington Post* did a sympathetic story on ferret recovery work, it called the stuffed badger model RoboBadger. The story also mentioned the rubber bands. An alert reader sent this story to columnist Dave Barry, who derived a staggering amount of humor out of the rubber bands, RoboBadger, and the Thanksgiving holiday season. (Barry, 1989). In the hallowed what-will-these-eggheads-come-up-with-next tradition, Barry wrote, "So we're talking about people who probably look perfectly normal, who have normal children and wear normal clothes and drive normal cars to a normal-looking building where they go inside and shoot rubber bands at ferrets. I bet they also argue over who gets to drive RoboBadger." Not so. In *Prairie Night*, a thoughtful exploration of the ferret project and endangered species recovery projects in general, the sole endnote for the chapter on predator avoidance training reads: "In answer to Dave Barry's speculation, we did not fight over who ran the badger—we took turns."

did raven calls. Moose who are "predator-savvy" are likely to associate ravens with carcasses, and carcasses with predators. When Berger lobbed his scent bombs their way, both scents troubled them strongly. (Berger did some of his research dressed as a cow moose so that he could approach moose without alarming them. This is probably not as much fun as it sounds.) They lowered their heads, retracted their ears, and their hackles stood on end.

This is the kind of danger that animals learn about fast. The area around Jackson Hole has recently been recolonized by wolves. The wolves went through the naive moose population like hot knives through butter. But the moose responded. Mother moose who had lost a calf to wolves were 500 percent more responsive to playbacks of wolf howls, spending much more time scanning their surroundings after they heard howls. It wasn't just that they were more sensitive because they had lost a calf, because mothers who had a calf hit by a car didn't increase their vigilance to howls.

Competitors

In this world it is also possible to meet individuals who wish you ill, without actually wanting to eat you. Maybe they want to move into your apartment, maybe they want to break up your marriage, or maybe they just want to eat your lunch. It's possible to learn who these vile individuals are. The Pacific gregory damselfish dwells on coral reefs, where it mostly eats algae. It defends its territory against other damselfish, against other algae-eaters, and, when it has eggs, against fish that might eat damselfish eggs. Biologist George Losey found that in the lab they generally left "predatory, bass-like fishes" alone, including a species it would not encounter in nature, the African tilapia, but that after a couple of weeks of watching tilapia gorge on algae, damselfish became testy. Losey tested this by training some tilapia to eat algae growing on bricks, and some to eat zooplankton from a dispenser. Damselfish housed with plankton-eating tilapia remained calm at the sight of them. Damselfish housed with algae-eating tilapia became increasingly hostile.

In a tank in the laboratory of John Todd, some small bullhead catfish were swimming about when a large bullhead catfish leaped in from a neighboring tank and attacked the little fish fiercely, mauling them so ruthlessly that all but two of them jumped out, fell on the floor, and died before help came. The big catfish was removed, and the two survivors began life anew. They established territories at opposite ends of the tank, which they patrolled vigilantly, and which they did not leave. Then Todd and his coworkers poured in some water from the aquarium of the big catfish, the Terror of the Tanks. Smelling the monster, the two little catfish fled their territories and huddled together, hiding until the scent had dissipated.

Listing your friends

It's also good to recognize friends. For many baby animals, there are critical periods during which it learns who friends and family are. When that period is over, the animal is much less likely to accept new individuals into its life. Human parents see this when an infant who was happy to be held by anyone and everyone suddenly becomes insultingly particular and wails if it's not in the arms of one of a very few people.

When David Macdonald's fox cub, Niff, was 10 weeks old, she abruptly lost her unquestioning friendliness to newcomers. Till then, she had loved everyone she met. Now she only trusted the ones she already knew, and was terrified by new people. If a stranger dropped by, Niff would frantically scrabble up the chimney.

The human being—friend, foe, snack?

A mother polar bear became alarmed when her year-old cub displayed interest in a Tundra Buggy, an armored vehicle containing humans. He walked up to the buggy, and when his mother tried to push him away, he wouldn't go. She walked away, then stopped, turned, and sat up on her haunches. She waved her paws and moaned. He gave in and went with her, and she led him a little farther away and nursed him before the two took off. She was letting him know that she viewed humans as a menace, a view it would be nice if all polar bears shared.

When the famous lion Elsa, subject of the book *Born Free*, had cubs, she wanted them to be part of her family, which included George and Joy Adamson. But by the time she introduced them, the cubs were old enough to be suspicious of humans. One day when the Adamsons had delivered a nourishing goat to the family, one of the cubs, Jespah, became annoyed at the sight of Joy Adamson handling their food and charged her. Elsa stepped between the two and whapped Jespah with her paw. Then she ostentatiously sat with Adamson in the Adamsons' tent, giving Jespah the cold shoulder.

Making sure animals think of humans in the right way is a perennial problem for those who rehabilitate wild animals. Biologist Anne Collet describes the behavior of an orphaned seal pup who was rescued and lovingly restored to perfect health by marine biologists and released into the sea off the coast of France. A week later, not having gotten the meals she was accustomed to being served, the pup swam onto the beach, spotted a human, flopped up to him, and fixed him with her big brown eyes, ready to accept food. He walked away. She followed. He kept walking, she kept following. He went to a phone to summon help and she sobbed so loudly outside the door that he felt compelled to let her in the phone booth.

The marine biologists came. Realizing their mistake, they now made a point of acting like jerks, until the little seal grew to dislike humans. "You can't believe how hard it is to be disagreeable to a young seal!" writes Collet. Subsequent baby seals had to be fed from hiding, chased, and generally treated in a cold, unloving manner so they wouldn't get too fond of our kind.

Many other wildlife rehabilitators have found that they must raise baby animals and tend injured adults in a hands-off way that does not reflect their concern for the animals, in order to keep them from becoming attached to or overly casual about humans. Thus Jeanne Lord, an expert on red fox rehabilitation, not only puts up screens so that she can stay out of view of the foxes she cares for, but avoids speaking, even in a whisper, where they can hear.

It has been suggested that tigers in Ranthambhore Park, who became tolerant of daily jeeps full of tourists, accordingly became easy targets for poachers.

Excessive familiarity with humans can play out in more than one way. In an experimental release of endangered Bali mynahs on the island of Pulau Menjangan, confiscated cage birds were used. Cage life had not filled them with love and trust for humans, but rather with contempt. The researchers complained of "antagonistic behavior to birdkeepers." Even after release the mynah LR04 would fearlessly fly down and peck observers on the ear. The scornful LR06 several times landed on an observer's head or attacked his boot. While both these birds succumbed eventually to a falcon, another was stolen by a poacher, perhaps in mid-sneer.

Chimpanzees released in Rubondo Island National Park in Tanzania, who came from German and Dutch zoos, also did not admire people. Of course, the reason the zoos were willing to yield up these particular chimps often had to do with their surly outlook. Most of the apes faded into the forest, but two of the males attacked people so fiercely that they were shot. The chimps were released in the late 1960s, but a few of the original zoo chimps were probably still around in the late 1990s, when one chimp was spotted stealing a blanket to wrap herself in and another was noted entering a store at the tourist camp and subsequently swigging from a bottle of liquor.

A lone bottlenose dolphin appeared in the coastal waters north of São Paulo and began socializing with humans. He followed ferries and small boats for a few months, and that was nice. Then he moved to the beach at Caraguatatuba and started mingling with bathers. People are a mixed bag. Bathers made contacts that included "simple touches . . . riding and jumping on his back, grabbing his fins, hitting him and even putting an ice-cream's stick into his blowhole." The dolphin began to respond sharply, sending 29 people to the hospital with injuries. Finally, he objected to a couple of drunken bathers, hitting them so hard that one went to the hospital with two broken ribs and the other died of internal bleeding, a first in dolphin-human relations. (The beach was patrolled for the rest of the season, and no one else was hurt.)

Sometimes not being afraid of humans is an advantage. Biologists reintroducing beavers to tributary streams of the Vistula River had both farm-raised and wild-trapped beavers at their disposal, and chose to release the farm-raised beavers in those sites "often troubled by man." The farm-raised beavers weren't scared off the way the wild beavers were.

In Monkey Mia, a cove in Shark Bay, in western Australia, some dolphins come into the shallow water and interact with people, who feed them fish. Most of the Shark Bay dolphins don't do this, and it's something dolphins learn. They learn where to come, and they learn the protocol. Biologist Rachel Smolker writes that there is a specific begging posture, in which the dolphins brace their pectoral fins against the sandy bottom, and hold their head up out of the water with open mouth. Smolker describes a dolphin called Joysfriend, who didn't usually come into the shallows. One day she visited, accompanying some dolphin regulars. "She stopped and looked around expectantly, as if to say, 'Okay, so where's the fish?' She moved back offshore and seemed to watch the other dolphins approach people and take fish. A while later she tried again, this time bringing her face up out of the water but failing to open her jaws. She was going through some of the motions, but again, she was several yards from the nearest person and still didn't have her jaw open. She looked awkward, like a person taking his or her first stab at stage acting."

What to do?

Creatures of other species who will feed you are few and far between. Meeting an enemy is more likely, and it's important to know how to act in such a case. How to act if you spot an enemy is such vital information that few animals that had to pause and learn the skills are still around. How to run, or freeze, or dive, or whatever it is that your species does, is generally built in. Practice can help, though.

When scientists kept baby guppies in a tank with adults, the adults chased the little guppies up to 300 times a day. Two days of this discipline was enough to give the young guppies an edge when they were put in another tank with bloodthirsty cichlid fish. Guppies who had not had to spend all their time dodging grouchy grown-ups were not nearly as good at getting away from cichlids.

Three-spined stickleback fathers guard their eggs, and when the tiny fish fry hatch out, the fathers guard them for another week and a half. When the little fish swim too far from the nest, the father chases them, catches them in his mouth, takes them to the nest, and spits them back into it. J. J. Tulley and F. A. Huntingford collected stickleback eggs and hatched them in the laboratory. The eggs were collected from Inverleith Pond, which is practically a stickleback sanctuary, with no fish-eating fish; or from the Burn of Mar, where fish-eating fish and birds abound. Half the eggs from each location had a father's tender care after hatching, and half grew up as orphans. When they were big enough, they were placed, one at a time, in an aquarium with "a realistic fibreglass model pike." After two minutes, the experimenters moved the realistic pike around "to mimic the stalking of a live pike" and scored the responses of the young sticklebacks. The young fish got points for retreating from the model and keeping their distance and lost points for feeding in open water without looking at the model predator.

Fish who had normal upbringings and whose parents came from the dangerous Burn of Mar were warier than fish from peaceful Inverleith Pond, showing a hereditary effect. But the Mar Burn fish were warier if they had a normal upbringing than if they had been orphans, suggesting that learning to dodge their father was useful practice for dodging pike.

You're not fooling anybody

If you are an edible individual, and you notice a predator sneaking around, it may be useful to let the predator know it's been spotted and doesn't have a chance of taking you by surprise. Many antelope stot or pronk: they run with short, high bounds, often flaring out white hair on their haunches. Although stotting makes antelopes conspicuous, it tells predators they have been seen, and seen by a creature that has the energy to stot. No point wearing yourself out chasing that one. Similarly, hares may stand on two legs, with their body and ears sticking up. Again the message is clear: I see you, creep.

Oh no! Any second now you'll grab me!

Grouse and ptarmigan are among ground-dwelling birds who perform distraction displays when they and their chicks are approached by a predator. The chicks scatter and sink down in the grass, where their streaked and speckled feathers are beautiful camouflage, and the adults flap up noisily and then stagger away as if crippled and mentally impaired, often dragging one wing as if injured.* Meanwhile the chicks are motionless and invisible. Often the predator will follow the

---

* The author was fooled by the first distraction display she witnessed, despite having read about distraction displays many times, and darted witlessly after a dramatically gifted killdeer, crying, "Look, it has a broken wing!" and pursuing it with intent to rescue and cherish. Only when the killdeer suddenly regained its powers and flew did the author realize what had happened and retrace her path to the spot where the killdeer had first seemed injured, there to find an insanely adorable killdeer chick squatting obediently on the gravel shoulder of the road. The author vowed not to be duped again.

adult, who flops just ahead until they are far from the chicks, where-upon the adult miraculously recovers, and flies off. Thus a naturalist doing a census of blue grouse on Vancouver Island was beguiled by the sight of a grouse hen acting like a maniac, clucking and staggering just out of reach of a black bear which ran after her until she had led it out of the valley where it had been nosing about.

A tame duck sitting on her nest spotted an otter (also tame) com-ing toward her with mischievous intent and, although she had no experience of otters and had never seen a distraction display per-formed, immediately stumbled off her nest and staggered through the water, flapping and quacking. The thrilled otter pursued her to the end of the pond, where the duck suddenly leapt into the air and flew off. "The otter's air of bewilderment was laughable," wrote Frances Pitt. Nor did it occur to the naive otter to go back and look for eggs.

Predators aren't always fooled. One summer day, Geir Sonerud saw a fox hunting in a Norwegian forest. When a black grouse started up and began doing its oh-no-I-can-barely-walk act, the fox ignored the grouse and instead began quartering the area, nose to the ground, keeping it up for 36 minutes and apparently locating and snapping up two chicks. After strolling on, the fox flushed a capercaillie (a large grouse). The fox spent 32 minutes again quartering the ground for chicks, catching two, and ignoring the capercaillie hen's desperate dis-traction displays.

Grouse do not perform distraction displays every time a predator approaches. Because grouse hens vary in the frequency with which they perform such displays for the benefit of hunters with dogs, and because the variation correlates with the number of inexperienced (and therefore perhaps easier to fool) foxes in the population, Sonerud spec-ulates that foxes learn to ignore the hen and look for the chicks, and that grouse learn how likely they are to be able to fool foxes and there-fore whether it's worth putting on their act.

Researchers banding ptarmigan chicks in Sweden made use of dogs who didn't get distracted. A knowledgeable person who sees a ptarmi-gan go into a display knows where to look for chicks but won't be as good at finding them as a dog. V. Marcström has described the diffi-culty of getting the right dog. It should have a good nose and lack culi-nary aspirations toward the chicks. It must also learn to ignore the histrionics of the mother or father ptarmigan, with their pathetic staggering and dragging wings, and focus on the search for chicks, keeping its nose to the ground and methodically sniffing the ground bit by bit—in other words, to be clever as a fox.

Cognitive ethologist Carolyn Ristau studied the distraction displays of piping plovers and Wilson plovers, birds that fake broken wings if predators get near their nests. The nests are on sandy ground and are mostly protected by the perfect camouflage of the eggs. A faking bird fans its tail and stumbles across the ground, peeping, fluttering, and dragging one or both wings. Ristau calls this a mixture of genetic elements and "more flexible behaviors." She investigated when plovers go into their act and how they employ it. The birds typically move closer to an intruder and start their displays in front of them. They don't randomly head in any direction—they head away from the nest. As they skitter away, they often look over their shoulders, enabling them to see what the intruder is doing. If the intruder doesn't follow, they often stop displaying and go back toward the intruder and start again. Ahem! I'm crippled here! Take a look!

Ristau ran experiments with beach-nesting plovers, having distinctively dressed persons either behave innocently, walking past without glancing at the ground, or wickedly snooping around as if hunting for nests. The next time they came by in the distinctive outfits, the plovers were apt to do a distraction display for the egg-hunting types but sit tight for the innocent-seeming beachgoers. Thus a plover is born with the impulse to fake a broken-wing display if its nest is threatened, but it is also able to assess how the display should be used and learn whether it's working.

Be careful, it might be some kind of a trap

One common method used by researchers studying wild animals is trap and release. Researchers record data on the trapped animals and let them go again. Kit foxes often become "trap-happy," gladly entering traps for the sake of the bait. Other kit foxes become trap-shy, and these two phenomena are maddening to statisticians trying to create equations that translate trapping records into population estimates. Before the phenomenon of trap happiness became well known, one study of kit foxes noted that one individual was trapped every night. The puzzled authors concluded that the fox must be a moron (not an actual quote). But a fox fancier who had actually lived with foxes felt sure that the animal in question had simply learned about a safe den for the night that offered free continental breakfast.

David Macdonald, author of *Running with the Fox*, did not find it easy to trap red foxes. His first traps, which he calls "monuments to my naïveté," didn't catch a single fox in three months. They didn't go unnoticed. Macdonald reports that the local foxes found his traps fascinating. "They circled them, urinated on them, dug under and climbed on top of them, defecated into them, even pawed through the mesh to extract the bait from within. The one thing that they never did was to enter."

Eventually, Macdonald writes, he figured out that the traps were "too small, badly designed, poorly sited and incorrectly baited" and with improved technique began catching foxes and putting radio collars on them. He captured foxes repeatedly, and he writes of the vixen Toothypeg that after a while she knew the routine, stopped biting, and waited placidly while he replaced her collar.

One rainy dawn when Macdonald checked his traps he found an uncollared vixen with unusual markings. She stared at him strangely as he put a collar on her. When he let her go, she jumped up, snatched the cap off his head, and raced off with it, discarding it in a nearby field. Pondering her strange actions and odd markings, Macdonald suddenly realized that the vixen was a fox named Sickly that he and his wife had raised from a cub and released. (Her old collar had fallen off.) That night his wife gave the old signal for Sickly: rattling a can of chocolate candy. Sure enough, Sickly arrived in a brand-new radio collar. He hadn't recognized her. Perhaps stealing his cap was a familiarity designed to remind him of their relationship.

In a patch of rain forest in Queensland, Australia, biologist William Laurance was trapping and releasing local small mammals. One day he set out some Elliot box traps, baited with oats. Elliot traps are designed to open out flat so they can be folded up when not in use. To do this requires pulling out a long wire that keeps the trap in the box position, a task Laurance needed pliers to perform. When Laurance returned one day, the bait was gone and all the traps had been opened, with the wires laid out neatly beside them. At first the baffled Laurance figured that local jokers, perhaps hostile to his research, had disassembled his traps, but it happened again and again.

He had to rule out humans since he didn't see any cars they could have come in, didn't think they would keep getting up in the middle of the night to disassemble his traps, and was sure none of them would take the trouble to lick the traps clean.

To find out what was going on, he interspersed the Elliot traps with cage traps. On his return the Elliot traps had been unfolded and cleaned out as before, and the cage traps held bush rats, fawn-footed melomys rats, and giant white-tailed rats. Laurance accordingly dusted each rat with nontoxic Day-Glo powder by species: the bush rats were blue, fawn-footed melomys were green, and giant white-tailed rats were red.

The next day the Elliot traps were opened, unfolded, and every speck of bait had been eaten, and they were speckled with red powder. The safecracker was a giant white-tailed rat—or rats, since, sadly, Laurance did not take the next step to determine whether there was a single rat at work or whether all the local giant white-tailed rats knew the trick.

Watch this!

The cunning of keas and their willingness to experiment made trapping them (for banding) a task that tested the ingenuity of all concerned. The researchers used drop nets baited with butter. Keas adore butter but they dislike being netted, so they tried to outwit the humans. One kea would go over to the net, which was poised to fall, grab the side, and shake it until it fell. Then the kea would clamber on top and eat butter through the mesh. Other keas waited until someone else had sprung the trap and then dashed out to snag the butter. Some specialized in what the researchers called the "fast run-through," in which they dashed under the net, snatched the butter, and tried to race out the other side before the startled biologists could drop the net. In order to keep the keas guessing, they had to keep moving the net and switching to different-colored nets. Thus scientific efforts to learn about animals are thwarted by animal learning.

Learning to avoid danger is one of the most clearly useful forms of learning. Innate reactions to some perils—huge birds with big talons—are valuable to some animals. But new menaces come along, and animals who can learn have the best chance of escaping them.

SEVEN

Invention, Innovation, and Tools: How to Do Something New, Possibly with a Stick

---

A hand-raised blue jay in a psychology laboratory, irked because some of the food pellets it was fed fell outside the cage and lay out of reach, devised a technique for dealing with this. The jay reached through the bottom of the cage, tore off bits of the newspaper lining, and industriously crumpled the paper. Then it stuck the paper through the wires at the front of the cage, where the pellets lay, and manipulated the paper until pellets came within reach. This was not what the researchers had planned to study, but they couldn't help being impressed. To see if the blue jay was set on using paper for this purpose, the fascinated psychologists offered the bird a feather, a paper clip, a plastic bag tie, and other items, and the jay used them all to rake in more pellets. Would the jay still try to rake in pellets if there were none to rake in? No, it would not. Would it try harder if it was hungrier? Why, yes.

When the researchers deprived the bird of pellets, the jay would sometimes dampen a piece of paper in its water dish and swab it in the food dish. Then it would either pick crumbs of food dust off the paper or simply eat the paper, dust and all.

The psychologists tested their eight other jays and found that five were tool users, two "displayed some components of the behavior," and one didn't seem to be a tool user at all, although it did like to play with paper. The researchers wrote that they suspected that one jay came up with the tool use thing "serendipitously," was encouraged when it accidentally retrieved a pellet, per-

fected the system through trial and error, and then the rest picked it up through imitation or observational learning.

ANIMALS COME UP WITH NEW STUFF all the time. A behavior might be so inevitable that animals invent it again and again. In fact, every animal of that species might invent it. Or it might be an innovation that's a little less obvious, which a few animals come up with from time to time. Or it might be so unusual that it is invented rarely or only once.

Sometimes innovation is a matter of applying well-established actions to new targets—eating a new food. That might seem trivial, but it can be pretty hard to get humans to eat new foods, and not only small children with suspicious taste buds. Whole populations refuse to consider eating fermented milk products, or jellyfish, or toasted locusts.

It's a cat-eat-monkey kind of world

It has frequently been suggested that man-eating tigers, lions, and leopards come into being when an injured or sick great cat, unable to catch wilier and more appealing prey, invents the idea of eating people. The famous tiger hunter Jim Corbett, who was often called to track and slay man-eaters, wrote that 9 out of 10 he shot had a serious injury, either from a gunshot or porcupine quills, and the tenth would be a case of extreme old age. "Human beings are not the natural prey of tigers, and it is only when tigers have been incapacitated through wounds or old age that, in order to live, they are compelled to take to a diet of human flesh."

A tiger in Muktesar unwisely tangled with a porcupine and got 50 quills driven into her face and foreleg, some right up to the bone. She lost an eye and was lying in thick grass licking her wounds, starving, when a woman cutting grass for cattle cut up to the spot where the tiger was lying. The tiger killed her with a single blow and limped away to lie under a fallen tree. Two days later a man cutting wood started to chip wood off that tree; the tiger killed him, and took a little taste. The next day she killed and ate a person who hadn't intruded on her sickbed or bothered her at all, and before she met her end she killed 24 people.

But in the Sundarbans region of India, an enormous delta system on the Bay of Bengal covered with mangrove forests, tigers regularly eat people, and it seems unlikely that they're all getting over porcupine fights. It has been pointed out that they don't eat very many of the people who enter the area, but individual humans are still reluctant to be among that small number. For a brief period, they didn't have to be. Tigers attack people from behind, an excellent tactic, particularly if the person is carrying a machete. Arum Ran of Calcutta suggested that people venturing into the forests of the Sundarbans (to fish, collect honey, or cut wood) should wear masks on the backs of their heads. That way the tiger would see a face on each side of the head and wouldn't see any place to attack.

In the first year this was tried, no one wearing a mask was attacked. "Some men reported that tigers would still follow them, sometimes for hours," writes Sy Montgomery. "Often the person would hear the tiger growling, as if it were frustrated that the Janus-man had somehow cheated it, yet it seemed unable to perpetrate a similar breach." But apparently the tigers learned that a face mask is not a face and resumed dining on even costumed humans.

The famous man-eaters of Tsavo, lions who killed and ate more than 135 people, were shot in 1899, and their remains went to the Field Museum of Natural History in Chicago. More than a century later, museum staff examined the skulls and found that one lion had a shocking mouth, with missing, broken, and displaced teeth and what must have been a painful dental abscess. The other lion would have kept a dentist busy, but his deficiencies were not incapacitating. Perhaps the first lion couldn't hunt anything but humans, and the other simply followed his lead.

Trendsetters

Who are the innovators? While many people prefer the animal Einstein notion, it turns out that you can actually collect data about which animals are likely to come up with new things. Hans Kummer notes that primates may be good innovators because they are "remarkably ill equipped with innate technologies." If they had a macaw's beak, they wouldn't see any advantage in cracking nuts between stones, and if they had an anteater's tongue, they'd never spend hours dipping straws into ants' nests.

Within a primate troop, Kummer expects social innovation among animals of low dominance rank, because the animals of high rank are happy with things just the way they are. In studies of captive macaques, lower-ranking animals were better at a lever-pressing task than dominant animals. The subordinates tried harder, for more irregular rewards. When dominant macaques lost status, they got better at the task.

Animals with lots of free time and energy are also more apt to come up with innovations, as in (the better) zoo colonies. Kummer and colleagues compared hamadryas baboons in the Zurich Zoo with wild colonies, and found that the captive animals had social signals that the wild animals did not have but that the reverse was not true.

Changes in the environment can spur chimpanzee thinking. In the early days of Goodall's research at Gombe she put out steel boxes which researchers could open remotely and which could be stocked with bananas. After four and a half months, three teenaged chimpanzees began trying to lever the lids open with sticks. While the boxes were too sturdy for this to work, "because a box was sometimes opened when a chimpanzee was working at it, the tool use was occasionally rewarded and, over the next year, the habit spread until almost all members of the community, including adult males, were seen using sticks this way." This would seem to be a case of superstitious tool use.

J. P. Cambefort hid tasty foods in the foraging areas of a troop of chacma baboons and a troop of vervets. The treats (banana slices for vervets, maize meal pellets for baboons) were marked with clues such as painted sticks, matchboxes, or bottle tops, so that the monkeys could learn to use the clues to find the foods. Juvenile baboons, who forage separately from the adults and are still learning how to find food, were quick to figure out the clues. The other baboons noticed this and propagation of the news was "instantaneous." In the vervet troop, the ones who figured it out first were just as likely to be adults as juveniles, but the news was slower to spread.

Who thinks this stuff up?

Researchers Kevin Laland and Simon Reader set up opportunities for guppies to show whether they were innovators. Fish had to swim through mazes to find unfamiliar but tasty food. Hungry guppies were more apt to explore and find the new food. Little guppies were more apt to do so. And female guppies were more apt to do so. The researchers, remarking that necessity is the mother of invention, take the view that needy guppies are more likely to innovate. The females need nourishment because they have to give birth, the small guppies need nourishment to become big guppies, and the hungry guppies— must we explain that they would kill for a snack?

In additional experiments in which they tried to control for motivational state, the researchers investigated whether there are "innovator fish." They found that fish that innovated before were more apt to innovate in a new test, which the researchers consider to indicate an innovative personality.

"At this stage it is not clear whether such differences reflect variation in mental abilities, . . . sociality, . . . boldness, . . . exploratory behaviour, . . . some other factor, or some combination of these factors," they write.* "None the less, it is interesting, and perhaps surprising, that we find evidence for innovative individuals in a species not particularly renowned for its intelligence or problem-solving capabilities."

In contrast, Dorothy Fragaszy and Elisabetta Visalberghi found no innovators among captive capuchin monkeys—capuchins that figured out how to sponge liquids weren't the same ones who figured out how to rinse sandy fruit.

Imo, all things to all thinkers

This brings us to the story of Imo, which has been told many times with different morals. Imo was a young Japanese macaque who lived on the small island of Koshima. Scientists had been observing the troop in which Imo lived for some time and putting out food to attract them to areas where they could be more easily seen. One food they used to lure them out of the forest and down to a beach was sweet potatoes. The potatoes got sandy, and the macaques brushed the sand off with their hands as well as they could, but this is why beach picnics are not unalloyed joy. In 1951, Imo, 18 months old, was seen taking her sweet potatoes to a stream and rinsing them.

Soon one of Imo's playmates was also seen washing her potatoes.

* See my forthcoming book, *The Seven Habits of Highly Effective Guppies.*

Imo's mother started washing her potatoes. Eventually most of the macaques washed their potatoes. Then Imo began to wash her potatoes in the ocean instead of the stream, perhaps for the salty taste.

The researchers started tossing wheat onto the beach, which was hard to eat without ingesting massive quantities of sand. At the age of four, Imo was seen gathering up handfuls of sandy wheat and putting it in puddles or in the ocean so that the sand sank and the wheat floated. She could then scoop the wheat off the surface of the water. Imo's little sisters Ego and Enoki picked up this method, as did Imo's (male) playmate Jugo. The habit also moved from offspring to their mothers, and generally from younger to older relatives.

Imo has been called an Einstein among macaques. A monkey genius! Psychologist Cecilia Heyes has suggested that it is also perfectly possible that Imo accidentally discovered the utility of putting sandy potatoes in water when she serendipitously dropped one there, and that other monkeys in the troop similarly learned by chance. Already in the habit of putting things in the water, the monkeys might have separately learned to wash sandy wheat. In this analysis, the fact that Imo did these things first merely makes her a genius of luck.*

That washing sand off potatoes is not such a bizarre and brilliant cognitive leap is shown by the fact that macaques on five other nearby islets also figured out that it was a good idea. "Imo did not exactly discover the monkey equivalent of the wheel," writes Frans de Waal.

Freakishly, pop science mystics spun the story that once a certain number of monkeys had learned to wash potatoes, suddenly all the monkeys on all the islands knew how. The knowledge had magically spread through macaque group consciousness. For fun, they picked the number 100. Thus, proposed the "Hundredth Monkey" advocates, if enough of us humans think beautiful thoughts, suddenly, when we reach the magic number, everybody will think beautiful thoughts. This is a relaxing, low-cost way to get things done, but sadly it's not true about the macaques and is unlikely to be true about us.

Meanwhile, back at Koshima, the monkeys only get sweet potatoes a few times a year, but they still wash them. Adults in the troop, scavenging the leavings of fishermen, learned to eat raw fish, and this spread to other members of the troop. The macaques are becoming

* Imo died at the age of 20 (normal for her species), leaving nine children.

increasingly littoral, spending more time on the beach, swimming and bathing in the sea, not only eating fish but prying limpets off rocks and catching octopuses and small fish in tide pools. Give them time, and they will become a great seafaring nation.

The Compleat Angler

Some green-backed herons practice bait fishing. In 1958, in an idle hour, biologists tossed a piece of bread to a heron, who surprised them by putting it in the water. He moved it closer when it drifted away, chased away coots that wanted to eat it, and caught small fish that came to sample it. When he noticed a lot of fish in an adjacent area, he carried the bread over there.

Herons not supplied with bread have been seen using twigs, leaves, berries, pieces of bark, moss, or Styrofoam for this purpose. Others have been reported to catch insects or worms for bait. If all a heron has is a twig, and the twig is too long, the heron may break it to a more alluring size, according to an observer in Kyushu, Japan—though American and African herons haven't been spotted doing this. A heron in Miami was photographed using fish food pellets to attract fish to their doom. His mother and little brother bait-fished too.

Juvenile green-backed herons aren't the best fishermen, as they often neglect to crouch down, and thus scare the fish away. "The young birds sometimes resort to eating the insects and earthworms themselves," reports zoologist Hiroyoshi Higuchi. Of three herons Higuchi observed, A and B were handicapped by inferior fishing territories but also by their tendency to use overly large bait. Heron C, in addition to having a territory that made it easy to crouch out of sight of the fish, selected his lures more judiciously.

Although green-backed herons are found all over the place, they've been spotted using this technique only in Japan, south Florida, and western Africa. People have tried to teach herons to do it, without success. There are scattered reports of a few other birds fishing with bait: some African pied kingfishers, a captive squacco heron, and a captive sun bittern. Maybe it is the sort of thing that a bird that spends hours staring at the water is inclined to invent once in a great while. And that other such birds are then inclined to pick up from them. Biologists James Gould and Carol Grant Gould suggest that it is "discovered by the Einsteins among the herons, and learned from them by only the brightest of their neighbors."

Smashing snails

European song thrushes are renowned for smashing snails on stones (called anvils). Some people thought they learned it from other thrushes, and some thought they were born knowing how. C. J. Henty raised song thrushes from the egg and offered them snails, pebbles, bits of wood, and anvils. The fledgling thrushes were enthusiastic about these things, pecking them, lifting them, carrying them, tossing them, shaking them, and hitting them. At first they focused on smashing pebbles and wood, which are easier to grip than snails. Once they smashed and ate a snail, their focus improved enormously. Although they still experimented with smashing new things, such as a collar stud Henty offered them, they no longer bothered with things they knew were inedible.

Henty notes that fledgling blackbirds—a species that doesn't regularly eat snails—did all the things that song thrushes do (pecking, lifting, etc.) but not with the same persistence. Trial and error combined with inherent tendencies seems to produce snail smashing in song thrushes. "Any animal that innately picks up large hard objects in the environment, . . . repeatedly . . . hits them on the ground, prefers the feedback from hard surfaces, and prefers doing the behaviour to items that provide food, is inevitably going to be a self-taught breaker of snails," writes Henty. (Henty neglects to mention that the animal should also enjoy eating raw snails.)

Catch the cunning sand lance

In the 1980s, humpback whales in the waters of southern New England were closely observed by researchers who went out on whale-watching cruises. Around this time the local humpbacks were adjusting to a crash in herring populations and were catching sand lance instead. A common way for whales to hunt sand lance is to blow a cloud of bubbles at them from underneath, then surge up with mouth agape and take a huge mouthful. (The water runs out through your baleen plates, and you eat the fish.) In the early 1980s whale watchers began to see a variation, which they called lobtail feeding. In lobtail feeding the whale makes a dive near the surface, so that its tail rises up out of the water. Just as the flukes at the end of the tail are about to slide under, the whale flexes its tail and violently slaps the water. Then it follows up with a bubble cloud and a feeding lunge. Whether this actually yields more sand lance is unknown.

In 1980 they hadn't seen any whales doing this. In 1982 they saw 8 whales (in a population of 250) lobtail feeding, and by 1989 more than half the whales were lobtail feeding. A lot of the younger whales were taking it up, and a few of the older whales were too. Two-year-old calves were seen off by themselves, practicing. They'd lobtail over and over, often without making a bubble cloud, and without making a feeding lunge. When they did make a bubble cloud, it was puny. By the time they were three, the same calves were like experts. Some of the lobtailing calves had lobtailing mothers, but based on the large number of calves and the small number of older females lobtailing, they couldn't all have been learning from their mothers.

Tool use

Tool use was for many years held up as a dividing line between the human and animal intellects. We used them, the others didn't. As various discoveries of animals using tools were made (some *bugs* use tools, darn it), the line was shifted to *making* tools. We make tools, they don't. Then various animals stepped over this line, and we began to hear less about the matchless intellectual prowess involved in toolcraft. But perhaps some of the early glorification of tool use (when we thought it was our strong point alone) was overblown, as hinted in the title of an 1987 article in *New Scientist* by Michael Hansell, "What's So Special about Using Tools?"

Hansell is unimpressed with the mental underpinnings he thinks exist when an Egyptian vulture drops a rock on an ostrich egg to break it; or when a woodpecker finch breaks off a stick and uses it to harry insects out of cracks in bark; or when a wasp closes the burrow in which she has laid her eggs by pounding the sand with a pebble she holds in her jaws. He concedes that some animals learn to be better with tools, but says that "there is no evidence here that there is something extraordinary about the learning process." Instead he says the last hope of proving that tool use is "something apart from other behaviour" is finding examples of insight, an animal equivalent of "James Watt looking at the boiling kettle and conceiving the steam engine." Sadly for this endeavor, few animals have invented the steam engine.

Nevertheless, along with language and social learning, tool use is one of the three areas where scientists look for evidence of cognitive skills. Indeed, many anthropologists suspect that the evolution of tool making and the evolution of language were closely meshed. This would be nice, because while spoken language leaves no traces in the physical record of the past, we can find some ancient tools, particularly those made of stone. (Wooden and fiber tools don't last, causing us to envision our forebears incessantly chipping, hacking, and flinging stones and not to envision their nets, snares, backpacks, furniture, and who knows what else. If there was a Basket Age, we'll have to wait for better archaeological techniques to find out about it.)

Intelligent tool use, it has been proposed, should not involve the use of the same tool behavior for every dilemma. It's said that if the only tool you have is a hammer, every problem looks like a nail. (Readers who are capuchin monkeys should think hard about this saying, as we shall see later.) That doesn't count as intelligent tool use. Parker and Gibson have very demandingly said that intelligent tool use "involves trial and error application of several complex object manipulation schemata such as aimed throwing, using a lever, banging with a tool, raking in with a stick, probing with a stick, in different contexts, such as opening objects, raking in out-of-reach objects, extracting objects from a container without opening it, using a variety of objects such as sticks, rocks, leaves. Intelligent tool use involves accommodation to the specific situation and exploring and manipulating physical causality."

Gorillas

Wild gorillas do not seem to be tool users. But captive gorillas use tools readily. Richard Byrne suggests that this paradox may be understood by the complex manipulations gorillas use to eat thorny, prickly, armored, sticky, and otherwise forbidding foods in their environment.

A captive gorilla child, Alafia, gazed at a moth fluttering around the gorilla enclosure, which repeatedly landed on the same wall, too high for her to reach. She swatted at the moth, but it really was too high. Turning away, she searched through a pile of hay in the enclosure until she found a short, stout stick, which she tucked under her arm as she went on searching. Finding a longer, skinnier stick, she went over to the wall and propped the short stick against it. She stood on the short stick, whacked the moth with the long stick, and when it fell, jumped down, grabbed it, and ate it. "She let out deep grunts of satisfaction as she mashed the moth around in her mouth and over her tongue, savoring what she must have considered a rare treat," writes gorilla ethnographer Dawn Prince-Hughes.

On another occasion, Prince-Hughes saw Alafia's 10-year-old brother Zuri limping around the enclosure, favoring his left foot. As she watched, Zuri collected short sticks into a pile. Then he pulled a thorn off a hawthorn branch, propped his left foot up on the stick pile, and prodded with the thorn at his big toe. Finally he pulled a splinter out with his thumb and forefinger.

Three eight-year-old gorillas in the San Diego Wild Animal Park used sticks to get leaves and seeds out of trees in their enclosure. These trees were protected with electrified wiring, so the gorillas couldn't climb them. Instead Milt, PD, and Penny tossed sticks up into the branches to knock down the mildly desirable foliage. If one stick didn't work, they selected a bigger stick. PD and Penny also used long sticks to draw branches down so they could grab them with their other hand. Penny was seen using a long stick to beat on a branch so foliage would fall off. The gorillas first threw sticks into the trees in 1996, when they were seven. By the next year Penny and PD had improved their accuracy, but Milt, who didn't practice, hadn't gotten better.

Bonobos

Wild bonobos, still little-studied, haven't been seen cracking nuts between stones. But they make rain hats. Living in the rain forest, they get rained on a lot. Rain hats are small leafy branches that the bonobo bends or places over its head and shoulders. This "may seem like a fairly simple, mundane sort of activity," writes researcher Ellen Ingmanson, but more is involved than one might think. First the animal needs to have the concept of constructing a covering. Not all bonobos do, and neither chimpanzees nor gorillas seem to. Then the ape has to select materials, arrange them effectively, and "behav[e] appropriately in conjunction with the rain hat."

About half of the bonobos Ingmanson watched at Wamba, Zaire, used rain hats. To the extent relationships were known, rain hat technology passed from mother to child. If a mother made rain hats, so did her children when they were old enough, and if she didn't, none of her children did.

Ingmanson watched seven-year-old Senta give it a go during one downpour. He was sitting in a small tree several yards up when rain began falling. He started to bend branches over, and he chose fine leafy branches, good for deflecting rain. But he bent branches at waist level instead of over his head and shoulders (probably because they were more convenient to reach), and after half an hour he had done a splendid job of covering his knees, but his head was still unprotected. "Senta . . . appeared somewhat perplexed." As it happened, his mother and four other females were sitting directly below him, benefiting from the rain-repellent lap blanket he had crafted. Three years later, Senta was seen making an excellent rain hat.

Orangutans

As with gorillas and bonobos, scientists have complained that it makes no sense that orangutans do not use tools in the wild and yet in captivity are adept. But in 1996, Carel van Schaik and E. A. Fox reported on a tool-using group of wild orangutans living in the Suaq Balimbang area of Sumatra, in freshwater and peat swamp forests. These orangutans are unusually easygoing and gregarious. They don't mind if others look over their shoulders and rip off their tool-using secrets. They strip slender branches to poke into the holes of termites, bees, and ants, to "prompt" them to exit and be eaten. Sometimes they bite the tip of the stick to flatten or fray it. Usually they hold these sticks in their mouths, manipulating them with their dexterous lips, but occasionally they hold them in their hands. They use sticks as chisels to chip off pieces of termite nests. They dip sticks into stingless bee nests to get honey. And they use sticks to deal with *Neesia* fruit. Although *Neesia* fruit contain delicious nutritious seeds, these are embedded in irritating hairs. A ripe *Neesia*, when it has cracked open, presents the problem of how to get the seeds without being stung by the hairs. The orangutan solution is to insert a stick into the crack in the fruit and scrape out the hairs. Then you blow the hairs off the stick, use the stick to push the seeds toward the apex of the *Neesia*, and scoop out the seeds with the stick or your finger.

(That tolerance and respect are relevant to the use of expertise is shown by the rhesus macaques of Cayo Santiago. They love coconut, but the only two monkeys who knew how to crack coconuts, by strategic flinging, gave up the habit. WK knew how, and so did his little brother, 436. When they lived in the troop where they were born, they were under the protection of their powerful mother, and WK cracked coconuts all the time. When he emigrated as a young man, he found that whenever he cracked a coconut, dominant animals in the new troop took it from him, so he stopped. 436 continued to crack coconuts, but only on rare occasions, when he was alone.)

Biruté Galdikas describes seeing the orangutan Cara react to rainfall by moving into a tree and breaking off two long leafy branches, which she held over her head as an umbrella. Her son Carl scooted over to take advantage of the shelter. On another occasion, Galdikas benefited by an orangutan's comprehension of the use of actual umbrellas. She was holding a juvenile orangutan on her lap when it began to rain. The young ape reached behind Galdikas, grabbed her umbrella, and passed it to her. Galdikas opened the umbrella, and several young orangutans joined her under its shelter. At the time, Galdikas was thinking about not getting wet. Only later, viewing a videotape of the incident, did she stop to be impressed by the young ape's understanding of the tool and its use.

Chantek

Chantek is an orangutan raised until the age of nine by Lyn Miles, who considers him her cross-fostered son. He learned a modified form of sign language, along with other things, such as the route to his favorite hamburger stand. Miles is fond of crafts and passed this along to her orange child. He wove crude potholders on little looms and made laced-up leather purses. As a child, Chantek played with large wooden beads strung on heavy string, a common toy. As Chantek's fine motor control improved, Miles brought him jewelry-making materials.

Today, when Miles visits Chantek at Zoo Atlanta, where he was transferred by the Yerkes Primate Center, she sometimes brings him jewelry kits in Ziploc plastic bags. Riveting video shows Chantek opening a bag and extracting a stiffened beading thread. He takes the bag of small beads, sticks out his lower lip like a steam shovel, and pours the beads into his mouth.

Taking the thread in his hand (Chantek is a lefty), he delicately extrudes a single bead from his lips onto the thread, and then another. If a bead displeases him, he sucks it back off the thread and extrudes another. The first time he made a necklace from one of these kits, the result was not what Chantek had in mind. "He didn't like it so he took it all apart, put it back in the bag, put the cord back in the bag, sealed it up, and gave it back to me," Miles said.

When he is finished, he has made a respectable beaded necklace that could be offered for sale anywhere.* His neck is so huge that he cannot wear it, although upon request he will put it on top of his head or on his beard, and he gives finished necklaces to Miles or occasionally to his companions. While Chantek works, his cagemates watch, their faces so close that they practically have their eyeballs on his lips.

Chantek became skilled with tools (hammer, pliers, wire-cutter, wrench, etc.) while he lived with Miles, and at the zoo he sometimes steals tools from the janitors, which he will only give back if asked nicely. In sign language. Which the janitors do not speak.†

Another literal tool user was Lucy, a chimpanzee raised in the Temerlin family. She knew how to use a screwdriver after seeing Maurice Temerlin use it once. She used it to dismantle a light fixture, learned from the experience how to dismantle light fixtures without getting shocked, and stole it whenever she could for such projects as taking the kitchen door off its hinges.

Chimpanzees

In 1960, when Jane Goodall was first observing wild chimpanzees at Gombe, Tanzania, the project was in peril. Goodall's mentor, Louis Leakey, had gotten funding for only six months, and Goodall hoped to

---

* I long for such a necklace, but Miles and Zoo Atlanta are not running a sweatshop, and very few exist. Yes, I realize they are covered with orangutan spit—that's the *point*.

† Complaints have been filed.

make discoveries that would draw more funding. The study was saved when she saw a chimpanzee, David Greybeard, modifying a stick into a tool for termite dipping by pulling off the leaves. "When I was at school we defined humans as 'man the toolmaker.' I sent Louis Leakey a telegram—I actually couldn't believe my eyes—and he sent back a famous telegram: 'Ha ha now we must redefine man, redefine tool, or accept chimpanzees as human.'"

The next snack sensation

David Greybeard was dipping a stick into a hole in a termite mound so enraged termites would seize the invading stick with their iron jaws, and he could pull them out and eat them. Sounds straightforward, but it takes years to become proficient. Here's what a chimpanzee child must learn to become a competent termite dipper. (We know because researchers tried it themselves.) First find a termite mound (and this is where I always fail), figure out where the tunnels are, and scratch a hole into one of them. Select a stick of the right elasticity, one that will bend with the bends in the tunnels but will not break. Remove leaves and projecting twigs and perhaps bite it to the right length. Then thread the stick into the passageway and wiggle it so that termites will detect it and grab on. Lastly, pull the stick out without knocking all the termites off.

Infant chimpanzees don't even try to fish for termites. They sit nearby while their mother fishes, watching her, playing with sticks, and occasionally eating a termite. Two-year-olds jab things in the termite mound randomly. Their sticks are the wrong size and the wrong kind. They insert sticks too clumsily and pull them out too soon. Three-year-olds do a little better, using longer sticks and showing more patience. Four-year-olds do better still, and they do catch termites, although they don't keep it up as long as grown-ups do. By five or six they do it as well as adults.

Ants

The chimpanzee children of Gombe take longer to master dipping for ants than dipping for termites because the ants bite. If an infant isn't clinging to its mother while she dips for ants it stays a safe distance away. Jane Goodall has described the chimpanzee Winkle dipping for safari ants while her four-year-old child Wonder sat apart with her own dipping stick, in the same pose as her mother, busily dipping for nothing at all.

So a five- or six-year-old ape that's skilled at termiting will still be a klutz at anting, will use poorly designed tools and wield them ineptly. At Gombe the accepted way to dip for ants is to select a fairly rigid stick, insert it into the ant hole, and stir. The ants will rush up the stick, plotting against you. When they are three-quarters of the way up, pull the stick out—act fast—and pull it through your other hand, crumpling the ants into a ball. Slap your hand to your mouth and chew quickly, before they bite you.

This seems to be a more efficient scheme, in terms of ants per minute, than the one used by the chimpanzees in the Taï Forest preserve in the Ivory Coast. There they simply put a stick in the ant hole, get a few ants on it, and nip them off the stick. They get a quarter as many ants. The chimpanzees at Gombe and Taï are eating the same species of ant, and primatologist Christophe Boesch has personally tried both techniques at Taï and found that both work.

Ants climbing up tree

The chimpanzees of Mahale, not far from Gombe, fish for wood-boring ants, usually in trees. Often a chimpanzee makes a tool and then travels to the tree, carrying the tool in its mouth or armpit or tucked between neck and shoulder or between thigh and abdomen. To get at the ants' nest, the chimpanzees may have to assume awkward and innovative positions.

Again, it takes time to learn this skill. As reported by anthropologists Toshisada Nishida and Mariko Hiraiwa, the two-year-old MA tried to get ants without a stick, got bitten on the foot, and hopped in agony. He also hung by his hands and tried to stick his finger in the entrance. Another young chimpanzee, Katabi, used a tool for the first time when he was almost three, spending six minutes preparing a branch tool. But it was too short and when he stuck it in the entrance hole it nearly vanished. When he was four, he climbed on the shoulder of an adult, Kamemanfu, who was ant fishing, and tried to fish from there, but Kamemanfu barked at him and Katabi's mother had to come get him. When he was nearly five he fished steadily for 23 minutes with a bark tool, until Kamemanfu chased him away. When he was over six, researchers still considered his technique clumsy and his persistence inadequate. He got bitten a lot.

Nuts

Chimpanzees in many areas crack nuts, and these can be a significant part of their diet. At Bossou, Guinea, their main foods are fruit, but during the middle part of the rainy season there's not much fruit, and they fall back on oil palm nuts, the pith of oil palms, and the fruit of the umbrella tree. Since tools are required to eat both the pith and the nuts of oil palms, primatologist Gen Yamakoshi argues that this population depends on tools for subsistence.

Researchers at Bossou set up an "outdoor laboratory," an open place where the presence of oil palm nuts and good stones for cracking nuts, both gathered by researchers, attracted chimpanzees. Observers crouched behind a screen. Researcher Tetsuro Matsuzawa examined the development of nut cracking in young apes. In the first stage, the infants hung out while their mothers cracked nuts, and manipulated one object at a time. One-year-olds would pick up a nut or roll it around on the ground or hold it in their mouths. They'd touch or hit or roll a stone. They would eat nut kernels their mothers gave them. Around two years old, the little apes would associate two objects: they'd push a nut against a stone or push a nut with a stone. They might put a nut on a stone. They began to attack the nuts, hitting them with hand or foot.

At three they'd put a nut on an anvil and hit it with their hand. Gradually they'd coordinate and string together their actions to crack nuts. By six or seven they were fully competent. At six and a half, Na used a third stone, a wedge, to level his anvil stone. Young chimpanzees whose mothers did not crack nuts (perhaps because of physical injuries) still learned to do it themselves.

Palm hearts, algae, and honey

At Bossou the chimpanzees have a clever and perhaps unique foraging activity—pestle pounding. This is a way of getting pith from the center of the stem of an oil palm. To do it you climb up into the top of an oil palm. Perhaps you pull out and eat the tender bases of the center fronds. Then you take a petiole of another frond and use it as a pestle to pound on and then excavate the soft juicy pulp in the crown of the palm. Also at Bossou, chimpanzees use sticks to skim algae from ponds. Chimpanzees in the Lossi Forest, Congo, use sticks to get honey.

Climbing shoes

Chimpanzees in Tenkere, Sierra Leone, like to eat the fruits and flowers of kapok trees. The kapok trees, as if foreseeing this eventuality, are studded with large sharp thorns. Apes move slowly and carefully through the branches. In Tenkere, but not in other places where chimpanzees feed in kapok trees, some of them use sticks to protect their feet and to sit on while they feed in these trees. Researcher Rosalind Alp first saw this in a teenager who twisted off a small thornless branch, put it in front of him on a thorny branch, and stepped on it, gripping it between his great and lesser toes. Standing on it, he picked fruit. The next day Alp saw a teenager (perhaps the same one) make "stepping sticks" for each foot. Sometimes when he wanted to move he moved the stick with his hand and stepped on it; other times he gripped it with his toes and moved his foot, carrying the stick with it like a sandal. Adults also stood and sat on sticks in kapok trees.

Like a Rock

An enclosure was designed for the purpose of holding and viewing a colony of captive chimpanzees. A paddock was built around an area with shade trees. Planks were fixed between the trees to form passages and climbing places for the apes. Above the planks, electrified wiring was wrapped around the tree trunks, because in a confined area animals often overuse and destroy plants. The 18-foot-high fence was chain-link topped with two sections of sheet metal, the top section of which leaned in at a 30-degree angle. An observation booth with large windows was built atop the fence. It was very nice, and carefully thought out, and then they added chimpanzees.

The eight infant chimpanzees played on the walkways, played with branches that fell off the trees and that were brought in so they could make nests, and were observed from the comfort of the booth. One of their games was to stand branches on end and swarm up as quickly as possible before they toppled over. A few years went by. One day six-year-old Rock, balancing on a pole near the observation booth, started to topple, stuck out his foot, and steadied himself. Aha! He could look into the booth! He could slap the glass! A week later Rock invented the ladder. The scientists arrived in the morning to find the observation booth a shambles. Rock had discovered that if you propped the branch against the wall under the booth, you could climb up, get in the observation booth, trash the booth, and escape. As reconstructed, in the night Rock may have been balancing on a pole next to the booth, toppled toward the wall, steadied himself there, and realized that he could now reach the booth. Rock's best friends promptly took up the same exciting activity, escaping with great frequency. Usually they returned of their own accord, but sometimes they had to be captured, often with lavish bribes of fruit. Since it was impossible to keep the apes out, the observation booth was boarded up.

The chimpanzees now propped branches on the walkways between the trees, to the portions of the trunks above the electrified wiring. (This required careful placement on the narrow planks of the walkways.) This allowed them to climb into the crowns of the trees, which they destroyed. Rock got hurt falling off a dead branch, and one of the trees toppled onto a walkway, so the dead trees were cut down, leaving only stumps, some of the posts of the walkway system, and a few planks. The industrious apes uprooted posts and stumps and wrenched loose the planks, and used them to escape. Posts, stumps, and planks were confiscated. Then they jammed short sticks into the cracks between sections of the fence to use as pitons, and escaped. Finally the entire colony was shipped to a monkey island in another state.

Stone tools? My folks used to make stone tools

A project to see if the bonobo Kanzi could learn to make stone tools like the ones made by prehistoric hominids proceeded in unexpected ways. The object was for Kanzi to flake flint into sharp-edged cutting tools. He needed to strike a piece of flint at the right spot with another rock. If he succeeded, he could use a sharp flake to cut a rope tied around a box containing some fabulous food item. Kanzi understood about the food, and he understood about cutting the rope, and, having watched demonstrations, he understood about hitting one rock with another to make a flake. But he had a hard time making flakes that were big enough, and after a few months of trying he was getting quite frustrated. One day he tried to get his friend, primatologist Sue Savage-Rumbaugh, to do it for him, but she wouldn't. He sat and thought, with rock in hand, and suddenly rose to his feet and dashed the rock to the ground with all his might. It shattered, and Kanzi uttered a glad cry, snatched up a nice big flake, and headed for the box. Since early humans probably did not flake flint this way, the researchers were disgruntled by Kanzi's method, and they carpeted the floor to thwart him. Kanzi tried hurling a rock at the floor, but nothing flaked off when it hit the cushioned surface. Undaunted, the brilliant bonobo examined the carpet till he found a join, peeled it back, and smashed the rock on the hard floor beneath. "We have assembled a videotape of the tool-making project, which I show to scientific and more general audiences. Whenever the tape reaches this incident there is always a tremendous roar of approval as Kanzi—the hero—outwits the humans yet again," writes Savage-Rumbaugh.

Capuchins and tools

Those who work with octopuses grumble about the hegemony of vertebrates; those who work with birds grumble about the hegemony of mammals; those who work with marsupials grumble about the hegemony of placental mammals; those who work with buffalo grumble about the hegemony of primates; those who work with monkeys grumble about the hegemony of apes; those who work with gibbons grumble about the hegemony of the great apes; those who work with orangutans grumble about the hegemony of the African great apes. And those who work with New World monkeys grumble about the hegemony of Old World monkeys and apes.*

Therefore all those who would strike a blow against hegemony† will be glad to hear that tufted capuchins, who are New World monkeys, have been spotted using tools to crack nuts, just like chimpanzees. Researchers

---

* I have met my quota and hope never to use the word "hegemony" again.

† This is really the last time.

Eduardo Ottoni and Massimo Mannu watched a group of tufted capuchins living in the Ecological Park of the Tietê River in Brazil. These are free-ranging former captives who had been living in the park for about seven years. This particular group was founded by capuchins who had been put on islands in the river but had swum to the mainland.

Capuchins had been reported cracking palm nuts by banging the nut against a tree or by smashing two nuts together, but residents in areas near where capuchins live also claimed that they used stone tools, and suggestive piles of stones and nutshells had been detected. Sure enough, these capuchins cracked palm nuts with a hammer stone and anvil stone. Sometimes the anvil was not a stone but a root or an old piece of pavement (not all parks are in pristine forest), and one hammer was a piece of wood. Capuchins aren't big monkeys and they usually held the hammer stone in both hands as they brought it down. Unlike chimps, capuchins can steady themselves with their tails as they perform this feat. Sometimes the monkey leapt clear off the ground with the effort.

Fifteen out of the 18 capuchins cracked nuts, and the 3 who didn't were infants. The most frequent nut crackers, by far, were the juvenile monkeys. Perhaps this is because they like fiddling with stuff, and perhaps it is because the adults chase them away from easier food sources.

Ottoni and Mannu classified some nut cracking as inept, such as when the monkey missed the nut entirely or didn't even put a nut on the anvil. Infants were by far the most inept, then juveniles, then subadults, then adults. They predict that it takes a capuchin years to become efficient.

Before you give a capuchin the Nobel Prize

But it so happens that capuchins absolutely love to pound on things. Many years ago the psychologist Elisabetta Visalberghi was passing the capuchins in the Rome Zoo and saw an adult male pounding on an unshelled peanut—with a boiled potato. "The fact that capuchins were doing something smart in a silly way, or something silly in a smart way, struck my interest," writes Visalberghi, who has since studied capuchins extensively.

Young capuchins love to pound, so if one capuchin is pounding on a nut and another one starts pounding things, it's not necessarily imitation. It could just be stimulus enhancement. It's pounding time! They seem to be enthusiastic manipulators of objects and terrible imitators. Of her capuchin Andy, Kathleen Gibson writes, "His skills in this regard are so poor that he has never even imitated my action of unfastening his leash." This although he is a clever fellow who rakes things into his cage, picks pockets, and wraps himself in a towel when watching scary war movies on television.

In another publication, Visalberghi and Luca Limongelli write, "During our 17 years of research experience with cebus [capuchin] monkeys . . . we have alternated between marvelling at their cognitive accomplishments and being plunged to the depths of despair over their inability or extraordinary reluctance to learn a variety of apparently simple tasks."

Captive capuchins were presented with a large box containing maple-flavored syrup. The box had holes in the top which led into plastic pipes which led down to the syrup. When they first saw the box, there were wooden dowels in the holes, and several of the capuchins pulled the dowels out and licked the syrup off, then reinserted the dowels to get more syrup. Later they were given the same box, but no dowels. The researchers had put sumac branches in the cage. The monkeys tried to get syrup by dipping in monkey biscuits, but that didn't work. On the second day the capuchin Nick bent a twig from a nearby branch into one of the holes, without actually removing the twig, and licked syrup off the end. Immediately Fanny, who had been watching Nick, removed a stick from a branch and dipped for syrup. The next day both Nick and Fanny broke off sticks to dip for syrup. Nick further covered himself in glory by snapping off a side branch of a stick he had broken off, "modifying an already-constructed tool by subtraction."

There were several baby monkeys in this colony, and they wanted syrup too. Alice's daughter Quincy, eight months old, explored the box extensively. She stuck her fingers in the holes. If monkeys would let her, she'd remove sticks they had inserted. Grasping dimly that sticks had something to do with getting syrup out of the box, Quincy took small twigs in her hand and slapped them against the side of the box. After 17 days, Alice and Quincy dipped a stick together. "Quincy's hand was literally on her mother's as the stick was inserted." Quincy got to pull the stick out herself and lick the syrup off. Within two days she was trying to do it herself, ineptly, holding a stick in two hands and stabbing it toward the box, missing every time. Two days later she succeeded in dipping a stick and two weeks later she was making her own sticks.

This was before the free-ranging capuchins in Brazil had been spotted cracking nuts, and the researchers puzzled over the fact that capuchins hadn't been seen to use tools in the wild. (The following year, a wild white-faced capuchin was sighted beating a poisonous snake to death with a branch. The author of the report added that capuchins often throw things at coatimundis, tayras, opossums, spider monkeys, and the author. On one occasion a capuchin who ran out of branches to throw at the author threw a squirrel monkey at her. I say it's tool use.)

Cause? Effect? Your point?

In experiments with capuchins trying to get food out of a clear plastic tube, Visalberghi and colleagues clarified what capuchins did and did not understand about tool use. The original apparatus was a clear plastic tube, mounted horizontally. In the middle of the tube was a peanut. To get the peanut out, the monkeys had to push it out by inserting a stick in one side or the other. They tried all kinds of methods, some of them fairly witless, but eventually succeeded. Visalberghi and Limongelli set up a variation in which the tube had a small (clear) trap on one side. If the capuchin pushed the peanut through from one side, the peanut would come out. If it pushed from the other side, the peanut would fall into the trap. Argh! If the peanut fell in the trap the capuchins bit the trap and shook it fiercely, but people who work with monkeys have learned to make the apparatus sturdy. Sometimes they watched the progress of the peanut as they pushed it, and held a hand under the tube, moving it as the peanut moved "as if by doing so they could prevent its fall" into the trap. That doesn't work either. They got nervous as the peanut approached the trap. But three out of four never improved, never showed any sign that they understood why the peanut might fall into the trap, only that it might. The fourth capuchin, a juvenile, did show improvement. Less and less often did she shove the peanut into the trap.

Alas, the successful capuchin, Rb, still didn't understand what she was doing. She had hit upon a strategy which is summed up as "insert the stick into the opening of the tube farthest from the reward," but didn't know why it worked. If they turned the tube so that the trap was on top, and there was no risk of the peanut falling in the trap, she was just as anxious, and peered into the tube just as many times to make sure she was using the end farthest from the reward.

Bowling for apples

Researchers who gave Japanese macaques a tube-and-stick problem got a different response. They fixed a large acrylic tube to a log at a site where they left food out to attract monkeys. To teach the monkeys to use a stick to get an apple out of the tube, they began by putting an apple in the middle of the tube, where the macaques could see it but couldn't reach it, with a hooked stick in the tube touching the apple. After various changes in the procedure, some monkeys could pull an apple out with a hooked stick, some could search for sticks to pull or push the apple out, and Tokei invented the idea of throwing rocks into the tube to knock the apple out.

Tokei had another method for getting apples: child labor. Babies under six months old were small enough to get in the tube. While the mothers Togura and Tomato would pull their babies out if they had gone in the tube, and take the apple from them, Tokei was more proactive. She'd take her baby and stuff it in the tube.

Birds with sticks

In Queensland, Australia, there's a species of palm cockatoo which makes even more hullabaloo than most cockatoos. They perch in tall trees and yell their heads off, and back this by stamping on tree trunks with their feet. If that isn't loud enough, they pound on trunks with nuts or sticks, striking up to 100 blows. Palm cockatoos have been seen snapping off branches and trimming them into drumsticks of their preferred size and shape. (Bird-watchers have been seen sneaking around and gathering discarded drumsticks.) A male palm cockatoo will perch atop a snag, stretch his wings wide, and pirouette while drumming on the tree with the drumstick held in one foot. Female palm cockatoos are thought to like this very much. One pair of palm cockatoos were both seen drumming on the day before their baby left the nest for the first time.

Otter not

Otters are skilled manipulators. The impulses that lead them to feel among the stones in a river or tide pool in search of lurking edibles such as crawfish or shrimp lead captive otters to juggle hazelnuts or slip their paws into pockets and fish out keys. Since we too like to manipulate things with our hands, this looks smarter to us than if they were doing it with, say, their lips.

Susie, a captive Alaskan sea otter, was given stones so she could display the famous propensity of sea otters to crack open clams. Susie also used them to pound on the edges of her concrete pool and smash the bolt holding the cover on the pool drain, which she presumably wished to investigate. They took the stones away.

Thumbs? I don't even need hands

A group of dolphins in Shark Bay, in western Australia, apparently use sponges as tools, but no one knows how. At first observers thought they were seeing a dolphin with a hideous growth on her face, but closer inspection showed that several female dolphins were in the habit of putting cone-shaped sponges on their snouts. Whatever they are doing with sponges on their faces, they do it underwater. When they dive after appearing at the surface with a sponge, the curve of the dive indicates a deep dive. They sometimes reuse a sponge, but they also get new sponges. The dolphins are occasionally seen to surface chewing something, and the amount of time they spend in this pursuit suggests they are using the sponges to forage in some way. The leading guesses at this point are that the dolphins use the sponges to protect their faces either from the spines or stingers of prey like lionfish or from abrasion by coral grit as they search for prey in the seafloor sand. Wild dolphins also smack yellowtail bream they've caught against the ocean floor to snap the heads off. (By many definitions this doesn't count as tool use, but it is a useful trick if you don't like eating bream heads.)

Captive bottlenose dolphins Frankie and Floyd tried to chase a moray eel out of a crevice. The eel wouldn't go. One of the dolphins killed a scorpionfish and held it by the stomach to avoid its poisonous dorsal spines. Then he prodded the beleaguered eel with the scorpionfish. The dolphins also liked to play with pelican feathers, positioning them over water jets and then chasing them. When feathers were scarce, the dolphins would swim up to pelicans and yank feathers out.*

The cheaper baby buggy

Gold-cheeked flag cichlids are attentive parents. A couple guard, clean, and fan their eggs, and when the fry hatch out they take them into their mouths and transport them to the places the parents feel the babies ought to be. The eggs are typically laid on a leaf, and the parents busily shuttle the leaf from one spot to another. In laboratory tests, they choose light small leaves that are more easily dragged in preference to large leaves to which some troublemaker has secretly glued small lead weights. A pair will move the leaf/baby buggy to deeper water with better cover if that is an option, and they move it more if there is a predator around, or even a minnow-shaped fishing lure painted to resemble a predatory pike cichlid. Also, if they see people approaching the tank, they drag the leaf to the back of the tank, usually behind a plastic plant or other object.

What have you done to your nose?

Kipling's fictional Elephant's Child uses his new trunk to break off a branch for a fly-whisk. Researchers observing Asian elephants in India often saw them switching flies with branches they had broken off. But since the elephants broke the branches off trees they were in the process of dining on, and since the elephants often ate the branches after switching them at flies, it could be argued that the elephants were not breaking branches for the purpose of switching flies, but just for eating. It would be as if you were eating celery, were bothered by a fly, waved your celery at the fly, and then ate the celery. Do not expect big tool-use kudos for that.

---

*This was hard on the pelicans, but they had it better than the moray eels, with whom the dolphins persistently attempted to mate, to the displeasure of the eels and the embarrassment of the staff.

The researchers focused on semicaptive elephants at a logging camp and a riding camp. Finding the elephants in camp, with flies about, the researchers presented them with enormous branches of *Butea* trees. *Butea* leaves don't taste good, so the branches were useless for eating. They were also too big for switching flies. Often the elephants picked the branches up, broke them to a useful size, and switched flies on their sides, backs, and bellies. They either put one foot on the branch and pulled a smaller piece off with their trunk, or they coiled their trunk around the branch and twisted a piece off.

An 18-month-old calf took a branch, snapped off a side branch, and switched flies with it. A 9-month-old calf tried but seemed too uncoordinated to actually succeed in switching flies. The researchers consider these cases of imitation and emulation, and hint strongly that elephants are just as good as primates any day.

Why not just whack it with a tire iron?

In 1974 a paper appeared in the ornithological journal *Western Birds* titled, "Do Crows Use Automobiles as Nutcrackers?" In 1978 a paper appeared in the ornithological journal *The Auk* titled, "Crows Use Automobiles as Nutcrackers." In 1997 a paper appeared in *The Auk* titled, "Crows Do Not Use Automobiles as Nutcrackers: Putting an Anecdote to the Test."

The first two papers each described a crow dropping a nut onto a road where it was run over by a car. The crow then hopped over and ate the meat of the nut. The first was a crow in Davis, California, which dropped walnuts on the road, and the second was a crow in Long Beach, California, which dropped a palm fruit on the road.* The third, 1997, paper described extensive observations of crows in Davis, on two roads lined with walnut trees. The authors concluded that the crows were dropping walnuts to crack them on the hard road surface. Cars might sometimes run over the walnuts, but the crows were no more likely to bring walnuts to the road or to drop walnuts on the road when a car was com-

---

* The first paper describes an observation of crow behavior made in the author's rearview mirror, since he was too busy to go back and observe further. In the 1970s car manufacturers did not provide warnings that "Actions in the mirror may be dumber than they appear."

ing than when there were no cars in sight. Moreover, 200 cars passed without running over a walnut. While the researchers could not rule out the possibility that crows ever harnessed the automobile in their culinary endeavors, they wrote that "their putative exploitation of moving cars is not adequately documented and should not be cited as an example of avian intelligence or adaptability."

More recent reports say that carrion crows in Sendai, in northern Japan, have been seen cracking walnuts under the wheels of cars. They don't drop the nuts, but place a nut on the road and then hop back until a car runs over it. This is said to be advantageous to younger crows. The idea put forward by researcher Yasuhiro Adachi is that for an adult crow it's just as easy and more reliable to simply drop the nut, but that inexperienced young crows tend to waste energy by flying too high when they drop nuts.* They save energy by using the car method.

Crows won't use this method unless they know how, and also have access to a suitable area with plenty of cars, a red light to make the cars stop from time to time, and plenty of nuts nearby. And adult crows have no reason to bother, so it's logical that the technique isn't more widespread.

Strange doings in New Caledonia

Crows and their relatives are frequently cited as nonhuman tool wielders. Heroes of this kind are the crows of New Caledonia, who display their prowess in the laboratory and in the wild, even when being filmed by the BBC. They use tools, they make tools, they make standardized tools, and they take their tools with them from place to place. One wild crow was seen arriving at a foraging site in the morning with a tool in its bill. It only remains to find that they hang their tools on walls with painted outlines of the tools to show where each one should go.

They make hooked-twig tools, from which they strip the bark and leaves, and use the hooked end to fish in holes and crevices for insects. They also make stepped-cut tools, which they cut from the edges of pandanus leaves. These are long and pointed and have small barbs on one side.

---

* Sadly, I do not know whether any Japanese ornithological journal has published an article entitled "Maybe American Crows Don't Use Automobiles as Nutcrackers, but Japanese Crows Do."

When foraging for the larvae of longhorn beetles in dead wood, crows make tools out of dry leaf stems, or out of twigs which they find on the ground or snap off nearby trees. The basic concept is that you annoy the larva by poking it, it unwisely grabs the tool in its jaws, and you pull it out and eat it. Sometimes it takes as much as ten minutes to dupe the larva into doing this.

A juvenile crow with this foraging party begged, and was given pieces of larvae by adults. The young crow looked into holes that adults were probing. (Usually this was tolerated, although once an adult shoved the kid away.) Sometimes the young bird picked up a tool that an adult had put down, and probed in the hole where the adult had been working. Twice, young birds tried incompetently to make tools, once using a piece of grass (too flimsy) and once holding the wrong part of a leaf stem it was snipping the leaf from, so the stem fell on the ground. "Adults did not behave like this, suggesting that the juvenile was inexperienced," writes researcher Gavin Hunt.

Crows who made hooked tools tended to hang on to them, and when Hunt and Russell Gray assembled a collection of 15 hooks, they got most of them by scaring off crows who had set their tools down for a moment. They also got two tools that crows had dropped and two that were left on a feeding table. The hooks are small but useful, and the crows consistently use the hooked end and not the straight end in their foraging (as evidenced by signs of wear).

Gavin Hunt, chronicler of the tool-using exploits of New Caledonian crows, says they appear to be "the only non-human species that manufacture and use hooked instruments." He goes on to demolish claims that chimpanzees have been seen to use hooked sticks to reach for figs—of those three chimps, only one used a stick with a hooked aspect to it, and half the time he held it by the wrong end! (This is the kind of statement that can only result in the discovery of chimps somewhere using hooks incessantly—to reach fruit, to scratch their backs, to crochet doilies with the motto Crows Are Birdbrains cross-stitched around the rim.)

Throwing the first stone

Two ornithologists were clambering around cliffs in eastern Oregon looking for ravens' nests. They climbed up to one and examined the six baby ravens they found within. When they exited, they were greeted by angry raven parents, who swooped at them furiously. As the ornithologists descended the cliff, a rock "the size of a golf ball" hurtled past. They thought a raven had accidentally kicked it loose until they looked up and saw a raven on top of the cliff with another rock in its bill. As they watched, the raven tossed it at them. They cowered against the cliff and watched in astonishment as the raven threw six more rocks, hitting an ornithologist on the leg. They returned later in the day, hoping to photograph this behavior, but the ravens were out of rocks and could only throw grit. The next year they came back to the site, but for some reason no ravens had nested in that spot.

The mad genius prisoner

A group of captive bald eagles got in the habit of throwing things, and it wasn't pretty. They threw a ring at a human, they threw rocks at crickets, they threw rocks at a turtle. I can understand that they might have blamed humans for their situation, and crickets could be edible, but surely the turtle was blameless? One of these eagles gripped a rock in his talons and used it to strike crickets and a scorpion; it also gripped a stick and used it to beat on that poor turtle.

I like eggs, but I don't know how to fix them

Egyptian vultures have become TV stars because of their habit of breaking open ostrich eggs by throwing stones at them. An Egyptian vulture who finds an unprotected egg in an area where there are no suitable rocks to throw will fly off and fetch one. Other vulture species in the area don't do this, although they are delighted to eat an ostrich egg that has already been broken.

Researchers kidnapped baby Egyptian vultures and raised them in captivity to find out how they learn to do this, specifically whether they copy other vultures who know how. A previous captive had been offered an ostrich egg, but instead of breaking it, had incubated it, leading to the suggestion that eating them was a cultural tradition. When the young vultures were presented with (fake, fiberglass) ostrich eggs, they were intrigued, but the thought of breaking the eggs or throwing things at the eggs did not occur to them. Researchers began giving them chicken eggs to eat, and these the young vultures liked. They soon figured out that they could pick up and throw these little eggs.

When they were a year old, they were offered fake ostrich eggs again, and they were interested, "suggesting that they were generalizing about shapes." But they didn't do much about it, even though there were stones handy. The researchers then gave one of the vultures a hen's egg cracked into a fake ostrich egg shell. Eureka! The next time the bird was given an ostrich egg, it immediately threw rocks at it. It didn't manage to hit the egg until its fifth throw, but after seven more tries it scored nine hits in a row.

The researchers concluded that wild vultures learn to break ostrich eggs, but not by copying other vultures. They argue that the critical thing is for the vulture to "experienc[e] an [o]strich egg as food." Only then will they throw stones at it. Egyptian vultures like to throw things: they throw small eggs to break them (in the wild, pelican and flamingo eggs), and one has been seen throwing stones at a tortoise. From throwing eggs to throwing things at eggs is not a great leap.

Nests, burrows, and dens

Building a place to live would seem to be related to tool use, since lots of object manipulation and material skills are involved. Why would it be tool use if you pick up a rock to throw at an egg, and not if you use it as part of the wall in a chic little two-room bungalow burrow you are building in the sand near a coral reef, like the yellowhead jawfish? But nest and burrow building is so obviously under so much genetic control that it has seldom been considered indicative of intelligence of any kind.

Gorillas, chimpanzees, bonobos, and orangutans all build nests to sleep in at night, and monkeys don't. (But among the lowly prosimians, bushbabies, mouse lemurs, and aye-ayes build sweet little nests.) Usually any behavior that great apes do and monkeys don't is considered to be a big deal, but there's a hitch here. We don't do it. Or do we—can we compare building beds and even houses with making a nest of branches?

Barbara Fruth and Gottfried Hohmann argue that nest building deserves closer examination. "With some exceptions. . . . nest construction at night and nest leaving in the morning became more the curfew times of behavioral observation than a topic of research itself." In other words, when the ape began to build a nest, weary scientists thought with relief of quitting for the day, not of compiling information about nest building.

Fruth and Hohmann studied nest building in chimpanzees and bonobos. They found that nests, which take a few minutes to make, include three elements: a solid frame, a central "mattress," and a lining made of leaves and twigs. It's not well understood how young chimpanzees and bonobos develop this behavior. They have an innate urge to make a nest, and hand-raised infants make nests out of blankets, sofa cushions, canvas bags, and whatever comes to hand. Observation of young wild chimpanzees hints that they develop their skills with practice. During the day, young chimpanzees and bonobos make play nests, and their technique is often poor.

Fruth and Hohmann propose the hypothesis that not only should nest building be considered tool use, "but it is also the original tool that led to the mental and physical ability to use the tool we see today." They postulate that the original ancestors of all great apes (including us) must have been nest builders. Perhaps nest building originated with feeding nests, when apes bend branches to make a convenient sitting place while they pillage a fruiting tree. Groups of nests, when apes build nests close together, might serve as an information center. Fruth and Hohmann also make the touching suggestion that nests may help apes sleep better.

In the Denver Zoo, the gorillas make ground nests. Ernie was a superb craftsman, and his work was much in demand by Bibi, who liked his handiwork so well that she regularly kicked him out and took his nest. Soon, whenever Ernie saw Bibi heading his way with that "I feel like a nap, buster" look in her eye, he would jump out of his nest, pick it up, and carry it to a safer spot.

I think I see what you were getting at

Nests have been referred to as frozen behavior, since they record a sequence of acts and decisions. Most of those acts arise innately, and most of those decisions arise when the animal compares what it is building to an innate template—a vision of the palace it was born to build. Birds raised by humans, birds who have never seen a nest, are for the most part perfectly able to build an excellent nest, given the proper materials. If not, they will improvise, like a peach-faced lovebird of my acquaintance who happily sliced a window blind into useful strips.

Nests are vital to the survival of baby birds. It makes sense that birds need to be born knowing how to perform such a complex, critical task. But learning creeps in. Nicholas and Elsie Collias have performed experiments illuminating the development of nest building in the village weaver.

Male village weavers build excellent nests, employing stitches and fastenings that include the loop tuck, the spiral coil, the alternately reversed winding, the half hitch, the overhand knot, and the slip knot. A nest is ovoid, roofed over, with a bottom entrance, an antechamber, and an egg chamber. It is suspended from a branch. The outside is made of long strips the bird tears from palm fronds or tall grass. Inside he installs a ceiling of short broad strips. The male, a good-looking black-and-orange individual, builds his nest in a day. To attract females, he hangs upside down alluringly from the bottom of his nest, flapping his wings and singing. Females like this. (It is a good bet that a guy who builds a house and hangs by his feet in the doorway singing in an attempt to attract a woman who will settle down with him is not a guy with commitment issues.) If a female likes the nest and the bird enough, she moves in and lines the nest chamber with fine soft materials. At this point the male adds a short hanging entrance tube.

Pulling off a project like this requires practice, and young males are eager to get started. Both female and male village weavers, from the moment they leave the nest, love to pick things up and manipulate them. They particularly love to poke and pull pieces of grass through holes. At about 10 weeks, females lose interest, but males remain obsessed.

Collias and Collias took village weavers from the nest at one week, before their eyes had opened, and raised them in bowls lined with cloth. At no time were they allowed nest-building materials to weave with. But they had each other. "It was not uncommon to see one of these deprived birds hold a protesting cage-mate's wing under one foot on a perch and attempt to 'weave' the wing feathers." If driven off, the obsessed birds would weave their own tail feathers.

Later, these birds with no experience were offered nest materials in a range of colors. They preferred green, and the more they thought about it the more they liked it. In the wild, this causes them to select fresh flexible grasses over brittle dry grasses that are hard to work with. If they were offered actual grasses when they were a year old, they were delighted, but they had a hard time tearing strips. What works is to perch on the base of the grass, bite through an edge of the grass leaf, and then, gripping the grass, fly away in the direction of the tip of the grass, tearing loose a long strip. The young birds couldn't tear a single strip on the first day. They'd perch in the wrong place, or they'd just grab the tip of the grass and try to fly off, or they'd tear in the wrong direction. They finally figured it out.

On the other hand, Phineas, a village weaver raised in isolation by Dr. Catherine Jacobs, never learned to tear strips. From the time he was six months old, Jacobs supplied him with pretorn strips, which he handled fairly competently. When, at the age of four, he was put in an aviary with other weavers, an aviary where fresh palm fronds to make into strips were delivered daily, he was still unable to tear strips. But Phineas was canny. He stole enough strips from other birds to make a crude nest, and then hijacked a half-built nest and completed it.

After tearing strips, the next step is to weave them into a sturdy vertical ring. The birds from the deprived background had a terrible time with this. In one experimental group, in the first week that they had access to decent materials, three males did not succeed in weaving a single stitch, while three males who had been allowed to mess with grasses to their heart's content did a fine job. In the second week the deprived birds started weaving strips into the wire frame of the aviary, neglecting a guava bush highly suitable for nesting.

Over the following winter, they had no chance to practice, but in the spring when they were given materials once more, the deprived birds did just as well as anybody. Except for LL, a deprived bird who was also bullied by everyone. Whenever he tore a strip of material, someone stole it before he could play with it. This was so constant that LL never learned to weave, let alone make a nest. He still liked to tear strips, and to play with them if he was permitted. He was transferred to an aviary where he was able to improve his social position and stand up to other males, but at the time of his death at the age of nine he had never built a nest.

In the wild, in their first year, young village weavers build crude sloppy things with loose ends and flapping loops, in little "play colonies" away from the adult colonies. Some nests in the play colonies have roofs but no floors.

What do young birds learn that eventually makes them good nest builders? Collias and Collias say it appears to be "what in subjective terminology one would call 'judgment.'" They learn to use materials of the right dimensions and flexibility; to hold a strip that they are starting to tear with their foot so it doesn't slip away; to persist in pulling on a strip until it is in place; when to let go of the strip to pull the next section into the right place; and to leave it alone once they have woven it into the right place. (Young birds are strongly tempted to pull it out and do it again.)

Collias and Collias, for some reason, induced one weaver to build a very long entrance tube on his nest. They did this by threading extra long strips around the entrance. He couldn't stand the dangling strips, so he wove them into an extra-long tube, 30 centimeters long instead of the usual 5 to 10 centimeters. Even though the researchers kept their hands off, all his subsequent nests had long entrance tubes, some even longer than 30 centimeters. "It would seem that the male develops a mental picture of the sort of nest to build, based on his experience," they write.

Old dogs and new tricks

While baby animals are particularly eager and able to learn, and while there are many things it's harder to learn as you get older, animals do keep learning. New neural connections are formed, new ideas are received, and new behaviors are exhibited. It's never too late to learn that a certain individual bites. You are never too old to learn that a particular food can make you sick.

Bluntnose minnows were taught to distinguish between water from Honey Creek and water from Otter Creek. Going to one kind of water brought a snack, going to the other produced an electric shock. Aged bluntnose minnows, nearly two years of age, "approaching senescence," learned this task as well as yearlings. Yearling minnows remembered the difference between the two creek waters for 15 weeks, however, whereas ancient minnows remembered for only 6.

Clarence, the house sparrow hand-reared by Clare Kipps, lived to a remarkable old age. At two, he had a stroke. He couldn't fly anymore, being unable to coordinate his wings. His balance was off, and he kept falling onto his back, calling "quite cheerfully" for Kipps to come and put him on his feet. Eventually he discovered that he could get back on his feet by doing a flip, "becoming so expert that before long he could leap instantaneously into the air from his inverted position, turn a complete somersault and come down the right way up—surely a feat for a small bird even in the first blush of youth!" (The adult chimpanzee MacGregor learned to progress in somersaults after he was crippled by polio.) The lack of coordination also made it hard for him to hop with both feet, as sparrows customarily do, and to the astonishment of ornithologists Clarence took up walking, right, left, right, left, just like the rest of us.

Innovation builds on learning. Sometimes an animal simply does everything it can think of with a stick and notices—learns—that one thing produces good results. At other times an animal learns in the same way what sticks can do and then thinks of a way to use that. If we could break intelligence in two, the ability to learn might be one part, and insight, leading sometimes to innovation, might be the second part.

EIGHT

How to Get Cultured

In Laguna, Brazil, people and bottlenose dolphins fish cooperatively. Fishermen with circular throw nets stand in a line, waist-deep in shallow water in the town's lagoon. The water is brackish and murky. A few dolphins drift slowly along the line a few meters away, facing seaward. Occasionally a dolphin signals by arching its back, slapping the water with its head, and then slapping the water with its tail as it dives. The dive roils the water, and the fishermen cast their nets into the space between dolphins and people, onto schools of mullet, croaker, or black drum. Those who catch fish carry their nets up to the beach and other fishermen take their place in line. Meanwhile the dolphins grab the fish who race out from under the net.

The fishermen can't see the fish, and only know when to cast nets by watching the dolphins, who herd fish toward the beach. As for the dive, "the timing of the dolphin's roll indicates that fish are present; the direction of the dolphin's movement indicates the location of the fish, and the vigor of the movement appears to indicate whether the school is large or small: the dolphin may show the head, back and dorsal fin, or just the head or blowhole." The falling nets presumably cause some fish to dash straight into the dolphins' jaws as they duck the nets. One dolphin may work the line for a couple of hours before leaving, and it is often replaced by another dolphin.

The fishermen do not feed, signal to, or train the dolphins. The dolphins—who can detect the fish—are in charge. If a dolphin moves to another part of the beach, fishermen hurry to follow. Sometimes people wait on shore, hoping a dolphin will show up. Dolphins spy-hop (stick their heads above water) to see if there are people ready to fish.

Researchers estimate the local dolphin population at 200, and 20 to 30 dolphins participate in the cooperative fishery. This is serious business, the pri-

mary source of income for about 100 families who sell the fish in nearby towns. On one day during fishing season, researchers saw six dolphin-human groups fishing in the lagoon, with up to 3 dolphins and from 4 to 40 men in each group.

Male and female dolphins both take part. Calves accompany their mothers. Some of the calves of cooperative fishers join in and some don't, but apparently no dolphins join who don't have a mother who was part of the fishery. At Imbé, 200 miles to the south, researchers saw the dolphin Geraldona and her four-month-old calf taking part in such a fishery. Twice, as Geraldona signaled to the fishermen, the calf followed behind her. The next three times the calf swam by her side. Finally Geraldona hung back as the calf went forward by himself and signaled the fishermen, who took his advice and netted mullet.

Rarely, a dolphin cheats by lifting or reaching under a net to grab entangled fish. The fishermen, alerted by mud clouds, throw sand or rocks at the dolphin to chase it away.

Laguna town records indicate that the dolphin-human fishery began in 1847, and some fishermen say their fathers and grandfathers fished with dolphins before them. That generations of dolphins participate is shown by Chinelle, a dolphin who has at least two daughters who are also active in the fishery, one accompanied by her calf. How the tradition started is a mystery. There are four such known fisheries in southern Brazil and scattered reports of similar fisheries around the world.

PEOPLE HAVE ELABORATE CULTURES, with enormous stores of learned information. In its everyday definition, culture includes language, religion, musical styles, ways of dressing, how to use tools and technology, rules about what parts of the body clothing should cover, styles of child-rearing, rules about who can marry whom, ideas about what people can eat and what they can't, and more. We use the word "culture" to refer to things as immediate as how closely you should stand to another person and as abstract as whether religion should be part of politics.

The essence of culture is learning from others. Animals, like people, behave badly often enough that it hardly seems likely that they would refrain from stealing each other's ideas. Yet there is, inevitably, debate over whether any animals really have culture. This begins with debate over what culture is exactly. (Ethologists traditionally speak of traditions, rather than culture.)

Some scientists define culture simply as the nongenetic transmission of behavior. In *The Ape and the Sushi Master*, Frans de Waal writes, "Culture is a way of life shared by the members of one group but not necessarily with the members of other groups of the same species. It covers knowledge, habits, and skills, including underlying tendencies and preferences, derived from exposure to and learning from others." Just how animals learn from one another is unimportant in de Waal's definition.

Other definitions are more exacting. Some cultural anthropologists take the view that there is no culture without language. Among the criteria that culture should show, as proposed by W. C. McGrew, are innovation, dissemination, durability, standardization, diffusion, tradition, nonsubsistence, and naturalness. Yet we can easily characterize some human cultures (other human cultures, naturally) as being anything but innovative. Using these criteria, it has been argued that chimpanzees show no cultural behaviors. Nor would the dolphin-human fishery described above qualify as a cultural behavior for humans or dolphins, since it involves subsistence—getting something to eat.*

Michael Tomasello is also strict about what he considers cultural behavior. In his definition, it must include imitation, and we know how long the arguments about *that* go on. With fairness, Tomasello comments, "Although I have expressed doubts about the facile use of the term 'culture' with respect to chimpanzees, in very few cases has the behavior of human children been examined with the skeptical scrutiny I have used in this analysis." Preferring to think that some animals and some humans do in fact show cultural behavior, I will, like de Waal, use a broader definition of culture espoused by many researchers, including pioneering Japanese primatologists like Shunzo Kawamura and Kinji Imanishi.

The wisdom of the ancients

Cultures pass on information that has been learned, not just to the children of the learner, but to other members of the group. Some social

* Primatologist Christophe Boesch writes that the nonsubsistence criterion "excludes food-oriented behaviors (difficult for a French person to understand!)" I can't understand it either—pass me a doughnut while we think this over.

animals benefit from having knowledgeable elders, who are not necessarily leaders. Primatologist Hans Kummer, following hamadryas baboons in Ethiopia, noticed that the aged and no longer dominant baboon Admiral knew the best routes across every obstacle and exactly where to dig for edible roots. Younger animals worked harder and found less. One afternoon the troop had not drunk all day. They were walking down a dry riverbed when Admiral, ahead of the troop, veered off to climb a bare hill with a patch of shrubbery on top. As he made his slow way up, he was passed by a juvenile and a female who went past the bushes. Admiral entered the bushes, and they turned back. In ones and twos, the entire troop went in. Afterward, the researchers found that in the middle of the bushes was a shaded hole in the granite, a yard deep, with a pool of rainwater at the bottom. Admiral knew it was there, but at least some adults did not.

In Amboseli National Park, Kenya, matriarchs are valuable assets to elephant herds. The older the oldest female, the better the herd's reproductive success. Perhaps she is better able to lead them to food and water. She also knows everybody. When elephants hear the contact call of a strange elephant, they bunch up protectively and wave their trunks around trying to sniff out the identity of the stranger. Scientists tested the reactions of herds in Amboseli whose matriarchs' ages ranged from 27 to 67 years old by playing them recordings of the calls of different elephants in the park. Herds with older matriarchs were far less likely to get panicky and go into defense mode, because the matriarchs knew the calls of more elephants. Herds with younger leaders spent more time fussing and worrying about invaders than in eating.

Short-finned pilot whales reach at least 80 years old but don't breed after 40. Jared Diamond reports seeing a 55-year-old killer whale with his 85-year-old mother. That females do not simply die when they become menopausal suggests to researchers that they are making valuable contributions, and perhaps part of that is knowledge.

Cultural hotbeds

In looking for conditions that might favor the development of culture or "proto-culture" in chimpanzees, Sue Taylor Parker and Anne Russon point to cognitive abilities such as imitation, the ability to teach by demonstration, and the social trait of female dispersal. If babies mostly learn from their mothers, then in a species in which females move from one group to another, they will carry traditions to the new group. (In some animals, females stay put and males move to new groups.) That chimpanzees use tools to get at recalcitrant foods such as termites, ants, and nutmeats also favors apprenticeship in tool use.

Duane Rumbaugh, Sue Savage-Rumbaugh, and Rose Sevcik, who have studied symbol and language acquisition in bonobos, chimpanzees, and children, note that young animals sometimes learn rapidly to do things their elders have laboriously struggled to master. Learning things in childhood may sometimes help develop "cognitive structures" in the brain of the young animal. Their improved performance could lead to a more focused learning experience for their own offspring. "Cultural gains might have been made indirectly and quietly as each gain, in turn, served to direct cognitive development of the ever-observant child to specific topics for reflection and refinement as he or she matures. Thus, the bedrock for geometric gains in building culture and technology across eons was laid in the minds of babes."

The arts, theater, and the dance

To some people culture means museums, symphonies, ballet. Some animals have musical and even artistic traditions. These can be passed on vertically, from parents to children, or vertically, horizontally, and obliquely. The striped-back wrens of Hato Masaguaral, Venezuela, have been studied since the 1970s by biologists who chart their genealogies. Striped-backed wrens live in family groups consisting of one breeding pair and as many as 12 of their children from previous years. Together the family maintains their nest, feeds new babies, and patrols their territory, an area of about an acre. Young males usually stay with the family for their whole lives, and may inherit the territory one day, although sometimes they strike out to find a new territory. Young females leave, often with their sisters, and look for a new family to join.

These are talkative birds. All the wrens utter nasal, raspy, clicky calls that sound to English speakers like a drawled "Where are you?" WAY, for short. They WAY back and forth all day. Males have a set of about 12 WAY calls, and females have about 4. Each family's males and each family's females have a set of WAY calls that are completely different from the WAY calls of the wrens in neighboring territories. Researchers suggest that during territorial disputes, WAY calls help birds keep track of which team is which.

Striped-backed wrens learn their WAY calls mostly vertically— young females learn from their mothers, and young males learn from their father and brothers. Young birds copy the songs of their elders with great fidelity. When researchers found two widely separated groups where the males had very similar WAY repertoires, they went to their genealogies and discovered that the two groups had a common great-grandfather, who had died two decades before, but whose songs his descendants still sang. He had left one family group to join or found another, and took the family songs with him.

Busy designer/architect/painter/musician/dancer wishes to meet females, LTR not desired

A male satin bowerbird doesn't fully develop his maddeningly attractive deep blue plumage until he is six or seven years old, but in the meantime a young bird is not just playing video games. He builds practice bowers—courtship sites—of increasing sophistication and grandeur. He goes around to the bowers of impressive older males and views their displays. Satin bowerbirds make very impressive bowers, even for bowerbirds. A male clears a space for his court and builds a broad platform, on top of which he builds an avenue of sticks. He decorates this with attractive objects, especially blue objects, since satin bowerbirds adore blue. (In addition to the male's indigo suit, both sexes have Elizabeth Taylor blue eyes.) Other favored accents include yellow leaves, yellow straws, blue and yellow flowers, snail shells, and shiny cicada shells. Parts of the bower are painted with paint the male mixes by chewing up vegetation with plenty of saliva. He also prunes the foliage above his bower so that shafts of light illuminate his artwork.

A male bowerbird displaying stands on his platform, often holding a priceless object of art in his bill, such as a blue feather. He faces his audience. He whirrs and prances. He fluffs his feathers and calls, flapping his wings to the beat of his call. He ornaments this song with chortling, buzzing, mimicry of other birds and environmental sounds, and tasteful interludes of silence.

Awestruck, a young male viewer goes back to his bower and practices his own song-and-dance routine. If he happens to visit the impressive bower of an adult male when that male is away, he may act as if it's his own bower: do a little painting, practice his display, or even put the moves on any female who drops by.*

A nice bower is key to social success. Some males do not get to mate all season, but one male observed by Gerald Borgia and colleagues mated with 33 females. (Sadly, photographs of his bower, which must have been stunning, do not seem to be available.) The researchers ranked bowers for quality and found that their ratings agreed with those of female bowerbirds: bowers should be symmetrical and neat, sticks should be densely packed, and the general appearance should be highly sculptured. Also, females seemed to feel bowers should have plenty of knickknacks, especially blue feathers.

Blue feathers are in short supply, and the most efficient way to get them is to steal them from other bowerbirds. A desirable blue celluloid band changed ownership several times a day until one bird wove it into the bower itself. Researchers got involved in this illicit traffic, but if they purloined feathers from popular, dominant males and put them in the bowers of nobodies, the feathers were stolen right back, and the status quo did not change. Thus, the most successful males are thieves who have too much clout to be stopped. This correlates with age. "Older males maintain bowers of better quality and decorate them more elaborately, and they are more successful in protecting their bowers from destruction. Moreover, older males give more refined courtship calls," writes Borgia.

Vogelkop bowerbirds are dowdier than satin bowerbirds, and they build elaborate decorated huts around saplings. They are hard to study because they live in remote mountain areas of New Guinea that are logistically and politically difficult to get to. Jared Diamond discovered that two populations of Vogelkop bowerbirds that look identical build distinctly different bowers. In the south Kumawa Mountains they glue together tall towers of sticks, which rest on circular mats of moss painted dead black, decorated with somber snail shells, stones, and acorns. In contrast to this gothic look, Vogelkop bowerbirds in the Wandamen Mountains make low, woven towers covered by a hut with

*Did we already see this movie?

a doorway, all resting on a mat of green moss adorned with colorful flowers, fruit, and butterfly wings. Diamond tested whether the birds in different areas decorated differently because they had different objects available to them. He offered the two populations identical arrays of poker chips in red, orange, yellow, blue, purple, lavender, and white, laying them on the mats of the bowers.

In the somber south Kumawa Mountains three out of five birds took the poker chips and dumped them in the woods. The other two birds, SK1 and SK5, were known slobs who never bothered to take fallen leaves off their mats, and they ignored the poker chips. In the gala Wandamens, birds dumped some chips and kept others, arranging them attractively in their bowers. While their preferences were not identical, all birds liked blue chips best and white chips least. Purple was the second most favored color, followed by orange. Meanwhile the adult birds busied themselves stealing each other's chips, first the blue, then the purple, then the orange. When they weren't stealing from each other, they went after Diamond's yellow film boxes, blue match-box, and black camera. W9 had a try at stealing another researcher's blue sock and brown shoelace while they were still on his foot.

What produces the style differences between Vogelkop bowerbirds in the two areas? It seems clear that learning is involved in making bowers and, for the females, in choosing the best bowers. "Females sometimes go about in groups to visit bowers, so that younger females might learn taste in bowers from older females," writes Diamond. Scientists comparing the mitochondrial DNA of Vogelkop bowerbirds from two populations, one of mimimalist spire builders like those in the south Kumawas, one of baroque hut builders like those in the Wandamens, found only slight genetic differences. It may be that the birds, driven by their artistic differences, will eventually become separate species.

No one understands my art

Then there is the mysterious case of stone handling. Japanese macaques living on Arashiyama, a mountain near Kyoto, have since 1979 been fooling around with stones every day for no clear reason. (I am including it here in case it's art.) The first monkey seen doing this was Glance-6476. She had carried some flat stones out of the forest into an open feeding area. She stacked them up, knocked them down, and scattered them with her hands. No other monkeys were seen doing this. But in 1983, when zoologist Michael Huffman resumed observing the troop, there were lots of monkeys doing it every day.

After feeding time (eating provisions put out for them), monkeys rest, play, and groom. Some of them fill their cheeks with grain and handle stones while slowly chewing. Although the forest is full of stones, monkeys may steal stones from each other. If these stones are more desirable, their excellence is based on criteria that humans haven't discerned.

In 1983, Huffman categorized eight ways of messing with stones: gathering them into a pile in front of you; picking then up; scattering them; rolling them in your hands; rubbing them together; clacking them together; carrying them somewhere; and cuddling them against you. By 1985 he added: repeatedly dropping the stones; rubbing them on a surface; striking two together in a "flinting" gesture; and picking up, rubbing, and clutching smaller stones or gravel. The younger monkeys came up with these.

Older monkeys don't handle stones as much as younger monkeys and they do fewer things with them. "Older individuals tended to become more conservative in their stone-handling behavior," says Huffman. Stodgy monkeys with their old-school stones.

The females Glance-6476, her cousins Glance-6775 and Glance-6774, and Blanche-596475 were the oldest stone-handling monkeys, the only ones in their generation to do it. The children of stone-handling mothers all handle stones, and Huffman suggests that baby monkeys find the sound of clacking stones familiar, since they were able to hear it in the womb. But monkeys whose mothers didn't handle stones also learned to do it, as in the case of Glance-69's baby, who tried to pick up stones when it was two weeks old. No monkey learned to do it after they were five. It seems that the custom at first spread within age groups and then was passed down to the children of handlers, who spread it in their own age groups.

If a young monkey is handling stones and its mother comes up to administer a grooming, the child will keep going while its mother grooms it. "When a stone handler is approached and solicited, for example, to play or copulate, he or she will sometimes abandon the stones and join in, but more often than not, the invitee will ignore the solicitor entirely." Oppress-7079, a macaque who gave birth to her first child, "did not at first appear to know what to do with her infant." She didn't seem comfortable when it tried to cling to her so this irresponsible, pleasure-mad young monkey handed her baby to a male, Deko-64, and went off and handled stones.

Fads

Some cultural phenomena spread obliquely through a group and then die away, passing novelties. One summer a pod of killer whales in Canadian waters went wild over headstands. For a killer whale this means treading water vertically while your tail is sticking straight up out of the water. If you do it right, your tail rises, majestically, higher and higher out of the water, but it takes practice not to wobble and fall over. By next summer headstands were over. No one was doing headstands.

At an oceanarium in Hawaii, where the rims of the tanks extended up to about human waist level, a fashion for leaning took hold among the dolphins one summer. Suddenly they all wanted to see how far they could lean out of the tank and balance on the edge of the wall. Sometimes a dolphin overdid it and fell out of the tank. Staff got annoyed, because picking up a 400-pound bottlenose dolphin and heaving it over a barrier is hard work, and what if a dolphin fell out at night when no one was around? Whenever staff saw a dolphin balancing on the wall they yelled and ran over and pushed it back in the water, but the dolphins seemed to find that even more entertaining. The staff were greatly relieved when the fad passed.

Humpback whales learn their famous songs, which change from year to year. (So, perhaps, do the less-studied bowhead whales.) All the males in an area sing the same song. Marine biologist Michael Noad has listened to thousand of hours of humpback song. In 1995 and 1996 he heard 2 out of 82 whales on the east coast of Australia singing the same kind of song west coast whales were singing. In 1997 many eastern whales began singing the western song. Most sang either the old eastern song or the new western song, but three whales had a song that combined the two. By 1998 every whale was doing the new song. Noad thinks the craze (which some popular commentators compared to the British Invasion of the 1960s and Noad compared to the advent of punk) stems from the fact that female whales get bored with the same old songs. He should know. Of his work, he remarked, "It's pretty soul-destroying in some ways, sitting there and listening over and over to whale songs."

Lions in Lake Manyara National Park were noted to spend a lot of time relaxing in trees. Whenever they had nothing to do, they did it in a tree. Lions in the Serengeti were far less likely to be found in trees. George Schaller examined possible reasons why the Manyara lions might be so arboreal. While lions in Ngorongoro Crater were reported to have retreated to trees during a plague of biting flies, there was no plague of biting flies going on in Manyara. Lions sometimes escape from hostile elephants or buffalo by climbing trees, but there was no plague of elephants. The trees were no different from Serengeti trees. Since the flies seemed no more pesky, the elephants no more savage, the buffalo no more intemperate, Schaller speculated that a prolonged epidemic of flies in the past might have gotten Manyara lions into the habit of retreating to the trees, and then, after the plague was over, the lions transmitted the custom culturally.

Learning to be a social animal—how we do it here

Katharine Milton observed a small colony of spider monkeys that had been created by placing 20 mostly juvenile animals on Barro Colorado Island, feeding them for a while, and then leaving them to their own devices. These animals had been bought in markets and probably had all been caught as infants from the wild (by means of shooting their mothers) and subsequently kept as pets. Fifteen vanished, and the remaining 5 juveniles survived and multiplied to a population of 12 at the time they were studied. Despite having had no wise elders, they had a normal spider monkey social structure and ate a fruit-supplemented-with-bugs diet such as wild spider monkeys customarily eat. (One cannot rule out the possibility that the 15 monkeys who vanished were eating the wrong stuff.) Milton takes from this the lesson that before one rhapsodizes about the plasticity of primate behavior and the "potential for innovation" they display, one should note the "monotony of much primate behavior" and the way that "species-typical behaviors are manifested with monotonous regularity and from study site to study site." She argues that environmental differences produce only "minor modifications" to this tedium.

Milton makes a good point, but it may also be that the environment in which they were released—one where spider monkeys had lived in the past before being exterminated—was like other spider monkey environments and helped produce standard spider monkey behavior. If they were in a genuinely different environment (spider monkeys in the desert! spider monkeys at the beach! spider monkeys on the moon!), it might produce a different social structure and dietary changes.

You're telling me that's normal?

Frans de Waal, who proposed the concept of reconciliation in nonhuman primates and pioneered its study, did an interesting experiment with two species of macaques. Stump-tailed macaques are large and rather peaceable for macaques. They are gentle and tolerant, less obsessed with dominance ranks than rhesus macaques, and they have a large repertoire of reassuring gestures. They use these when they reconcile after fights, and they reconcile three times as often as rhesus monkeys do if they have a fight. Don't they sound nice? But their tails are mere stubs.

With student Denise Johanowicz, de Waal set up a mixed colony of juvenile macaques, consisting of some two-year-old rhesus and some two-and-a-half-year-old stump-tails. They chose older stump-tails in the hope that this would make them more influential.

The rhesus were frightened by the big stump-tails. While the stump-tails wandered about looking at their new accommodations, the rhesus all clung to the ceiling in a worried bunch. After a few minutes the braver rhesus, though still hanging on to the ceiling, grunted threats at the stump-tails. A dominant rhesus would have answered back and a subordinate would have run away, but the stump-tails ignored it.

Eventually the rhesus came down from the ceiling and the species mingled, although at night they slept in two piles, one of stump-tails and one of rhesus. Again and again the rhesus found that the big, confident stump-tails, clearly dominant, were unaggressive. The two species played together and had the usual quarrels, but the stump-tails were amazingly—amazing to the rhesus—eager to reconcile with their opponents. The stump-tails were also anxious to groom the rhesus, a friendly act that was probably accelerated by the stump-tails' fascination with the lovely long tails of the rhesus. By the end of the experiment, all the monkeys slept in one big furry pile.

More important, the rhesus monkeys had changed an aspect of their behavior. They had not adopted stump-tail gestures or calls, but they had increased their rate of reconciliation dramatically. They now reconciled just as often as stump-tails did. After five months, the stump-tails were taken out, and the rhesus monkeys went right on reconciling like civilized, mature primates. They still acted like rhesus, but now they acted like really nice rhesus. Rhesus monkeys make a soft, pleasant sound, "girning," to show friendly intent, and these rhesus were big girners.

It may be that with an inverted experimental design, with older rhesus and younger stump-tails, the stump-tails could have been taught not to be so ready to reconcile, could have learned to be touchier. But if the researchers had to pick one experiment, they picked the right one, for it is more impressive to learn that animals can be encouraged to be sweeter than to learn that they can be encouraged to be meaner.

If you don't like it, why don't you leave?

The fact that the aggression level of primates varies among groups can be seen in some large colonies of chimpanzees. The captives have tons of free time to spend on their social relations, and they can't get away from each other, so they have stronger reason to seek harmony than wild chimpanzees. This may produce different rules for acceptable conduct, particularly limits on male violence against females and infants. In places like the Arnhem Zoo, the Netherlands, these limits are enforced by coalitions of adult females of a kind that do not seem to exist in wild groups.

In some zoos, there are no such coalitions. In a private zoo, two chimpanzees lived in a small, old-fashioned cage, an iron-barred, concrete-floored box 5 meters by 10 meters. There was a 15-year-old male, Bonz, and a 33-year-old female, Jingles. Both had been raised by humans. Jingles's only pastime was begging for food from visitors, but Bonz had two pastimes, begging for food and terrorizing Jingles.

Her life was miserable and had been for years—and then the zoo opened a new exhibit, an enclosure of almost an acre, with grass, trees, and a stream. There's a fake termite mound where chimps can dip for peanut butter. They sent Bonz to another zoo and put Jingles in the new exhibit with two juveniles who had been raised by their mothers. The juveniles explored with enthusiasm, chasing toads and chipmunks. Jingles did not seem to grasp that all this space was available to her. She stayed within 100 feet of the entrance to the sleeping quarters. She seemed "almost agoraphobic." When the young chimpanzees first groomed and hugged her, Jingles would start with terror. "I don't think anyone had hugged her before," reports Kathleen Morgan, who with her students studied the effects of changed environments on captive primates. Jingles eventually learned to groom and hug. One day she even crossed the stream.

Paternalistic culture

In the 1950s, Japanese primatologists watching macaques noticed that, in some troops, adult males cared for young monkeys during the four-month-long season when the females were giving birth. They called this phenomenon "paternal care," although it is unlikely that the males could know which infants were their own. The "paternal" monkeys were tender and vigilant, and did everything for the one- and two-year-olds that their mothers did, except for nursing them. Each male developed a relationship with a particular infant. This was first noticed in the Takasakiyama troop. A survey of other troops showed no sign of paternal care in eight troops, one or a few examples of it in seven troops, and a strong tradition in two troops. The paternal males clasped the infants, carried them on their backs, groomed them, played with them, and threatened tourists and primatologists who got too close to them. The babies were even permitted to grab food that the males were preparing to eat. This behavior started abruptly when the first baby of the season was born. Sometimes a baby was cared for by a male both during the delivery season when it was around one year old (8 to 16 months) and in the following year. Thus the infant Cob, who could be identified by her limp, was cared for by the male Pan two years running.

Cob was a big crybaby, and Pan was aggressive. "When he is protecting an infant, he becomes especially furious," writes primatologist Junichiro Itani. One day Cob got frightened when tourists tossed peanuts noisily, and she ran away crying "Ki. Ki. Ki." Pan raced over, clasped her, and threatened the tourists. When Cob ran off to play, Pan tried to attack the tourists.

Uzen, a sociable and ambitious but unaggressive macaque, was very fond of children and was so often seen gathering half a dozen and playing with them that researchers nicknamed him "Kindergarten Teacher." Observers felt that this gave him improved access to the central part of the troop, where the highest-ranking macaques spend their time. He almost always had an infant clasped to him when he entered the central area, perhaps "playing the part of a passport for the central part of the group." This may have contributed to his rise in status.

It would be lovely to think that the male macaques of Takasakiyama got together and decided that mothers needed help with child care when they had new babies, but that doesn't really seem like macaques. And it would be sociobiologically interesting to hypothesize that in a troop where males take care of the older infants at this difficult time, more babies survive, and so more of the males' genes are passed on—or that males who care for infants are more likely to rise in rank and conceivably to leave more descendants as a result—but there's no reason to suppose that the difference in this behavior among troops is genetic. The primatologists concluded that it was a behavior elicited when males saw other males caring for infants, and passed on culturally.

Were you born in a barn?

Biruté Galdikas describes social gaffes committed by young orangutans who are being reintroduced to the wild after having been partly reared in captivity. They do not know how to keep a proper social distance from adults and are apt to offend by walking too close to them or trying to climb into their nest. Since the orangutan response to an orangutan faux pas is sometimes to bite big holes in the other orangutan, learning manners is very important.

Mandara, a three-year-old gorilla raised by humans, was placed in a zoo exhibit with Tomoka, a large, gentle silverback. Tomoka had lived with other gorillas for years, although not with baby gorillas. He knew how a male gorilla starts a conversation with a stranger—by showing how impressive he is. He strutted around, pounding on his chest. The only gorilla Mandara knew was a younger playmate, and so she assumed that Tomoka was inviting her to play and she chased him. The more fiercely and impressively he strutted, the more she frolicked. When she simply wouldn't get the message, Tomoka snatched her up, put her on the ground facedown, opened his jaws wide, and pressed his enormous teeth against her back. Then he walked away. Mandara, shocked but unhurt, got up and sat in the corner. She was a quick learner, and when she was introduced to the other adults, there were no such crossed signals.

Lion lore

Elizabeth Marshall Thomas has described a change in the culture of lions in the Kalahari Desert. When she and her family lived there in the 1950s, the local people, the Ju/wasi, showed them how to behave around lions and how to expect lions to behave around them. The attitude was one of cautious respect. The message to be conveyed by each party was "We won't bother you if you don't bother us." When lions and people met in the bush, for example, the thing to do was to "walk purposefully away at an oblique angle without exciting the lion or stimulating a chase." And this was what the lions did, too. No one loses face or shows weakness by retreating, and no one provokes anger or alarm by advancing. Simultaneously, lions and people recalled business elsewhere. (I saw this courtly routine performed when bird-watchers and a black bear family met on an Arizona mountain trail. The bird-watchers, going uphill, suddenly preferred to look for birds on their right, and the mother bear, coming downhill with her cubs, remembered something that needed to be investigated off to *their* right.)

There were disputes, since people and lions stole kills from each other, but they did so in a quiet, look-we-outnumber-you-do-the-math kind of way, not with hostility. "I thought we were finding out about lion nature," Thomas wrote.

Lions and people had then lived together in the Kalahari for a long time. In the 1960s the government evicted the Ju/wasi from the part of the Kalahari that became Etosha National Park. In the 1980s, Thomas revisited the area, and found that the lions had changed. Her first hint of this was when two teenaged lions charged and chased the van in which she was driving. Thirty years without human inhabitants had broken a chain of cultural transmission. Young lions had grown up without ever seeing their elders respond to humans. They did not use the dance of the oblique departure. Humans (other than park rangers in jeeps wielding guns with anesthetic darts) were a novel phenomenon crying out to be investigated. Do they scare easily? How fast can they move? And perhaps even—are they tasty? Deprived of the observational learning experiences they would have had if they had met people while with the elders of the pride, they were ready to learn by experiment.

Anubis versus hamadryas

The closely related hamadryas and anubis baboons have very different male-female relationships. Female and male anubis do not have enduring relationships other than friendship. Mating takes place when the female is in estrus. Male hamadryas, however, control one or more females in an arrangement Hans Kummer calls a marriage. The male herds the females together, keeps an eye on them, and is deeply domineering. He does this at all times, not just when the females are in estrus (which isn't often). Kummer and colleagues performed the experiment with wild baboons of putting hamadryas females in anubis troops and anubis females in hamadryas troops. When an anubis female was placed near a hamadryas troop, a male hamadryas immediately claimed her, and was then vexed by her reluctance to follow him. Communication was not the problem: she had no difficulty understanding what he wanted, but she found his wishes bizarre. He attacked her every time she strayed, more harshly each time. Within half an hour, an anubis female knew that when the male stared at her, she was too far away and was supposed to come closer. But even knowing that, she didn't like it, and as the days went by and the male had to keep rounding up his exotic wife, he gave up. "The experiment had shown that hamadryas males can enforce a marital relationship with threats but cannot maintain it permanently," Kummer wrote.

Meanwhile a hamadryas female placed in an anubis troop would attach herself to a male and groom him devotedly. Soon she would notice that he made no efforts to herd her, and that he did not protect her from other baboons, and she would lose interest in him. One hamadryas female found a nice male anubis who did protect her, and although he did not herd her, he followed her around. "Not a bad choice, it would seem," wrote Kummer. "But she finally left this male too and lived unmarried."

Landmarks, paths, and migration routes

Since the natural landscape changes over time, a lot of important information about where to go and how to know when you're on the right track to get there cannot be entrusted to the genes. Cultural transmission is a great way to pass on this vital data in many species.

Teenaged French grunts spend their days schooling with other juveniles in favored spots on the coral reef. At twilight the juvenile fish swim together along "featureless (to human observers)" routes to feed in sea grasses, and at dawn they migrate back to their usual coral head. The routes can be more than half a mile long. Because grunt generations overlap, it seemed possible that they were learning the routes from each other, not simply taking the route that would be obvious to any French grunt. Curious naturalists kidnapped juvenile grunts and added them to schools in other parts of the reef. When twilight came, the new kids followed the locals to the sea grass beds, and at dawn they went back with them. When the scientists went one step further and removed the locals, the new kids kept taking the route the locals had shown them. French grunts who were slapped down in a new area from which the residents had been removed, and who didn't have anybody to follow, worked out routes for themselves. But the routes they chose were different from the ones the previous residents had laid down.

Buffalo shuffle

Along the shores of Lake Manyara, in Tanzania, buffalo herds rest in the late afternoon. The buffalo lie down, some on their sides, and there is little activity. Most buffalo do not even chew their cuds, the bulls leave the cows alone, and the calves do not pester their mothers. "It seems that nothing is happening," writes researcher Herbert Prins. "However, this is not the case."

After Prins had been studying buffalo for two years, he realized that what he had thought was a random activity of no importance—a cow getting up, shuffling around, and then lying down again—was actually meaningful. He had thought they were just stretching their legs, but one day it occurred to him that the cows who did this adopted a particular stance, held their heads higher than usual, but not so high as to indicate alarm, and gazed in one direction for about a minute. Each time a cow got up, she'd gaze in the same direction, and this direction indicated where the herd would go when they ended their rest period. In a herd of 950 buffalo, normally about 5 cows might be standing and gazing at any one time. Prins calls this the "voting posture."

Around six o'clock, or when the shadow of the escarpment moves over them, the herd begins scrambling to their feet. Calves get up and nudge their mothers to do the same. Within a few minutes, Prins reports, "they start trekking, at the beginning independently of each other, **in the same direction.**" Such herds do not seem to have leaders, and an animal that is in front at one moment will not be there for long. Often bulls may be seen at the front of the herd, but if they stop and graze, the cows walk on past them, and the bulls follow.

The herd heads to an area where they will graze by night and the following morning. Once Prins had figured out what the staring cows signified, he experimented with trying to head the herd off with his car, chasing them away, or splitting the herd, but in every case, the buffalo went around, doubled back, or regrouped, and went where they had planned to go all along.

Sometimes the herd does not achieve consensus in the direction of their voting postures, and then they may split into two herds.* "Voting" is done by adult cows, with a very few teenagers (subadults) of both sexes taking part. Bulls, who are transient members of the herds, do not vote, and when they get up during the rest period, the direction in which they look, walk, or face while grazing is random.

Prins argues that these buffalo live in an area of fine-grained patches, where at any given time the grazing is too low in protein in many spots (less than 7 percent). They compete for grazing with other herbivores, particularly elephants. The decisions about where to graze are important to their survival and are based on local knowledge of the

* Sometimes the idea of reaching consensus by means of fixed staring is attractive.

state of the grazing in many locations. (So when I said earlier that eating grass probably didn't require as much learning as eating grass-eaters did, that was too glib.)

Going nowhere

Many species of migratory creatures also have sedentary populations. This isn't genetic, as it is possible to indoctrinate a bird from a migratory population into stay-at-home ways. Whooping cranes follow their parents in migration and learn the route. But if their parents do not migrate, neither do they. Conservationists have been able to establish a nonmigratory population of cranes in Florida, the children of captive cranes.

Young greylag geese at a research station in the Alm Valley of Austria, where the water does not all freeze over in winter, show the migratory urge on stormy days in fall. Having been raised by humans or by other geese who are not accustomed to migrating, they have no place to go. They fly for hours, in classic V formations. In the evening they land, back at home.

It's a bird, and a plane, and a guy in a plane in a bird suit

In addition to the sedentary population of whooping cranes, the International Crane Foundation (ICF) is also trying to establish a new migratory population with a different route. Instead of migrating between Canada and Texas, these cranes will migrate from Necedah National Wildlife Refuge in Wisconsin to Chassahowitzka National Wildlife Refuge in Florida. There are no adult cranes who know the way, so instead the ICF is leading cranes on the first flights in ultralight planes. To avoid imprinting, the people around the young cranes feed them with realistic whooping crane hand puppets and wear whooping crane outfits (and never speak), so the pilots must also wear crane suits. Enormous, implausible crane suits, but it seems to be working.

Leading cranes in an ultralight is dangerous work, especially when predators appear, and worried young birds scramble to fly next to their parental figure, the plane.

The young cranes were first taught to follow the planes on local flights. Then the pilots led them south in easy stages of 20 miles a day.

Often bad weather grounded them. Cranes led by real bird parents can soar on thermals and save energy, whereas cranes flapping behind a slow plane work much harder. In the first group, the Class of 2001, one crane dropped out over Kentucky. A search was launched for the missing bird, Number 6. When he fell behind, Number 6 returned to the place where they had spent the previous night, an airstrip in Adair County. A resident looked out her window, saw lonesome Number 6, and alerted the search team. By the time they got there, he was gone. Fortunately he had not gone far, so they were able to get radio signals from a transmitter he was wearing. They followed the signal to a river valley and, figuring he might be near the river, began broadcasting crane calls. The young bird answered and appeared flying above them. Desperately hoping to call him down to a suitable landing spot, team member Dan Sprague, clad in his crane suit, sprinted to a clearing at the top of a hill, blasting crane calls on his crane vocalizer. No sooner had he reached the clearing than Number 6 landed beside him, peeping pathetically.

The cranes made it to Florida, and the following spring they flew back to Wisconsin by themselves, with no one leading them. They remembered the way and made excellent time. The Classes of 2002 and 2003 also did well. The ultimate test is whether the birds will breed. If they do, they should lead their children on the new route that the airplanes showed them.

The mountains are up, you fools

Reindeer and caribou migrate yearly, using traditional routes, although, in the case of caribou, not always the same routes each year. Reindeer do not like to travel alone or in small groups, and prefer the safety of the largest possible herd, but they also like to be with reindeer they know and to migrate together with them. Although their urge to migrate is innate and they may have an innate compass that tells direction, maps are not innate. Young reindeer who travel with the herd may learn landmarks along the way, thus acquiring a map. In the 1960s, some Swedish reindeer farmers, worried that grazing conditions along the migration route wouldn't support all the reindeer, decided to lend a hand. In the autumn, they rounded up reindeer, loaded them into trucks, and drove them to the traditional wintering grounds. So far so good.

But in the spring, when the reindeer migrate back to the mountains, many stayed behind on the winter range. "This has caused much trouble to the reindeer owners," zoologist Yngve Espmark reported. The problem may have been partly that many animals did not know the route and partly that their social structure had been disrupted when they were put on trucks. Forced to scatter on the winter grounds in order to get enough to eat, they may not have known the landmarks or anybody they felt safe about following in the spring migration, so they stayed put.

The big chimpanzee survey

The strongest case for animal culture comes from a survey of wild chimpanzee groups across Africa. Comparing chimpanzees in Guinea, Ivory Coast, Uganda, and three locations in Tanzania, the sites of the seven longest-running studies, researchers counted 39 behaviors that took place in some sites and not others or that were done in different ways. They looked at the impressive stuff, like nut cracking, and at less serious stuff, like the custom of using large leaves as a seat cushion or using a leafy branch to fan away flies or doing rain dances. "The profiles of each community . . . are distinctively different, each with a pattern comprising many behavioural variants," the study's nine authors write.

The Sassandra River, enemy to knowledge

A dramatic example of differing cultures comes from West Africa, where there are chimpanzees living on both sides of the Sassandra River. There are coula (*Coula edulis*) nuts on both sides of the Sassandra. And there are stones on both sides of the Sassandra. Yet the chimpanzees on the west side of the river crack coula nuts between stones and the chimpanzees on the east side apparently do not. The reason may be historical and cultural. About 17,000 years ago serious drought conditions allowed a tongue of the desert to reach the sea from the north, cutting the West African forest lands in two. During this period, it is guessed, the chimpanzees in the western forests learned to crack coula nuts. When the drought ended, the forests returned and the river swelled. So there was forest on each side of the river, but chimpanzees could not cross. Further north the tropical forest changes to semideciduous forest, poor in nut trees, and so although chimpanzees can cross the river there, the nut-cracking culture doesn't reach so far north. Thus it's suspected that the culturally transmitted invention of nut cracking would come as intriguing news to the chimpanzees east of the Sassandra.

Rehabilitation and culture

An animal raised by people may have a culture, but it won't be the same as the one its wild peers have. If the plan is to release the animal into the wild, this may create problems. The released animal may not know where to go (because it has never seen others go there), or what to eat or how to eat it. It may also (like the rehabilitated orangutans with bad manners described earlier) have a hard time fitting into the existing culture, if there is one.

It may be possible to save a species, through captive breeding, but to lose its culture when no wild individuals are able to pass what they have learned to subsequent generations. Whatever a tiger learns in an empty cage, it's not likely to be of much use if it is released in the woods. If there are no animals left in the wild, then the young animals must invent their culture from scratch or be supplied with some of its elements—like crane migration routes—by the rehabilitators.

A tropical island

In 1987, Govindasamy Agoramoorthy and Minna Hsu released some captive chimpanzees on a swampy island off the coast of Liberia. These chimps came from a laboratory that used them for vaccine research, and some had been pets before that. (People who get baby chimpanzees as pets usually betray their darlings in one way or another once they become moody, mischievous teenagers with enormous teeth, muscles like steel bands, and minimal impulse control. It is fortunate for our species that human teenagers cannot thus be given or sold to labs.)

Before release the chimpanzees were kept in large outdoor enclosures on the island in social groups. They were released only if they were in good health, and only younger chimpanzees were selected, since the researchers had found in the past that older apes were harder to rehabilitate. Thirty were released, with the younger and lower-ranking apes being given first go at settling in, followed by the older, bossier chimps.

Once released, a few chimps preferred at first to sleep on top of or in the holding cage, but most of them made nests in trees like wild chimps, and soon all of them did this. Seven of the chimps, who had been pets most of their lives, couldn't handle the change, partly for social reasons. There was fighting, and three chimps had to be temporarily taken away for medical treatment. The first two babies born did not survive.

Many of these chimps had shown stereotypical captive behavior before release—they rocked, pinched themselves repeatedly, or ate their own feces. After release, this stopped. Although Agoramoorthy and Hsu provided food, it proved unnecessary. Most of the chimps started gathering wild food within a few days, climbing palm trees and eating nuts, fruits, plants, and insects. The chimps immediately began using tools, employing rocks to smash nuts they placed on tree trunks. It's unknown whether the chimps had seen or done anything like this in their previous lives, or whether it just seemed like the obvious thing to do with really hard-shelled nuts. I myself have reinvented this technique on camping trips when cones full of pine nuts were lying around, and I didn't feel all that smart.* One week after his release, a male chimpanzee was seen tackling a weaver ant nest, eating ants and their eggs. (This is something that never occurs to me to do when I am camping.)

"Younger chimpanzees watched and learned feeding strategies from older animals," Agoramoorthy and Hsu write. Though the older chimps knew more, the younger ones "learned social and survival skills more quickly than older animals." In 1989, civil war in Liberia forced the researchers to abandon the study. In the late 1990s, there were still chimpanzees living on the island, and more babies had been born to them.

Uncultured youth

When captive-reared California condors were released, they showed a distressing tendency to hang around human habitations in gangs, vandalizing property, flying into power lines, and getting into trouble. The condor recovery team has concluded that simply raising chicks with

---

*Smart would have been bringing a nutcracker.

condor puppets is not enough. Mike Wallace, of the California Fish and Wildlife Service, and head of the team, puts some of the blame on the kibbutzlike rearing system in which the young birds are raised together in large cages. While this reminds them that they are condors, not human beings, the atmosphere is one of a "big rumpus." In nature a condor chick is an only child, whose sole companions are its parents, and the tone is more staid. "It produces a more timid, cautious bird, one more likely to survive," Wallace told *The Economist*. The answer may be to release the young birds with older mentors, who will encourage serious study of life skills as opposed to hanging out in rowdy groups in parking lots daring each other to eat car parts.

With this in mind, the condor species recovery team released AC-8 in April 2000. AC-8 (short for Adult Condor 8) was born in the wild—no one is sure when—where she grew up. In 1986, with the species on what looked like its last legs, AC-8 was captured. (By 1987 there were no condors left in the wild. All living birds were in captivity. Releases of captive-reared birds began in 1992.*) Under the auspices of the U.S. Fish and Wildlife Service, AC-8 produced 12 chicks in the next 14 years, many more than she could have raised in the wild, since condors do not breed every year. She also spent time with the young condors raised in captivity, since staffers hoped she would be a mentor to youth and that they would imprint on her. (The condor keepers didn't call her AC-8, but Grandma.) Eventually she stopped laying fertile eggs. At the time of her release, her age was guessed to be between 28 and 40. She was released together with two captive-reared 10-month-olds, in Ventura County. She flew 80 miles to her old foraging range and settled back into the old life. Whether any of the young condors modeled their ways on hers is unclear. Just three years later, a participant in the Tejon Ranch Company's annual Pig-O-Rama pig hunt spotted AC-8 sitting in a tree and shot her.

Orangutan lessons

The issue of rehabilitation and culture is clearest with orangutans. While the biggest threat to the species is loss of habitat, the large red-

* Crazed bird-watchers do not count sightings of captive-reared condors as real sightings. AC-8, having been born in the wild, counted.

dish apes are easy to shoot from trees. There is a lucrative illegal trade in baby orangutans (for pets), which are obtained by shooting their mothers and hoping the baby is not killed in the subsequent fall. As a result, there are two kinds of orangutans needing rehabilitation: infants and older apes who have been confiscated from the black market by law enforcement; and former pets who have grown too big, too old, too destructive, and, in some cases, too obviously lustful for the families who originally purchased them. Some of the former pets miss their favorite television shows. Others probably do not miss being chained under the house for years.

In rehabilitation programs run by scientists Willie Smits or Biruté Galdikas, orangutans have a chance to learn survival skills that will fit them for life in the wild. "Most people think that if you let an orangutan loose in the forest, it will know what to do naturally," scientist Robert Shumaker told the *Smithsonian*. "That is wrong. Orangutan juveniles aren't genetically programmed—they must be taught about their environment, just like humans." Shumaker studies cognition in the orangutans at the National Zoo, and hopes that understanding how orangutan minds work will help rehabilitators give the apes the skills they need. Ordinarily, they learn while spending years in the company of their mothers. That it can be done by humans is shown by the case of Uce, the first infant Smits rehabilitated. She was offered for sale in a Bornean market, but there were no takers because she was visibly ill. Smits found her crate discarded on a garbage heap, with the dying infant inside. He nursed her back to health and spent years with her while she learned to live in the forest. At last report Uce was living in the Sungai Wain Forest, and when visited by Smits, she climbed down from the trees to show him her baby, Bintang.

As researcher Anne Russon notes, rehabilitant apes must not only learn a set of skills for wild life, they must also unlearn things that were useful to them in captivity, such as getting food from people. Teaching orangutans to locate, prepare, and eat forest foods is a challenge for people who don't eat those foods themselves. Knowledgeable orangutans who are willing to be role models make much better demonstrators. With no one to show them a better way, some of the naive orangutans adopted inefficient or useless foraging techniques. Judi ate ants one at a time, pinching them up between thumb and forefinger, whereas wiser animals mopped up a bunch of ants on the back of their hand and then licked or sucked them all off. Tono, aware that termite nests (these are spongy nests built in trees, unlike the earth nests that chimpanzees exploit) contain savory termites, tried to open them by banging them on hard surfaces, which worked exactly never. Yet he persisted for months, once trying to break one open on Aming's head. The nest didn't break, and Aming took it badly.

Adapting to the forest is hard, and some orangutans in rehab have been known to wait at bus stops on roads through the forest and climb aboard looking for an easier life. Others tough it out: Judi eventually learned to eat ants by the handful.

Enculturation

Animals raised by people, particularly great apes, are sometimes called enculturated animals. They have been raised in our culture, not their own. Anne Russon considers some of the orangutans at Galdikas's Camp Leakey bicultural—they can survive in the forest and are also at ease in camp among humans.

Some scientists are interested in enculturation as a process that may change what great apes can do cognitively, and others also focus on how enculturated apes are entitled to be treated. Lyn Miles calls Chantek her foster son. She rues the fact that "our society isn't prepared for dual-culture creatures" and hopes that someday Chantek can have a home with other enculturated apes, an environment "featuring agency and choice," one where he would have companions he could sign with. Perhaps he will be able to go to an ape preserve in Maui, Hawaii, now under construction by the Gorilla Foundation as a place where Koko and other gorillas can live.

In one experiment focusing on enculturation, human two-year-olds, chimpanzees raised like human children and "exposed to a languagelike system of communication" (lexigrams), and chimpanzees raised by chimp mothers were given a chance to imitate a demonstrator. The human toddlers and the chimp raised by humans both copied the human demonstrator, and the chimpanzees raised by their mothers scarcely did at all. Based on this, Michael Tomasello asks which group is likely to have intellectual abilities most like wild apes—the captives raised by their mothers or the enculturated chimpanzees? His hypothesis is that "a humanlike sociocultural environment is essential to the development of humanlike social-cognitive and imitative learning skills."

Frans de Waal takes a different view. The experiment shows that chimpanzees can imitate as well as human children when the model is a species they have grown up with, he suggests. He suspects that imitation is important to wild chimpanzees, and that this ability is useful in their natural lives. It's not that enculturated chimpanzees have become smarter, better imitators from being around us, but that they accept us as part of their community. "Rather than having been lifted to unprecedented cognitive levels, apes raised by people have become ideal test subjects simply because they are willing to pay attention to psychologists."

The chimpanzee Jun, in the Tama Zoological Park in Tokyo, was insultingly uninterested in a human demonstrating how to crack walnuts with a stone hammer and anvil, ignoring him through dozens of sessions. But she was instantly riveted by the sight of Sachiko, another chimpanzee, doing the same thing, and took it up herself. Orangutans, too, are far more apt to imitate relatives and friends than they are to imitate strangers.

Wild culture

Christophe Boesch notes that every time the environment of captive primates is enriched, the animals seem to get smarter. But neither a featureless barred cage nor a house with lots of toys and visiting graduate students map directly to the lives of wild chimpanzees. "Not only space and opportunity to explore and play but also the absence of large and complex social interactions have a negative influence on the psychological development of the captive individuals," writes Boesch. "In my opinion, rules in the wild are more stringent, with survival and reproductive success acting as 'teachers,' so that the cognitive development of wild chimpanzees may be more advanced than that of their captive counterparts."

The shortcut that is culture allows animals to learn things their predecessors have learned without having to take as much time or run as many risks. If they elaborate on that, then, building on the discoveries of earlier generations, animals who are no smarter than their ancestors can live smarter lives.

NINE

Parenting and Teaching: How to Pass It On

At the San Diego Wild Animal Park, zookeepers worried about an unmotherly gorilla, Dolly. Dolly had been captured in West Africa when she was nine months old and raised at the zoo with other gorilla children and no adults. Her first baby was born when she was 10 years old. Dolly wanted nothing to do with him, although he was a fine healthy baby who tried to cling to her, whereupon she would peel him off. Unhappy zookeepers took him away to raise and named him Jim.

They did not blame Dolly, noting that she had had little socialization, and that "all imported individuals and over 80 percent of infants born in captivity . . . passed through the critical period of juvenile development in the care of humans. Under such conditions, these infants have been unable to observe and imitate adults of their own species engaging in normal patterns of social interaction."*

When Dolly got pregnant again, the zoo created a training program "with the specific aim of inducing maternal behavior." The first part of the program was showing Dolly films "depicting gorilla mother/infant [behavior]." This did not work. Like so many youth, Dolly was unimpressed by educational films and more intrigued by the projector. This is a reasonable intellectual choice, but unless Dolly's next child turned out to be a film projector, not useful. The second part involved giving her a doll (called a "surrogate

---

* Yes, yes, I am willing to believe that your own family may have been pretty dysfunctional. Perhaps you do not feel that they engaged in normal patterns of social interaction, but I'll bet they were the same species as you.

infant" in the scientific literature and a "baby" when speaking to Dolly) and teaching her four commands about handling it. Dolly knew the word "baby" from occasions when zoo staff brought Jim over for her to view through the bars. She liked the doll, and when it was offered, she drew it in carefully through the bars of her cage.

Researcher Steven Joines taught Dolly to respond to "Pick up the baby, Dolly"; "Show me the baby, Dolly"; "Be nice to the baby, Dolly"; and "Turn the baby around, Dolly." This last was aimed at getting Dolly to hold her baby in a suitable position for nursing.

When Dolly's next baby, Binti, was born, Dolly was an "exemplary" mother, cradling her immediately. The only need for the commands came when Binti, unlike a doll, started to cry. "Dolly did not recognise, and displayed confusion at, the infant's crying and appeared unable to deal effectively with the situation on her own. However, the command 'Be nice to the baby, Dolly,' delivered by a familiar human, soon resolved the confusion." On command, she cradled Binti to her, and Binti hushed. "Within a week, she no longer required prompting." Dolly did so well that they gave Jim back to her, and she was a good mother to both.

ANIMALS OF EVERY KIND would become extinct if some of them didn't produce the next generation. Raising children is literally vital. Like other essential behaviors it is both instinctive and learned.

People sometimes wrestle with the question of whether they can be good parents if their own parents were unsatisfactory, and while it's clear that a lot of parental behavior is learned at an early age, resulting in the phenomenon of your mother's or father's words coming out of your mouth, Dolly's case is one example of parental behavior learned later in life.

Getting together

Sexual selection, in which animals choose their mates, is a powerful force in evolution. When animals go beyond mating and actually pair off for a season or a lifetime to raise their children, the choice is doubly important. A mate should have good genes, but a spouse should also have good skills.

Many papers have been written about strategies that animals might use to pick a mate. In a paper on mate selection in barnacle geese, Sharmila Choudhury and Jeffrey Black review some strategies. The "random mating" strategy is just what it sounds like—the animal goes for the first possible mate. The "fixed threshold" strategy is to select the first mate who "meets some minimum requirements." In the "one-step-decision" strategy, the animal decides at each encounter with a potential mate whether to stop looking or keep searching. In the "sequential comparison" strategy the animal also samples prospects but compares each new candidate with the one before. And in the "best-of-N" strategy, the animal compares N potential mates, ranks them somehow, and picks the best one. Peahens use this last method, and I am sure we can all think of people who have employed each of these strategies.

Barnacle geese mate for life, and they split child care, so the mate choice matters. They take their time. In wild flocks, some geese find each other right away, and others go through several trial liaisons before they meet their dream goose. Choudhury and Black raised geese in captive flocks and monitored their romantic lives carefully. In their second winter, the age when wild geese start pairing up, these young captive-reared geese were moved into large enclosures where they could mingle and ask each other searching questions. A male barnacle goose hints that he likes a female by herding her with loud calls and neck stretches. She may encourage this if she likes him, or she may reject him. If things are going well, they'll stroll, share food, and do triumph ceremonies together. If love derails, it's easy to tell. "One partner would start to direct threats at the other, respond negatively to or ignore any courtship advances, and/or start courting a new trial mate."

About half the birds settled down with the first goose they courted, but others went through up to six test geese. Trial marriages lasted from one day to nine months, but a few days was usual. How they made their choices was not wholly clear, although the researchers did their best to figure out what makes a barnacle goose desirable. They rated them on social dominance, weight, facial patterns, and vigilance toward predators. I am happy to state that vigilance was measured by mounting a stuffed fox on a stick and walking it past the enclosure, which caused every goose in the flock to put its head up and stare keenly. Geese who kept their heads up longer were scored as more vigilant. Hoping to get more data, they repeated this test, but the geese had habituated. (It's the stuffed fox on the stick again. So, are you see-ing anyone?) The researchers could not identify any trait that any one goose consistently liked in another goose.

Barnacle geese did not always dump the goose they were courting when they met another goose they really liked. This is referred to as the "partner-hold" tactic and it led to trios of geese strolling around together. This usually didn't last for more than a few days, but five males were able to form lasting relationships with more than one female. The only noticeable characteristics of these polygamous males was high marks for vigilance. Perhaps they could have protected more than one female from a stuffed fox on a stick.

How did these geese learn who was right for them? Their trial mar-riages were indistinguishable to observers from marriages that lasted, so what makes one goose lovable to and compatible with another remains a mystery. They did not seem to be using a fixed threshold cri-terion. Since they never went back to a goose they had previously been with, they were not methodically collecting rankings in a best-of-N strategy. Perhaps 40 percent of the time the geese appeared to be searching by the one-step-decision strategy, trying one goose after another till they found the right one, but about 60 percent of the time they seemed to be using "a modified version of the 'sequential compar-ison rule'" by means of the partner-hold tactic. By pairing with two geese at once, they could compare and contrast in a manner that seems obnoxious to me, but if it's okay with the geese, who am I to interfere?

Make sure she's a real girl

Speaking of stuffed specimens, a vital step in mating is to make sure that you are directing flirtation at a suitable mate and not a robot or dummy. You thought that went without saying? Ask the lovelorn male corn crakes who were netted while courting a stuffed female. The ornithologists who set up the dummy attracted males by imitating the call of the crake ("crake crake"). Some males who approached the dummy attacked and some courted it, craking, doing an attractive spread-wing display and then trying to mate. One crake made 23 attempts to mate with the dummy, then went away and returned with a green caterpillar, which it politely offered. When the dummy did not take the caterpillar, the crake ate it himself and tried displaying again. The good news for him is that the ornithologists were not netting crakes that day. The bad news is that they were taking pictures.

Learning to get along, or why cockatiels hate arranged marriages

Cockatiels are in the mating-for-life camp. Unromantic researchers compared the breeding success of cockatiels who were allowed to pick their own spouses with cockatiels who were assigned a mate by researchers. Cockatiels who were allowed to marry for love raised more fledglings. All couples did okay at the first stage of family life, nest inspection. But the "force-paired" couples were less likely to get as far as building a nest in the nest box, and less likely to lay eggs, sit on the eggs, or have the eggs hatch out.

However, if the birds in arranged marriages spent time together before the breeding season, they did better than birds who had been thrown together more recently. This suggests that spending time together helped them get to know one another's quirks and synchronize their activities better. They still didn't have the breeding success of birds who chose each other. Pairs of cockatiels who had been sundered by researchers, assigned to other mates, and then reunited after three months had not lost the magic, and were as reproductively successful as they had been before the terrible mix-up.

Marital harmony

Ngaio, an inexperienced young female black robin, paired up with Crunch, an inexperienced male. Crunch sensed that when you court a girl, you should feed her dainty insect items, but it took Ngaio a month to accept them. Perhaps this was because Crunch was rather aggressive, chasing Ngaio when she was feeding and even when she was trying to mate with him. And perhaps this is why Ngaio dumped him for Mertie the next year.

A young male named Pani had a different problem: he'd start to feed his mate, Baby Blue, and then grab the bug back.* Black robins are a species in which both male and female feed the young, and the

---

* Women hate that.

male feeds the female while she is incubating the eggs, but males often take a while to figure this out, leaving their brides to feed the whole family themselves. Pani did such a poor job of feeding Baby Blue when she was sitting on their eggs that social services/researchers had to step in and feed her.

Lovebirds mate for life, and they're decisive. They fall in love just like that. "It takes no more than a few hours to establish lifelong pairs." But it takes a lot longer to get things working smoothly. At the beginning they're awkward. They misunderstand each other. They take offense. Often a male makes "many mistakes" in attempting to mate, and is "threatened and thwarted" by the female. (For one thing, he needs to notice whether her head plumage is fluffed up—a good sign—or sleeked down—a bad bad sign.) By the time they've raised a few broods they understand each other better, fight less, and mate more often. The male is less often seen showing frustration by squeak-twittering, displacement scratching, and switch-sidling.

Mating with your mate

Sex is where the next generation comes from, so I feel obliged to point out that learning is involved here, too. Animals often display a strong, not to say profound, interest in sex along with complete confusion about what it actually is.

Researchers studying endangered black-footed ferrets note that while an experienced male ferret needs little direction from a mate and in most cases "simply grabs the female by the neck and mounts," things may not be so simple for beginners. A young male ferret without experience is likely to hang back, and a female may encourage him with "soliciting" behavior, approaching in a low crouch with her head and neck extended toward him. She will pause before moving forward a little more. She may chuckle invitingly. Once she gets close she may put her head under his chin. (Hint, hint.) If he doesn't react, she may rub against him. (Yes, you.) "In one extreme case, an older female, having exhausted all the procedural niceties without response from an inexperienced male, finally seized him by the neck, pulled him into the nest box, and crawled beneath him. Her efforts were finally successful in spurring the young male into action."

Tigers need to learn about mating, and inexperienced tigers may have awkward moments. "If you have two tigers that have never mated before, they have a pretty difficult time," says tiger expert Ron Tilson. When tigers mate, the male seizes the female by the back of the neck. "It's a huge irritation to the female," and when sex is over the female usually turns and belts the male with a savage clout. Males learn to anticipate this and jump back, but it can be a shock for an inexperienced young male tiger who thought things were going rather well up until then. "He doesn't expect this to happen, and he really gets clobbered, and he gets very angry," Tilson explains. He contrasts this with one older male who knows what's coming and at the right moment leaps so high that the female's swat misses him altogether.

Breeding captive tigers seems to be a learning experience for zookeepers as well as tigers. Female tigers in estrus are extremely fierce. "When a female is in the wild she needs to be able to handle these very aggressive males. She has to be pretty tough to deal with them. The male has to hang out with her for about three days till she's ready to ovulate, and during these three days the male is trying to copulate with her and she doesn't want to, and there is an extraordinary amount of growling—it scares the devil out of keepers." Whenever Tilson (a coordinator of the tiger Species Survival Plan) recommends that a tiger breeding take place at a zoo where the process is unfamiliar, he knows that eventually he'll get a horrified phone call: "'Gee, Ron, we put them in together—we thought she was ready, she was calling and rolling and rubbing against the walls—and my God, there was *a big fight*!'" Tilson tells them to be brave and put the tigers back together to work things out.

You may also have been wondering about the sex lives of chimpanzees who've been raised in boxes for the first three years of their lives. Well, it's a problem. If for whatever reason you took chimpanzees from their mothers within 12 hours of birth and reared them "in closed boxes in isolation from humans and other chimpanzees," you might, years down the line, wonder about their capacity for normal sexual behavior. (This is old research. They don't do this to chimpanzees anymore.)

It turns out that this kind of upbringing handicaps chimpanzees sexually as well as making them "withdrawn, fearful, socially eccentric animals." Male chimps raised this way and then put in a small cage with a female chimpanzee were unable to make headway with girls their own age, whether the females were raised in a box or wild-born. Only with older, "sophisticated," wild-born females did anything happen. "Copulations were the result of the skill and persistence of the females, who frequently effected intromission by trapping the male in a corner and backing onto the erect penis," the authors note, impressed.

Box-reared females were not active participants in sex, either, but after a month or two allowed experienced males to mate with them. Things went somewhat better for the box-reared chimpanzees when they were placed in a group with other chimpanzees who had slightly more normal backgrounds, and some of them mated regularly, though the observers scorned their technique ("brief, perfunctory"). Of the box-reared females, only one out of seven avoided sex entirely. Of the five males, one was not tested, three "exhibited sex behavior which, if not completely normal, was successful to the extent that intromission was accomplished," and the fifth, while avoiding other chimpanzees, had occasional liaisons with a 55-gallon drum.

Wait, so that's where babies come from?

Before birds can cherish their nestlings, they must cherish their eggs. Eggs are demanding and need to be kept warm almost constantly. They need to be humid, but not too humid. They need to be turned daily. Birds generally have strong innate wishes to tend their eggs, but that in itself can lead to problems.

Inexperienced California condor parents must work out a routine. Condors are egalitarian, and mother and father take turns incubating the egg while the other one goes out for some refreshing carrion. It's not a sacrifice for the parent who sits on the egg. Condors like to do it. It's satisfying. Sometimes when the other parent arrives to take over, the first parent isn't ready to leave.

CCM and CCF, wild condors observed in the 1980s, had not worked this out. CCM hated to leave his egg. He would sit on it for as long as 10 days and when CCF showed up, he'd attack her. It sometimes took CCF a couple of days before CCM—who had to have been hungry by then—would let her take a turn. Their fights were so prolonged that in 1983 concerned condor watchers decided to seize the egg if the next fight lasted for more than three hours, because they worried that while CCM was busy kicking CCF out of the nest cave, the egg was getting cool. The next day, a grieved condor watcher recorded "senseless hysteria" and domestic violence at the residence, and authorities took the egg after four hours. (It hatched in captivity.) They replaced it with a lovable dummy egg.

CCF, who had always been "submissive toward the male," and who only got to take a turn by being meekly persistent, eventually got fed up. One day she arrived to incubate the egg, CCM got ugly, and she bit him. She bit him until he let her sit on the egg. After that, any time he refused to yield, she bit him, and he behaved himself for the rest of the season.

In later years CCM died, and CCF was trapped for captive breeding. She was so savage to all males that biologists did a blood test, wondering if such an aggressive creature could be female. "It is tempting to attribute the aggressiveness of this bird to her history of coping with her hyperaggressive former mate," write Noel and Helen Snyder. It wasn't until 1995 that CCF met a guy she could live with and resumed breeding.

Picking the right neighborhood

Even when birds know how to make a nest, they may need to learn about selecting its location. Pinyon jays learn from experience about dangers to their nests. Researchers studying pinyon jays in ponderosa pine forests found that the principal perils came from predators and cold snaps in the weather. If a jay's nest was in too shaded a position, and a cold snap came on, snow would remain on the nest and it would often be abandoned by the chilled parents. Nests in more open positions got more sun, but they were more apt to be robbed by crows or ravens, who cruise over the treetops looking for nests. Researcher John Marzluff looked at how male jays (the male selects the nest site) picked sites after a nest failure.

If a nest was preyed on, jays tended to build their next nest lower and in a more concealed spot. However, in two cases where jays built nests low down and had the nests robbed by cats, they then built nests near the treetops. If a nest failed due to cold, they tended to build the next one higher up. When jays successfully raised young, they built their next nest in a similar position.

Jays are long-lived and accumulate experience over the years. Young jays are more apt to make the mistake of choosing a lovely sunny nest site without worrying about the eating habits of the neighbors.

Marzluff suspects that in the past, jay nests were most likely to fail due to snowfall, but notes that the jays in this population have been subject to increasing predation as the raven population has boomed. The fact that pinyon jays can change their nesting practices quickly as their environment changes enhances their survival.

I like the look of natural wood

The Mauritius kestrel also changed its nesting practices, rather late in the game. This species was down to just four wild birds in 1972. Things didn't look good. While the use of organochloride pesticides had a lot to do with its decline, the final insult was that some idiot had introduced macaques onto the island of Mauritius, and the macaques liked to climb trees, reach into kestrel nests, and eat kestrel eggs. Cats, which some fool had introduced, didn't help, nor did mongooses.* Before the kestrel population plummeted, museum specimens had been collected. By looking at these specimens, ornithologist Stanley Temple was able to tell whether the bird had nested in a tree cavity or on a cliff. Kestrels who nested on cliffs had tail feathers worn down by the basalt. Apparently kestrels had mostly nested on cliffs until about 1900, when for some reason, they began nesting in tree cavities. They went on nesting in trees and providing macaque snacks as their population dwindled. In 1974, the last remaining pair thrilled well-wishers by nesting on a cliff and raising three chicks. Since that worked well, they went on nesting on the cliff, and their children, who had grown up on cliffs, also nested on cliffs.

Since then the population has been augmented by captive-bred birds, and macaque-proof nest boxes have been supplied. Pesticide bans were enacted. There are now hundreds of Mauritius kestrels.

Adoption

Among the various ways animals produce families, there are nonstandard options. There's a surprising amount of willingness to raise other animals' kids out there. Common eider ducks nesting in Manitoba frequently brood clutches of eggs that include the eggs of other eiders and then raise the resultant ducklings. One or two extra eggs or ducklings

* Introduced by some cretin.

are no more trouble to these tolerant mothers. DNA testing has shown that some polar bears in the Canadian Arctic are raising cubs who aren't their own. How this happens isn't known, but some speculate that the mothers don't know or don't care whose cubs are whose. Since polar bears are solitary, this theory holds, they don't need to distinguish their cubs as individuals. If there's a cub around you, it must be your cub. But when gathering at bonanzas such as a whale carcass or a garbage dump, you might take the wrong cub home.

Wildlife rehabilitators Jan White and Cheryl Millham explain how to get a mother merganser (handsome fish-eating waterfowl) to adopt orphaned chicks. The ideal target is a merganser with chicks the same size as, or smaller than, the ones you want her to take. First take the orphans out and let them peep loudly to attract the attention of the mother. "Maneuver as close as you can to her and gently underhand toss the chick towards her," they pleasingly advise. Then back off. With any luck, the mother merganser will call her chicks, the adoptees will rush toward her along with the others, and she will accept them as more of the gang, thus saving you many weeks of care during which the chicks would be gobbling 100 goldfish a day. (These are the smallest, "comet" size of goldfish. When the chicks are bigger you could trap minnows, and a mere 30 to 60 a day per chick should suffice.)

In Kenya, a crowned eaglet was brought to Simon Thomsett. The downy eaglet was about a month old and uninjured, but her home had apparently been destroyed by logging. Thomsett, knowing a bit about crowned eagles, and having "a high regard for their intelligence," decided to see if he could place the young bird for adoption with a pair of wild eagles. They had nested earlier in the season in their territory along the Amboni River, but had abandoned the nest after two months when their egg didn't hatch. Although five weeks had passed, Thomsett enlisted the aid of the park warden of Aberdare National Park, Phil Snyder, who ascended the forbidding nest tree, a task which took two and a half hours, and fooled around in the nest a little (he could walk in it—it was two meters across!) before hauling up the eaglet in a sack and leaving her there.

The next day the chick sat in the nest peeping of her hunger and solitude to any raptor that flew over. The male eagle who had nested there weeks before flew past doing a territorial display. "Suddenly he stopped the display and I felt sure he had seen the chick," Thomsett reported. The next morning the male and his mate returned to the site. The female perched in an adjacent tree and peeped to the eaglet. The eaglet peeped back. The female eagle flew to the nest, where the chick bowed submissively. "Immediately the [female] showed signs of parental behavior." Soon the male joined in. When Thomsett returned the next day the female was on the nest, the chick was snacking on a young antelope her new parents had brought her, and the male was off hunting for a family of three.

(It is very tempting to supply dialogue for the parent eagles. "Honey, remember our egg? It never hatched, did it?" "No, it didn't." "You're sure?" "Of course I'm sure. I brooded that egg every day for two months, remember?" "Yes, that's what I thought." "Why do you ask?" "Well, I think there's a baby in our nest." "That's ridiculous. We haven't been back there in five weeks. How could there be a baby in our nest?" "I know, I know. . . . I think it's hungry.")

Does your mom think I'm cute?

During the reintroduction of wolves to Yellowstone National Park in the 1990s, a female called either Natasha or Number 9 was widowed shortly after giving birth to nine pups. When the male was shot, the pack fell apart, and there was no one to help her feed the puppies. So biologists recaptured the family and put them in a pen until the pups were less needy. When the pups were about three months old, a windstorm knocked a tree onto the pen, making a hole just large enough for two of the pups to get through. Biologists worried that either coyotes or a grizzly bear would kill the pups. Just then a hero arrived, Number 8, a young male wolf from another pack. (Actually, he had been noticing Number 9 for a while.) Now he ran the coyotes off and started regurgitating food for the pups. He hung around, protecting, feeding, and playing with the pups until the biologists finally let Number 9 and the rest of her children out. Sure enough, she had been admiring him too, and they became a pair.

Running away from home

Herring gull chicks who have lost one or both parents and are at risk of starving will try to get adopted. This is risky, since adult herring gulls frequently attack strange chicks, and may kill and even eat them. The hungry little orphans get as close as they can to the nest where they hope to live, and crouch down. If the adult pecks at them, they creep even closer, since gulls have a reluctance to peck at chicks in their own nest. Usually, within a day or two the chick is adopted. White stork chicks in Spain sometimes join other stork families, but not because they are starving orphans. They are "greedy eldest chicks from the largest broods," and by moving in with a stork family with younger chicks, they can be fed for a longer period before they are forced to earn their own way.

It's a girl? What's a girl?

Animals giving birth for the first time may have no idea what's going on. Equine veterinarian Sarah McCarthy describes an eight-year-old mare named Merganser foaling for the first time. After the foal emerged, Merganser got up and walked away. The humans chased her, brought her back, and showed her the bay filly, to Merganser's obvious delight—she just hadn't realized there was a baby horse around until they showed her. One of McCarthy's own horses, Holly, "wanted a foal very much," as evidenced by her extreme interest in foals and her tender nickering when she saw them. (Nickering is a friendly sound made to foals as well as in other situations.) When Holly foaled for the first time, at age 12, she failed to grasp the implications of labor. After the delivery Holly lay in the straw of her stall, feeling sorry for herself and "calling for cold cloths." McCarthy grabbed Holly's head and brought it around to where she could see and smell her chestnut colt, Tristan. Holly was ecstatic—at last a foal of her own!—and began to nicker. When she had her next foal, she knew what to expect.

Dearie, a knowledgeable thoroughbred brood mare, had foaled 10 times. For her eleventh, she was sent to a veterinary clinic because of worries that she might have problems. She was in an outrageously advanced stage of pregnancy and staff were keeping an eye out for signs of labor. One staffer, checking on Dearie, lifted and peered under her tail—and Dearie began to nicker hopefully. (She foaled successfully, but not that very second.)

It's hard to see how animals could know they're about to give birth the first time, unless they could figure it out from seeing other animals give birth. In many species laboring mothers sneak away for a little privacy, making it unlikely that other females will witness the process. So the first-time mother will often sneak off herself because she feels unwell, go through delivery on her own, and then get the surprise of her life when confronted by her progeny, whether that consists of a foal, 14 puppies, or a big blue egg. Hard-wired parental doting will usually take care of matters from there. It is not completely unknown for humans who don't realize they're pregnant to go into labor and not figure out what's happening until they spot an infant loitering about the premises, so it ill behooves our species to sneer.

Midwifery

In the wild, when an elephant gives birth, she is closely attended by the females in her herd, who remove the placenta, help the calf up, and caress it. Female dolphins cluster around a female giving birth and help the calf to the surface to take its first breath.

The Rodrigues fruit bat, or golden bat, is an endangered species. At a small captive breeding colony, biologists noticed that a pregnant bat was in labor and in difficulties. An unrelated female was hanging next to her, paying close attention. The pregnant bat did not seem to understand that although bats usually hang head-down, in this case she needed to hang feet-down to let gravity assist the baby in being born. Failing that, hanging horizontally from the top of the flight cage in a "cradle" position would have been better.

The helper licked the pregnant bat and folded her in her wings. Four times the helper left her place next to the laboring bat, moved into full view in front of her, and hung feet-down. When she did this, the mother also turned feet-down, and the helper moved back and wrapped her wings around her. The bat pup slowly began to emerge, with one foot and one wing protruding, and the helper groomed the baby and fanned the mother. "The helper continued to alternate between 'tutoring' the mother in a feet-down posture, grooming the emerging pup and mother, and resting." The pup was finally born an hour and three quarters after labor began.

Since these bats always hang head-down except to urinate or defecate or to give birth, it seems that the helper hung feet-down to give the first-time mother a hint. Had she only been doing so out of the sort of unconscious sympathetic movement that makes us flinch when we see someone take a blow, it's hard to see why she would have moved into the mother's view to do so.

Meanwhile, two males in the colony didn't get involved in the delivery, but when one of the people watching tried to take close-up photos, they moved in front of the females and threatened the photographer, spreading their wings.

Keeping track of the baby

Since other animals may kidnap babies, mothers must learn not to let others borrow their baby unless they're sure they can get it back. Lorel, the biological mother of the loquacious bonobo Kanzi, had never had a baby before. Raised by humans, she didn't know that she was entitled not to let others hold her child. When the bonobo Matata, an experienced mother with a child of her own, reached out for the 30-minute-old Kanzi, Lorel let her take him. Matata never gave him back.

Nursing

Suckling the young, that amazing female mammal specialty, is indispensable for the survival of babies (unless the setting is a zoo, where keepers can step in with bottles and formula). Many hand-reared apes don't suckle their babies, because they've never seen it done. But they can learn. Human nursing mothers (who happened to be zookeepers) were enlisted to educate an orangutan at the Seattle Zoo. Melati, a captive-born ape with a complicated childhood, had had essentially no experience with a mother-infant situation. She ignored her first baby, and it was taken away and hand-reared. When she got pregnant again, they brought in the breast-feeding demonstrators. Melati watched with interest and "autonursed," sucking on her own nipple, but didn't try to imitate them by nursing an orangutan doll or, when she had her second infant, by nursing it. Instead, when the baby cried, she kissed it tenderly, and the baby sucked on her tongue. She did carry this baby nicely. Discouraged, keepers took the baby away on the third day and fed it in the nursery. Three days later, when they saw that Melati had a good milk supply, they gave the baby back. Melati had a wacky new scheme: she would autonurse, fill her mouth with milk, and then transfer the milk to the baby's mouth. On the ninth day they finally worked it out: Melati held the baby to her bosom, the baby nursed, and the zookeepers sighed with relief.

Large sea mammal mothers in captivity must keep their babies—who can swim vigorously—from bashing against the walls of the tank. Alexandra Morton describes the tragic outcome when Corky, a captive killer whale, carefully interposed her head between her new calf and the tank wall every time it approached. To nurse, killer whale calves must locate the mammary slit, which is marked by a white spot under the whale's tail. But killer whales (also called orcas) have white spots behind their eyes, too. The calf kept nuzzling Corky's eyespot. "This led us to believe that baby orcas are led by instinct toward a white spot, much as a seagull chick pecks at the red spot on its mother's beak. Because Corky constantly had to use her head to push the baby away from the circular tank edge, however, the baby had fixated on the wrong white spot—the one behind her eye," writes Morton. The calf never succeeded in suckling and was removed, but humans failed to raise it. In the wild the other females in the pod would probably have guided the baby to the right spot, but when a diver bravely entered the tank to try to perform this service, Corky was so worried by the sight of a human gripping her child that she perpetually swung her face, with its alluring white spot, toward the calf.

Weaning

Young mammals must eventually be weaned, but they often disapprove. Sociobiologists have produced theoretical analyses of how the parent's interests conflict with the offspring's interests, but it all boils down to the offspring seeing no reason why a good thing should end and the mothers being sick and tired of nursing these great louts.

Biruté Galdikas describes the incessant tantrums of Mel, a five-year-old orangutan of formerly sunny temperament, when his mother Maud was weaning him. He would add emphasis to his discontent by hanging upside down from a branch by his feet while screaming at the top of his lungs and punching the air wildly.

Primatologist Toshisada Nishida describes a chimpanzee mother when weaning. "She shakes off her youngster when it tries to ride on her back, and prevents it from suckling by folding her arms in front of her breasts or by lying prone on the ground." Five-year-old Katabi was being weaned by Chausiku. His display of hysterics included hanging by one foot and banging his head against the tree trunk, shaking and screaming. Nishida noticed the seemingly distraught Katabi sneaking glances at Chausiku to see if she was relenting.

Faustino, a chimpanzee child at Gombe, felt deprived of attention when his mother Fifi had another baby. He threw constant tantrums. One day Fifi climbed high into a tree with her two children and tossed Faustino out of the tree—except that she held on to his ankle so that he dangled, screaming. He did not have any tantrums for the rest of the day. Jane Goodall saw it happen again on another day, and again Faustino was subdued for the rest of the day.

Not all ape mothers are so emphatic in their discipline. Sue Savage-Rumbaugh describes the bonobo Matata being called to intervene in the trials of her son Kanzi. Savage-Rumbaugh and Kanzi might be squabbling over whether Kanzi should pick up her papers, which he had just grabbed and flung on the floor. Kanzi would yell for Matata. She would "lumber over with a suspicious look on her face and attempt to determine what was going on between me and Kanzi." Both primates would appeal to Matata to see the justice of their position. Often Matata did not see things as Savage-Rumbaugh did. Toilet training, for example: Matata, who was not toilet-trained, and who had never known a bonobo who was, could not see what Savage-Rumbaugh was getting at, "apparently assuming that sitting on the potty was an unusual form of punishment that I had invented for her son." When even his mother felt that Kanzi needed to be reined in, Matata would take his hand or foot in her mouth and slowly increase her pressure until he realized that she was serious, and behaved himself.

Becoming better parents

One of the main responsibilities of parents is safety—keeping kids from being eaten, keeping them from falling off cliffs and out of trees, and keeping them from eating poison. Chimpanzee mothers and older siblings do not like to see babies putting strange things in their mouths, and if they see them with a novel food item, they will snatch or flick it away from them. The chimpanzee mother FT thought it was a bad idea when she saw nine-month-old PN reaching for ficus leaves, so she moved PN's hand away. Stubborn PN grabbed more leaves, so FT took the leaves out of PN's hand, picked all the ficus leaves in reach, and dropped them to the ground.

Keeping children from being eaten is one of the most pressing concerns for many animals. Young animals are easy to catch, tender, and bite-sized, so predators focus on them. Bonnie and Clyde, whooping cranes who had triumphed over their own irregular childhoods (they were raised by biologists in crane outfits) and built a happy home next to a lake in Florida, successfully hatched two chicks. One day, both first-time parents had stepped away from the nest, and a bald eagle swooped down and grabbed a chick. When it came back for the second chick, Bonnie covered it with her wings, and Clyde leapt at the eagle and frightened it off. Observers promptly named the chick Lucky. Two months later, when Lucky had even more meat on his bones, a pair of eagles came by, and the incensed Bonnie and Clyde went after them so fiercely that one eagle had to spend two weeks in rehab.

Mares also become better parents as they get older. In wild populations, a higher proportion of the foals born to older mares survive. Studies of feral horses living in the Kaimanawa ranges of New Zealand found that older mares didn't work any harder at it than younger mares, but they targeted their care more successfully. The older mares were more attentive and doting when their foals were very young (under three weeks) and less so when the foals were older. Older mares let their very young foals nurse longer and put more effort into maintaining contact with them. This seemed to be the result of experience—younger mares who had lost a foal changed their behavior with the next foal and acted more like the older mares.

Whether older parents are more successful parents has been studied in monogamous birds. There may be two effects here: individuals in a pair may be better at raising a family, or they may be better *as a pair*. The effects are clearest in swans, which are long-lived. Although they mate for life, mortality is high, so old and experienced animals who have been widowed are found in new pairs with inexperienced young birds. By studying enough pairs, researchers found that older Bewick's swans are better parents, even when mated to younger swans. But they also found that the longer a pair had been together, the better they did. By their calculations, Bewick's swan pairs get better for 11 years and then, apparently, they are the best parents Bewick's swans can possibly be.

Barnacle geese—remember them and their trial marriages?—also improve their teamwork over the first seven years they are together. Perhaps they get better at coordinating their incubation and foraging duties, and perhaps they are better at presenting a united front to other goose couples and grabbing the best nesting sites.

In Chilkat Pass, willow ptarmigans who stay together for three years hatch out more chicks than newlyweds. On offshore Pacific islands, Cassin's auklets who've been together forever (okay, seven years) do no better at laying eggs early, hatching out chicks, or producing big chicks, but excel at keeping chicks alive. Red-billed gulls that have been together for at least five years manage affairs so smoothly that their eggs are laid an average of nine days earlier in the season. They get better, drier nest sites, are feeding their babies when food's most abundant, and even have time to nest again if anything bad happens to the first eggs.

As for Adelie penguins, they do better if they stick with their old mate—but what if he or she doesn't show up at the breeding ground? It's important to start promptly, and you can't be waiting around, dressed in your tux, looking at your watch, while all the best mates and nest sites are being snatched up. Thus ornithologists calculate that the best plan, if you are an Adelie, is to show up on the breeding ground as early as possible, go to the spot where you and your mate were so happy last year, and do your best to line up a nice mate at once. If last year's mate shows up, dump the new one, and rekindle the old flame—you'll raise more chicks if you do.

Older men, older women

In many bird and animal species older mates are in higher demand than novices. (Avital and Jablonka prefer to speak of a preference for middle-aged mates, since they feel that few creatures are attracted to actual dotards.) European sparrowhawks will not choose yearlings as mates unless there is no one else available, and it's true that they will fledge more chicks with an older bird. Jane Goodall has written of the powerful allure the chimpanzee matriarch Flo held for male chimpanzees despite her grizzled hair, worn teeth, and tattered ears. Many male chimps came to Goodall's camp for the first time when Flo was in estrus and they couldn't bear to let her out of their sight. One researcher looked at the choices of adult male chimpanzees when they encountered two females who were each in this desirable condition. In 30 out of 38 times, the male approached the older female.

Helpers

Some young Seychelles warblers stuck around to help their parents take care of the next clutch; they got helping experience and had better reproductive success when they started their own families. Both males and females did better if they had helping or breeding experience. Females do most of the nest building in this species, and inexperienced females had a fatal tendency to build nests that were anchored to a branch and a thin leaf stalk, as opposed to the stout branch forks selected by experienced types. Nests not in forks were destroyed by gusty winds and heavy downpours. Male birds with experience did a better job of guarding the nest from skinks and fodies.

In some pinyon jay families, teenagers, especially male teenagers, help out with their younger brothers and sisters, and in some they do not. The choice seems to be learned, for a pinyon jay from a nonhelping lineage fostered into a helpful-lineage nest became a delightfully helpful young bird.

Several related species of voles have very different family lives. Meadow voles, for example, are promiscuous, do not form pairs, and leave all child care to the mother. Prairie voles pair up and cleave unto one another, and the fathers do everything that the mothers do for the children except nurse them. They huddle over them, groom them, and retrieve them if they wander off. Older children stay in the family and help raise their younger brothers and sisters. The more devoted the father vole, the more time the teenagers spend helping out. These prairie voles who help with their younger siblings are excellent parents when they breed and spend more time with their kids. Their kids grow faster and are bigger when they are weaned.

If you take baby meadow voles (the promiscuous children of single mothers, remember) and have them raised by prairie voles, they will grow up to form pairs and the males will be devoted fathers. Perhaps they act this way because they have learned to act this way; perhaps they act this way because the presence of a father in their childhood affected their hormone levels; and perhaps both things are true.

Teaching

Whether animals teach things to each other (and this mostly means teaching the children) is another of those quarrelsome areas in animal behavior. Some take the view that animals do teach in certain ways, and others take the view that what those animals are doing isn't really teaching.

What is it to teach? When is teaching others simply enabling them to learn? If you let others watch you perform a difficult task, is that teaching? If you set up the task for them to perform and let them try, is that teaching? If you set up the task in simplified form for them to try, is that teaching? If you tell them what to do, but don't demonstrate, is that teaching? If you physically orient their body so they can do it better, is that teaching? If you give them a book that explains what to do and yell at them until they read it, is that teaching?

An influential article by T. M. Caro and M. D. Hauser, "Is There Teaching in Nonhuman Animals?" takes an inclusive view of teaching, describing three forms: opportunity teaching, coaching, and active teaching. This is their basic definition of teaching.

> An individual actor A can be said to teach if it modifies its behavior only in the presence of a naive observer, B, at some cost or at least without obtaining an immediate benefit for itself. A's behavior thereby encourages or punishes B's behavior, or provides B with experience, or sets an example for B. As a result, B acquires knowledge or learns a skill earlier in life or more rapidly or efficiently than it might otherwise do, or that it would not learn at all.

By this definition, all of the above examples except simply allowing someone to watch you do something you were going to do anyway qualify as teaching.

Opportunity teaching is giving others a chance they wouldn't otherwise have to learn something. When a mother cheetah brings a gazelle fawn to her cubs and releases it for them to catch, that's an opportunity they wouldn't otherwise have.

In coaching, the coach watches what the other animal is doing and encourages the right moves and discourages the wrong ones. Young vervet monkeys learn to improve the precision of their innate alarm calls, so they eventually give an eagle call only for an eagle or hawk, and not for a dove or a falling leaf. When they give the correct alarm call they may be encouraged by the fact that other vervets follow up by giving alarm calls. Perhaps these vervets would have given that call anyway, so they're not really coaching. But sometimes a little vervet gives an alarm call for the wrong thing and frightens its mother. When its mother sees that she hid from a dove and not an eagle, she is apt to slap the baby—that's coaching.

As always, learning can produce error. One day a baby vervet saw some stampeding elephants and gave a leopard alarm call. It so happened that at almost that very moment, an adult male in the troop, looking in the opposite direction, saw a leopard and gave the leopard alarm call too. For months thereafter, that little monkey gave leopard alarm calls for elephants.

In active teaching, one animal wants another to learn something and changes its own behavior according to how the pupil is doing. Some take the view that animals, or certainly most animals, cannot really teach, because they do not have a theory of mind. That is, they cannot put themselves in the mental place of another animal and perceive what it knows and what it does not know. If the adult animal doesn't know that the baby doesn't know, how can it teach? Others argue that this knowledge may be necessary for some forms of teaching, but not for all.*

Walk this way

The more you read about learning, the more elusive teaching becomes. We don't teach babies to talk or walk. It is impossible to keep them from walking and talking. We may be able to coach them to walk and talk sooner. Possibly this is a delusion, but just as it is impossible to stop babies from striving to walk and talk, it is impossible to stop parents from urging them on.

* Most of the teachers I have had did not have any idea what was going on in my mind, and that's how I preferred it.

When a mother or father primate encourages a baby to crawl or walk, Dario Maestripieri argues, this doesn't mean that the parents are aware of the infant's mental state. They may not have an idea about what goes on in the infant's head; they may just want to see the child walk, and have learned that infants walk sooner if gently encouraged.

We are not the only primates who teach, or coach, infants to crawl and walk. Some captive pigtailed macaque mothers set their infants down, back a few steps away, look at the infant and pucker their lips and raise their eyebrows, enticing the baby to come. If it can't support itself on all fours, she may support it with a hand. A wild Barbary macaque in Gibraltar placed a week-old infant (probably his child) on the ground, backed about two feet away, and leaned in to look at and chatter to the baby. The baby chattered back and crawled ineptly toward the male. When it got six inches away, the father backed up a little more and chattered to the baby again.

Researchers studying captive rhesus found that some mothers encouraged early independence by breaking contact with their infants in the first weeks of life, while others did not break contact and in some cases restrained their infants when they eventually tried to move away. The mothers who broke contact accompanied this with encouragement for the infants to follow, by walking backward, looking at their infants, and smacking their lips. Their children became independent slightly earlier, breaking away from and returning to their mothers at a slightly younger age than the infants of the clingy mothers.

Spider monkey mothers encourage their older children to lead the way along foraging routes, and put a lot of time into it. The mother will start out along the route and then sit down on the branch. Eventually the little monkey gets tired of the inactivity and walks a little way along the route through the tree, and the mother at once falls in behind, thus encouraging her child to lead and learn to travel independently, not simply follow.

Bat rehabilitator Patricia Winters received 15 orphaned pallid bats one year. Their Latin name is *Myotis yumanensis*, and bat enthusiasts call them yummies. When Winters had gotten them through infancy she put them in a flight cage with two adult male California bats. She hoped the sight of the flying adults would encourage the yummies to try flying. Winters was a little worried because yummies are twice the size of California bats. When she heard a racket from the cage, she feared the yummies were beating up the adults, "because they can be very rowdy." She peeked in and beheld "flight school." The males would land a short distance away from the yummies, turn to face them, and call to them. The apprehensive yummies "acted like they were going to get crushed to death on the rocks below," but finally they flew the tiny distance to the males. The males repeated this at greater and greater distances. "Within two nights the babies were flying like champs."

On the prowl

If the only role of parents were to keep the young alive while they mature, there would be no reason for predators to take their cubs along on hunts. Cubs and pups are clumsy and noisy and scare the prey, but they've got to learn sometime. When they are a little more skilled, mother and children may actually increase each other's success.

Mother grizzly bears who swim out to islands off the Alaskan coast to pillage seabird colonies, digging up nest burrows and eating eggs, chicks, and all, are followed by their cubs. The cubs learn where the islands are (some are as much as 10 miles offshore) and that delicious food can be found there, but no teaching is necessary. The cubs can see and taste for themselves.

Kids want limits, don't you, honey?

If you're going to take the kids with you, they're going to have to learn to behave themselves, and animal parents and elders do not always take the view that children who misbehave can be sweetly reasoned with. Zoologist Maurice Burton describes what he takes to be maternal discipline in the dusky wood rat. A mother wood rat struck her son, knocking him off a log. The observers concluded that she didn't recognize her child, but Burton argues that she was disciplining him for not following behind her in single file as young wood rats usually do. He immediately got in line. She may also have been in a terrible mood because they had only recently been released from a wood rat trap set by those same observers.

Burton describes a mother water vole swimming peacefully, followed by her child. When the young vole veered off obliquely, she turned and "there was a flurry of water and every sign of chastisement after which the two swam on, the young one following directly behind the mother."

If a river otter cub races ahead of its mother, she may nip it on the nose. Chastised, the cub falls flat on the ground. If the mother is very angry, the cub may lie there "until the mother returns to it and caresses it as if telling the cub that everything is all right, but just to behave."

Dolphin mothers usually feel that their calves should stick close. In one oceanarium the calves Delphi and Pan saw no reason for this caution. If a calf was persistently disobedient, his mother would push him to the bottom of the pool and hold him there for a few seconds. Then she'd let him go and he'd behave. For a while. When the calf got bigger, the mother had a harder time holding him down. Instead she'd swim below the calf and lift him *out* of the water.

Teenaged chimpanzees who accompany adults on boundary patrols get hints on how to behave during this dangerous activity. If they utter a sound, they get shushed by older chimpanzees who touch them or make threatening faces at them.

Greylag goslings follow their parents wherever they go. When goslings are raised by humans, who walk faster and cheat by wearing shoes, it is easy to tire the goslings out. The goslings keep following through the scary woods and over the hard roads because they don't want to be left behind. But they learn that their parents are unreasonable taskmasters, and refuse to go on subsequent expeditions.

Apprenticeship

Humans do teach actively but not nearly as much as we think. (This fact may be partly obscured by the fact that most of the people studying this subject support themselves as teachers.) Most tool use is not actively taught. Beginners learn through apprenticeship, which involves observational learning, trial and error, opportunity, and occasional coaching. Anthropologists who have studied the human acquisition of complex skills have found very little direct instruction. Thus a novice on a commercial fishing trawler learning the process of purse seining "does not learn by simply internalizing a body of knowledge presented to him by others. He is, in fact, given little or no information at all." He is not taught the names of parts and he is not told why he is supposed to do tasks he is assigned. He learns one thing at a time, in no particular order, and gradually assembles them into sequences of actions. Traditional Navajo silversmiths and weavers don't tell their apprentices what to do, but simply let them watch and try. Papuan craftsmen who make flaked stone tools don't teach explicitly, but by demonstration.

Okay, that counts

The canonical examples of active teaching by animal parents come from chimpanzee nut cracking. Even those who define teaching narrowly accept these, although they say how rare it is.

Wild chimpanzees of the Taï National Park in Ivory Coast crack nuts on stones that serve as anvils, striking them with stones or branches that act as hammers. This skill takes years to master, and chimpanzee mothers encourage the development of the skill in several ways. Christophe Boesch divides the behavior of these chimpanzee mothers into stimulation, facilitation, and active teaching.

They stimulate nut cracking by leaving nuts or hammers near anvils, something they do often when their babies start to show an interest in cracking nuts, at around three years of age. On several occasions mother chimpanzees placed a nut in perfect position on the anvil, placed the hammer next to it, and then left it there, making things as easy as possible for the inexperienced young chimps. Males and females without offspring don't leave their stuff lying around. Someone might take it.

Mothers also help out by supplying nuts or better hammers to their infants. Typically this involves letting the kid snatch the hammer. A chimpanzee named Ella let her five-year-old son Gérald take four hammers from her in a row. Using hijacked hammers, Gérald was able to crack open 36 nuts during a 40-minute period in which Ella cracked only 8 because she was kept so busy searching for new hammers.

On several occasions researcher Christophe Boesch also saw chimpanzees actively teach a child how to crack nuts more effectively. Watched by her six-year-old, Sartre, the chimpanzee mother Salomé was cracking panda nuts, which have particularly hard shells and which require careful positioning to extract the three separate kernels. She cracked 18 nuts and Sartre ate 17 of them. Then she watched as he took up her hammer and tried to crack nuts. He succeeded in getting the first kernel out of a nut, and then replaced it "haphazardly" on the anvil. Salomé took the nut off the anvil, cleaned off the anvil, and carefully repositioned the nut. She watched as Sartre successfully cracked the nut further and got the second kernel.

Nina, a five-year-old, was trying to crack nuts with an irregularly shaped hammer. It wasn't working, so she kept changing her position, her grip, and the position of the nut—still no success. After nearly 10 minutes, her mother, Ricci, got up from where she had been resting and came over. Nina gave her mother the hammer, and watched as Ricci very very slowly rotated the hammer into an effective position. She did this so deliberately that it took her a full minute, Boesch reports. As Nina watched, Ricci cracked 10 nuts. She gave Nina 6, and bits of the other 4. Then she left the hammer and anvil to Nina, who cracked 4 nuts in the next quarter of an hour, all the while holding the hammer exactly as her mother had, even when she changed positions. Ricci and Salomé are generally conceded to have been teaching their children, even in the strictest sense of the word.

Creating learning situations

The safety provided by an animal parent keeps children alive and also gives them time and confidence to learn. Eytan Avital and Eva Jablonka write, "The presence of their parents allows young animals to explore their environment much more readily, since, by providing a protective and comforting atmosphere, the parents act as stress minimisers. . . . There is an obvious, and usually adaptive, association between emotions and learning in both animals and man."

That young ravens learn what is safe by exploring the world in the presence of their parents became clear to Bernd Heinrich when he raised a family of four young ravens in the Maine woods. He took them on walks through the woods and made a point of touching many things. They touched and examined everything he touched, but they also investigated almost everything else they saw. By the time they were four months old, however, they were showing signs of neophobia—fear of the new. If they had seen an object, say a film canister, in the first four months of their lives, it didn't worry them. But if it was a novelty, it was scary.

The sequence of the indulgent parent making the young work a little harder is common. In an Israeli olive orchard, great tits raised a family of chicks. When the chicks were newly hatched, the parents brought insects and put them in the babies' mouths. When they were two weeks old, almost fully feathered, the parents would perch in the nest entrance and the babies would have to leap to get the insect. Soon the parents would perch in the entrance for a second, displaying the tantalizing food, and then fly away, encouraging the young birds to fly after them.

In the Kalahari, after bat-eared fox cubs have been weaned, their mother goes out hunting at night while their father stays with them. When they are a little older, he takes them out foraging. If the father catches a sunspider, a terrible-looking creature that bat-eared foxes consider edible, he gives a low whistle to call the cubs to his side. Then he releases the sunspider and the cubs try to catch it. If they bungle the task, he catches it again and lets it go again.

Modeling behavior

For some British red foxes, earthworms are a big food item. One evening David Macdonald witnessed the old vixen Toothypeg foraging with her son. Toothypeg walked silently back and forth, inspecting the damp grass with watchful eyes and cocked ears. When she detected a worm she'd stab her nose into the grass and grab it. Meanwhile her cub was dancing around in the grass, performing the classic "mouse leap," in which a canid leaps high in the air and comes down with both forepaws on some savory rodent sneaking through the stems. This didn't work. Worms slithered away between his toes, leaving him biting at the dirt.

The cub turned to see what his mother was doing. She had caught a worm that had its tail anchored in a burrow. Toothypeg was expert at pulling such worms loose, but this time she stretched the worm out and held it in place. She waited, and the cub took hold of it. He pulled too hard, and the worm broke in two, so he only got half. Toothypeg caught and held another, and let the cub take it. Again he snapped it. By this time the cub was deeply interested. Toothypeg caught a third worm, this one with most of its body anchored in its burrow. She stretched it, and tapped its taut body with a forepaw, gently teasing it out, and the cub— and Macdonald—watched in fascination. A worm-catching artist was at work. Slowly she eased the worm out of the soil, and let the cub have it.

He began trying to catch worms using her quiet careful examination of the turf instead of his previous pouncing folly.

Two days later Macdonald saw them catching worms again. Toothypeg was catching four times as many worms as her son—but he was catching them. A month later he was as quick as his mother. (Which makes it all the luckier that Macdonald caught the brief, minutes-long tutorial performed by Toothypeg.)

Teaching beaching

Killer whales in the southern Indian Ocean, observed by researchers stationed on Possession Island in the Crozet Archipelago, taught their calves a diabolical, dangerous way of catching elephant seal pups. Elephant seals would be basking on the beach in apparent safety when a killer whale would come shooting out of the water, strand herself on the beach, grab a young seal, and let the next wave carry her back into the water.*

Christophe Guinet and Jérome Bouvier were watching a group of killer whales: A1, an adult male; A2, A3, and A6, adult females; and A4 and A5, female calves. A4 was A2's child, and A5 was A3's child, but the female A6 was also involved in the care of the calves. The calves A4 and A5 frequently beached themselves when no elephant seals were present. Occasionally they beached together. In 1988 young A4 beached herself and couldn't get back into the water—observers returned the calf to the water (after taking advantage of the opportunity to measure her). Without their help, she would probably have died. Yet in ensuing weeks she continued to practice beaching. A4 mostly practiced stranding with female A6, not with her mother. Between the 1989 and 1990 field seasons A6 vanished, and in 1990 observers did not see A4 beaching herself at all. In 1991 A4 resumed her beaching practice, mostly with A3 rather than with her mother. That year A4 almost got stuck on the beach again. She couldn't seem to get back to the water. As she was struggling there, her mother, A2, swam away from the beach about 50 meters, then turned and acceler-

---

*Yes, I'm sorry, this is a grim tale of baby animals eating baby animals. We do it ourselves when we feed, say, egg salad to toddlers, but that is of course different, for reasons I am unable to state with a straight face.

ated toward the beach. When she was 5 meters away she wheeled suddenly, creating a large wave that lifted A4. Buoyed by the wave, A4 was able to turn around and swim back to deep water.

Meanwhile A5, a slightly younger calf, was accompanying her mother on attempts to actually get a seal while beaching. In 1991 she was videotaped in the act. A5 was swimming along the shore with her mother, A3, when they spotted an elephant seal pup on the beach near the water's edge. They lined up side by side and shot onto the beach, where A5 snatched at the seal. Her mother, right behind her, gave her a push toward the seal, and A5 grabbed it by the side. The mother then used her head and the front end of her body to push A5 and the seal back into the ocean.

Between 1963 and 1995 observers on Possession Island found six hopelessly stranded killer whales, five juveniles and one adult. (I'm tempted to guess that the reason A4 did all her practicing with adults other than her mother was that her mother didn't want to encourage A4 to do anything so dangerous, but I have no evidence.)

Killer whales also catch elephant seals by the beaching technique in Argentina, on the coast of Patagonia. Diana and Juan Carlos Lopez saw an adult male stranding side by side with a juvenile. The adult would grab a seal and fling it over to the juvenile.

Ornithologists watching royal terns in a wintering area in coastal Peru were surprised to see adults calling to young birds, leading them away from the flock and giving them fish. Sometimes the adult would, in an instructive manner, drop a fish for the young bird to catch or at least pick up off the water. The young terns sometimes follow their parents, uttering "the squeaky begging call." These are not tiny fuzzy chicks. By the time they get to Peru at seven months old, they have migrated 3,000 miles from their breeding grounds. They can still use a little help.

In 1954, in the ornithological journal *The Ibis,* Col. Richard Meinertzhagen gave an account along similar lines, but with more detail, of osprey school. Osprey are handsome fish-eating raptors, and in the family Meinertzhagen observed, the parents were ready to stop bringing fish to their no longer helpless chicks. The two babies were ready to fledge, and the adults brought fish and perched nearby, while the kids screamed that they were starving. The adults repeatedly flew off with a fish in their talons, as if hoping the frantic children would follow. At first the fledglings stayed in the nest, but the next day they ventured to a nearby rock in the lake, and the parents fed them there. After another day or two, the parents were able to lead the young birds flying over the lake. A parent caught a fish, flew toward a fledgling with it, and then dropped it in midair. Each time the parent swooped down and snagged the fish before it hit the water. They did this many times, until one of the fledglings grabbed a dropped fish in the air and took it to the rock to eat. The other fledglings came over and they quarreled until "an infuriated parent" came over and shoved one of the kids off the rock. (We are told that the one who got to keep the fish was the one who caught it. I hope so, because otherwise it wouldn't be fair.) The parents went back to dropping and catching fish until the second fledgling got one.

The next day the parents dropped fish into the water and retrieved them, over and over, until the hungry young birds started grabbing the floating fish. By the next day, they were diving for living fish.*

Classroom time

In a more scholastic setting, a bird was noted modifying his behavior in an instructive way. When teaching young gray parrots using the model-rival method, one of the participants was the older and more skilled parrot Alex. Psychologist Irene Pepperberg noticed that during such sessions, Alex spoke with unusual clarity and would say the word

*In *The Thinking Ape,* Richard Byrne asks why no one has reported seeing osprey doing this since 1954, and notes that Meinertzhagen was "the soldier in charge of covert operations in Palestine during the First World War" (although he discovered and collected the giant forest hog *Hylochoerus meinertzhageni*), and that "scientific fraud is not unknown!" Since Byrne wrote, it has become clear that Meinertzhagen was also covert about natural history. He had a fine, comprehensive collection of 25,000 bird specimens, and it seems that he got such a complete set in part by stealing specimens from museums and relabeling them as birds he'd collected himself. As related in Scott Weidensaul's *The Ghost with Trembling Wings,* Meinertzhagen's lies to conceal his thefts inhibited rediscovery of the mysterious forest owlet of India—scientists had been searching in the wrong place and habitat. Clearly we must be suspicious of information from Meinertzhagen. Still, he really did discover the giant forest hog for Linnaean science. And the osprey behavior he relates is rather like royal tern behavior. So please keep an eye on your local osprey nest. If you see them acting pedagogical, get witnesses. Try to deploy a video camera. And leave the giant forest hog out of it.

being taught repeatedly, with admirable diction, which was not, apparently, his usual habit. "The intentionality of the behavior, however, cannot be proved," cautions Pepperberg, and she is right, for although it seems probable to me that Alex was speaking slowly and clearly to help the younger birds understand, it might also be that he was doing it to drive Pepperberg crazy.

Training

People have used the learning abilities of animals to teach them or train them to do all sorts of things. As many early attempts to teach animals elements of language showed, we don't always go about it in the best way. Drilling and artificial reward structures often don't work. Animal students hate rote teaching just as human students do, and if Koko, the famous signing gorilla, is made to practice a sign after she has lost interest, "she becomes restless, moody, and intransigent. Often she'll walk away." She also gives deliberately silly replies. When Koko was joined by the gorilla Michael, she tried to teach him signs by molding his hands into the proper shapes, as hers had been molded, but Michael wouldn't allow it.

Dogs

Dogs are taught to do all kinds of things. They are trained to sniff things out, including drugs, termites, people who are lost or hiding, snakes, bodies, explosives, missing pets, truffles, land mines, and contraband shellfish. One dog was inadvertently trained to sniff out plastic bags like the ones the police department stored drug samples in. Dogs are trained to herd sheep, cattle, and geese.* In one fish hatchery, they are trained to jump in the water and herd trout into a net. They menace malefactors, rescue drowning people, and pull sleds. As service dogs they guide blind people, alert deaf people to sounds such as telephones and smoke alarms, and let epileptic people know when a seizure is coming, as well as fetching things, hitting switches, and car-

* When you herd sheep and cattle you want to hang on to every last one. When you herd geese you want every last one gone from the golf course or the airport landing field.

rying things. All these training regimens involve refining and channeling natural dog behavior patterns so that their innate tendencies are modified by learning.

Learn to herd

Dog that herd animals are using instincts that were originally selected for because they're useful in hunting. This is why sheep often get the horrors when they're herded: the behavior of the sheepdogs makes them think a murderer is stalking them. But instead of cutting off the path of a fleeing animal and ripping its throat out, the herding dog cuts off the path of the fleeing animal and sends it racing back to the herd.

I once watched a pet border collie, who had never seen a sheep, taken to a sheepdog training arena. The dog, whom I'll call Fido, glanced into the corral and saw some sheep. Then she looked at her owner and wagged her tail. The sheep shifted nervously and Fido licked her owner's hand. She picked up a stick and brought it to her owner. "Look at the sheep, Fido!" said her owner, and Fido gazed fondly at her owner's face. She lay down and chewed on something dubious.

The trainer went into the pen and the sheep fled. Fido showed no interest. The trainer gave her own highly trained collie a few commands, and the dog sent the sheep hurtling around the pen, selected one, and penned it in a corner. Fido was electrified. She stared. She shook with excitement. She suddenly knew what she was born to do.

After a while they took Fido into the pen. All she knew was that she wanted to chase sheep, but it was clear that she had the instincts to perform more sophisticated maneuvers. She knew, for example, how to give the sheep the eye, a dreadful where-shall-I-bite-them-first stare that chills sheep to the marrow. Her owner had merely been idly curious about whether her border collie had the right stuff. Now she had a prodigy on her hands.

Heard what? Heard whom?

Under the auspices, in part, of the Office of Naval Research, researchers trained a bottlenose dolphin to attack and herd sharks on command. (You can see what they were thinking: "Sharks! Flipper, help me!") First the dolphin was trained to head-butt small dead sharks around the gills. Then big dead sharks. Then they opened a flume between the dolphin tank and a shark tank and let in a small live sandbar shark. The dolphin hesitated but then, on command, butted the shark, chased the shark, and herded the shark back into the flume.

Good! Then they let a small bull shark into the tank and the dolphin got upset and started barking and chirping and echolocating like a fiend and refusing to respond to commands. Bull sharks have been known to eat dolphins. They had to take the shark out and go back to training with dead sharks and then a living sandbar shark before they had the dolphin responding properly again. Then they let in another bull shark, and the dolphin got worked up again, following the shark, carrying on like crazy, but refusing to attack.

They tried the dolphin with nurse sharks and lemon sharks. In each case the dolphin pursued the shark and made "soft contact" with it. This sounds rather like a dolphin that has learned that if you agree to attack a little dead shark, they will ask you to attack a big dead shark. And if you agree to attack dead sharks, they will ask you to attack living sharks. And if you agree to attack timid living sharks, they will ask you to attack scary living sharks. You have to draw the line somewhere or you'll find yourself wrestling gators on a daily basis.

You'll thank me one day.
No, I wasn't talking to you, Mickey

Of orphaned owls raised by wildlife rehabilitators, Kay McKeever writes, "The gravest consequence of separation from natural parents is the loss of the whole period of gradually acquired experience in what to look for, where to look for it, and how to catch it *and* kill it." The job of the parents is to keep the young owl fed while it is learning, "usually pretty clumsily," to do these things. McKeever, having raised young owls on a diet of dead white mice, begins by putting the mice in different parts of the flight cage so the owl has to search for them. Then she gives the owls live white mice. "Astonishment, instinctive excitement, and fear will probably be the reactions from the owls." The mice, who don't know about owls, scamper freely, and it can take so long for an owl to make its first attempt on a mouse that McKeever says rehabilitators may need to put out food and water—for the mice. But as soon as the owl makes its first lunge, the mice, quick learners, go to cover. As the owls get hungrier, and think about how delicious mice are, and glimpse mice dashing for shelter, they listen to the mice rustling in the leaves on the bottom of the cage, and begin to associate the sound with the visual images they already have of mice, McKeever writes.

"Once the full implication is familiar—usually pretty fast—the owl does not wait for sight of prey but begins plunging into the leaf litter at every slightest sound." Now that the owls are learning to listen as well as look for mice, the rehabilitator can switch from white to dark mice and put them in the cage at night. The sign of success is "the appearance of decapitated dark mice stored in nooks and crevices around the cage," McKeever says, and getting to this point can take four weeks. Exactly how well this procedure prepares owls for hunting in the wild isn't known, but bands have been recovered years later from owls who were rehabbed this way.

Training as communication

For bioacoustical research, Diana Reiss was training a dolphin called Circe, using conditioning techniques. Circe was rewarded with bits of fish, which Reiss made by cutting fish into three sections, a head, a middle, and a tail. Circe kept spitting out the tail sections, but Reiss noticed that if she cut the fins off the tail section, Circe ate them.

The negative reinforcement, or punishment, was a time-out. Since Circe liked training sessions and interacting with people, any time the dolphin misbehaved, Reiss would step back from the pool and wait for a minute, just looking at Circe. This was very effective.

One day Reiss tossed Circe a piece of fish. It was a tail, and Reiss had forgotten to cut the fins off. Circe spat it out, swam to the other end of the pool, and looked at Reiss. "After a few seconds, I got the strong feeling that she was giving *me* a time-out." Several days later, Reiss experimentally slipped another untrimmed tail section into the routine, and Circe promptly gave her a time-out. Not only had Circe learned the tasks Reiss was training her to do, and the signals that Reiss used, she had made one of the signals her own. The next step really ought to have been to rig up a device with which Circe could reward Reiss with chocolate or grapes so that we could find out what Circe would have trained Reiss to do. Surely with the right behavioral tools, Circe could have trained her tasks beyond "Cut the fins off, dummy."

Not that old trick again!

Clarence, a house sparrow hand-reared by Clare Kipps, had a repertoire of tricks he did on request. During the London blitz, when he was young, Clarence was in demand as a performer at children's wards, rest centers, posts, and private homes "where there were nervous people." With Kipps as his foil, he played tug-of-war, selected cards from a deck, and twirled a card in his beak.* As his smash finale, Kipps cupped her hands, and when she called "Siren's gone!" —the signal to take refuge in a bomb shelter— he hid in her hands. After a moment he peeked out, "as if enquiring if the All-Clear had yet sounded," which always brought down the house. When the war ended, Clarence gave up show business. Years later, when Clarence reached the venerable age of 12, he had a stroke, which Kipps nursed him through. Realizing that she had no record of his stage career, she arranged for two photographers to document his act. Although he had not practiced his routine in six years, he remembered it perfectly.

Trained mothers

As in the case of Dolly, great apes raised by humans may not show mothering skills. In 1991 at the National Zoo, the first infant gorilla in many years was born to Mandara. Zoo staff were worried about how Mandara, who was only seven, would react to the baby, but she handled him perfectly. Almost a year later Holoko had a baby, her fourth. At the zoo where she had lived before, she had had a stillbirth and two other babies, which she ignored and which were raised by humans. Staffers hoped she would do better this time. But the first time they saw Holoko's baby, it was in Mandara's arms along with 11-month-old Kejana. Holoko was not interested.

Mandara cared for both infants perfectly, but after a few weeks she started weaning Kejana. Kejana perhaps did not agree that he was getting the care he was entitled to. After a few months he decided that Holoko should fill the gap and pursued her relentlessly until she began acting as his mother. Unlike a newborn, he could follow Holoko and climb into her arms, and she must have seen that it was useless to resist.

* Nervous people were more easily amused in those days.

Parenting classes

Zookeepers at Apeldoorn, the Netherlands, were concerned when one of their young gorillas, Mouila, became pregnant. (As a birth control measure, they had separated the females from the dominant male, Balu, but apparently the younger male, Bongo, was more capable than they realized.) "As we considered that Mouila had come into captivity too young to have learnt the techniques of infant care . . . we felt some teaching was necessary." They scheduled viewings of humans cuddling a young black spider monkey. Every few days, Mouila spent 20 minutes with a zookeeper and the baby monkey. She was allowed to touch and sniff the baby. At the first session Mouila looked interested but stayed several meters away. At the next session she drew quite close until the baby "gave a little scream," and she hurried away. In subsequent sessions, she showed more interest, touching the baby gently. She no longer rushed off if the baby made noise. By the tenth session, she was hinting that she wanted to hold the baby, but that wasn't allowed, in case she refused to give it back. The zookeepers felt that Mouila was becoming "familiar with a living small creature, a 'sound-producing, moving, brittle thing.'"

Concerned that Mouila might not know how to nurse her infant, the zoo enlisted two nursing mothers to demonstrate (on the other side of a fence). The mothers made sure Mouila saw the sequence of a baby crying and then being soothed by nursing. "Mouila showed satisfying interest." When Mouila's baby was born, she handled it with affection and confidence and suckled it competently, moving the little gorilla into the right position.

I call her Aunt Gigi, but she's not really my aunt

Extreme ignorance resulting in terrible maternal technique is often seen in captive animals of species that do a lot of learning, but is unexpected in the wild. Jane Goodall was surprised to witness the dreadful mothering of Patti, an ape in the Kasekala troop. Patti had transferred into the group, and her own childhood's story isn't known. "We know from studies of wild Japanese monkeys that females who lost their mothers during infancy, but survived, have turned out to be abusive mothers. Their own first-born infants have a much higher percentage of deaths than do infants born to females who did not lose their mothers while they were young. I suspect Patti may have been an orphan," Goodall wrote.

Patti was so slapdash and unfeeling that her first baby was dead within a week. She ignored him when he cried and carried him by one leg so that his head bumped the ground. Tapit was her second, and she did not carry him by one leg, although she often carried him back-to-front. If he was climbing near her and she wanted to move on, instead of gently drawing him toward her like the other mothers, she'd grab an arm or leg and jerk, causing Tapit to clutch the tree and scream. She'd jerk him loose, slap him on her abdomen, in all likelihood backward, and trundle off with Tapit screaming his head off. If he got lost and started to weep and then shriek, Patti would sit and watch, making no move to get him or let him know where she was. Once she left him alone and screaming in the forest. Goodall was following the chimpanzee Melissa and her daughter Gremlin when they heard the screams. Gremlin carried Tapit until they found Patti.

Luckily for Tapit, a childless female, Gigi, took a loving interest in him. She went everywhere with Patti and Tapit until he was four, and he spent half his time, or more, with Gigi. She was also there when Patti had her third child. But by this time Patti knew her stuff and she was a fond, skillful, attentive mother.

Animal children learn about life, and if they survive, they may become animal parents. Aided by instinctive hints, they learn to be better parents. They may even learn to help their children learn.

TEN

What Learning Tells Us About Intelligence

The crows of New Caledonia are inveterate tool users who make hooks and rakes to extract uncooperative insect larvae from cracks and dead wood. Two New Caledonian crows in an Oxford laboratory, Abel and Betty, were being tested to see if they could pick the right tool for the job. The task was to get a little bucket with meat in it out of a tall upright pipe that was fixed in place. The pipe was so tall and the bucket so small that the crows couldn't reach it with their beaks. They were offered two pieces of wire, one straight and one with a hook, to see if they had the sense to use the one with the hook.

Both Abel and Betty apparently understood that the hooked wire was better for the purpose. In fact, the fifth time around, when Betty selected the hooked wire, Abel stole it from her for his own use. Undaunted, Betty took the straight wire and bent a hook in it.

The impressed researchers changed the experiment in response to this unexpected result. They gave the crows only straight wire. Betty tried to get the food with the straight wire and, when that didn't work, she bent it into a more useful shape. She did this either by wedging one end of the wire in the tape the experimenters used to fasten various things in place or, if there was no tape, by holding one end down with her feet, and then bending it with her beak. Then she fished the bucket of meat out of the pipe with the bent wire, and Abel promptly stole meat from her. (Why didn't they test Abel and Betty separately? They don't like to be apart and are "less motivated to participate in experiments" when separated. Oh, Betty.)

When a paper in Science *reported these experiments newspaper stories celebrated Betty's brilliance. That Betty invented a technique for making a hook to get food does seems pretty smart. Although Abel didn't make hooks, he had already devised the effective foraging technique of stealing from Betty. Maybe that's just as smart. If Betty's a crow Einstein, Abel is a crow Machiavelli. Suppose the researchers keep putting crow food in places that are harder and harder to reach. Eventually, I suppose, Betty will invent the pneumatic drill in order to obtain crow snacks embedded in concrete—and Abel will invent municipal noise ordinances, a partisan police force, and a system of fines, and will use them to obtain crow snacks from Betty. Unless Betty invents Mace first.*

ALTHOUGH NO ANIMALS are clever enough to present a threat to us, there are always some people who want to emphasize the us-them divide as strongly as possible, and intelligence is often where they choose to locate the gulf. That learning is not the same thing as intelligence is shown by all the things people and animals can learn to do without understanding. Our immune systems learn to respond to pathogens without being smart.* On the other hand, an animal that knows quite a lot but can learn almost nothing does not impress us as smart.

Fans of IQ tests might like to know that Koko the gorilla is reported to score consistently between 70 and 90 on age-graded IQ tests (the human average is 100). More interesting are examples of what Koko can (and cannot) do. She can use maps. When she lost a toy outside her enclosure, and her human friends made a map of her enclosure, she was able to point out the spot on the map, just outside the enclosure, where the toy had fallen.

Despite our first-place status in the intelligence Olympics, we can't win every single event. At the task of looking at two shapes and figuring out which is the mirror image of a third shape, a task that often appears on intelligence tests, pigeons and college students were equally accurate. But pigeons were *faster*. The researchers suggest that pigeons use some different, automatic process, and that they need it more than we do, because they fly around and look down on things that are oriented arbitrarily, whereas the things we look at are more consistently oriented. In other words, when we do it, it's smart, and when they do it, it's not. I'm not convinced that the world around me is so consis-

---

* Which is why pollen makes people sneeze. Get a clue, immune system, it's only pollen.

tent, but I agree that navigating in three dimensions must be harder than navigating in two. So we shouldn't feel bad about being inferior to pigeons at mental rotation. But we should avoid going on game shows where we would face teams of pigeons at mental rotation tasks, because that really would be embarrassing.

What can we say about the repeated failure of attempts to set up "Man is the only animal who . . ." rules? Common sense tells us that even if some animals make tools, or laugh, or invent forms of communication, we really are a very different animal. Our mistake may be in trying to find some unique quality in ourselves, when it may be that what's different about us is quantity. We're unusually smart, we're really chatty, and we've taken the tool thing to ridiculous extremes.

Dumb brutes

It is easy to find examples of animal stupidities. Fortunately you can love animals without admiring their intellects. Lieutenant Colonel Locke described the inflexibility of tigers trying to drag away dead prey that has gotten stuck. "Once a tiger has made up his mind to do anything, he will often go on trying to do the same thing over and over again, although several simple alternatives may be available to him." A tiger trying to drag a cow over a high fence is one example he cites, as well as several cases of tigers trying to drag an animal whose horns have caught on a raised root. The tiger keeps pulling in the same direction, when all it needs to do to disengage the horns is to pull a short distance in another direction. "Whether this obstinacy is due to lack of intelligence or to stubborn pride I cannot say."

A peninsula?
Is that like an isthmus?

Greylag geese like to nest on little islands in lakes, where they are safe from foxes. Konrad Lorenz writes that they do not understand the difference between an island and a peninsula, however. While it is doubtful that geese ever learn this distinction, they do learn that nesting at a particular site didn't work and nest elsewhere next time.

Superstition

Like people, animals are superstitious. Spinner dolphins in an oceanarium all had "lucky corners," which they associated with getting a fish reward. They'd do a group leap and, expecting to be rewarded for the leap, would scatter to their lucky corners. That's where they'd be when they got fish, so the association was strengthened, when in fact they would have gotten fish anywhere.

My dogs believe that barking drives away the package delivery truck. Over the years their hysteria at its every appearance has increased. I suspect they are so vociferous because they anticipate success. The truck has come before, they barked like crazy, and it went away. Hurray! We saved us! They never do the control experiment: if they don't bark, would the truck still go away, or would the emboldened truck driver break into our house and steal their kibble?

What is smart?

It is possible to think of intelligence as one thing, a "g factor" that can be assigned a single number, or as a collection of attributes. Richard Byrne writes, "Intelligence certainly means more than flexible learning: terms like 'thinking clearly,' 'solving difficult problems,' and 'reasoning well' recur in attempts to define the ability. The scope of intelligence is quite wide, including learning an unrestricted range of information; applying this information in other and perhaps novel situations; profiting from the skills of others; and thinking, reasoning, or planning novel tactics. (Which should remind us not to expect that intelligence is a single 'thing,' but a bag of devices and processes, endowments and aptitudes, that together produce behaviour we see as 'intelligent.')"

Being very good at just one thing doesn't impress much. Many of those who exalt imitation as a crucial, rare, perhaps uniquely human, ability do not count vocal imitation by birds because it is *only* vocal imitation. (But don't forget Okíchoro, the parrot who waves bye-bye.)

"Learning occurs in all animal species and encompasses a variety of levels of adaptation, including such low levels as habituation and associative learning, . . . and is often highly specialized, inflexible, and limited in scope, as well as species-specific," write Sue Taylor Parker and Patricia Poti. "Intelligence, which occurs only in a few long-lived, large-brained species, is, in contrast, quite generalized, flexible, and broad in scope, as well as species-specific." Richard Byrne argues that we should expect to see intelligence in animals that live in uncertain conditions, that "intelligence should most benefit extreme generalists, species adapted to exploit continually changing environments since they must daily cope with novelty in order to survive."

If learning doesn't equal intelligence, because even dopes learn some things, but intelligence includes the ability to learn, the missing ingredient is understanding what you've learned. Byrne writes of understanding situations and responding appropriately. But people and animals often come to understand situations only after they have learned how to respond to them. (This may produce the "prelearning dip" described in chapter 1.) Then the question is whether we can generalize, whether similar situations in the future will be met with understanding, or whether a whole new learning process will have to take place.

Insight

Some researchers have sought intelligence as shown by understanding. Bernd Heinrich, in his work examining raven intelligence, is interested in things they don't learn to do, but instead do instantly, through a flash of insight.

One Maine winter Heinrich put a chunk of suet out to attract birds, and many species came to hammer at the frozen fat with their bills, chipping off flecks of suet to eat. One day Heinrich went out to replenish the food and scared a wild raven away from the suet. Instead of chipping little bits off, the raven had been in the process of chiseling a groove across one corner of the suet chunk. Had Heinrich not interrupted, the bird would have been able to carry off quite a large chunk of suet once it had carved the groove all the way through. "This was a raven Einstein," concluded the stunned Heinrich, photographing the suet. "Intelligence is doing the right thing under a novel situation, precisely as this bird had done."

Scientific journals showed no enthusiasm for the suet story, however, dismissing it as a mere anecdote. Heinrich needed an experimental situation for ravens to show insight. He came across a story in *Ranger Rick* suggesting that kids put out bits of food for chickadees on the ends of strings, so they could watch the chickadees pull up the strings. "I could not believe that chickadees could actually do that. If they did, it seemed to me, they would have to be trained," wrote Heinrich. Ravens, however, might just be able to grasp the pull-on-the-string concept.

Heinrich presented the string problem to some ravens he had raised in aviaries, ravens who had never seen string in their lives. They *had* seen hard salami, and they strongly approved. Heinrich tied pieces of salami to long strings which he tied to high perches. The strings were so long that the salami couldn't be pulled up in one or two moves.

The ravens were interested. They jumped on the perch and pecked the top of the string. They peered at it. They lost interest. The second time Heinrich presented this setup, one raven, Matt, flew up to the perch, reached down, pulled up on the string, placed his foot on the loop he had pulled up, pulled up more of the string, placed his foot on that, and so on, until he had pulled up the salami, which he ate. Eureka!

In the five groups of ravens Heinrich tested, several birds knew how to pull up the salami right away, many eventually figured it out, and a few never did. In further tests of what the ravens learned and what they knew, Heinrich discovered that if he hung a sheep's head from a string—and a sheep's head is a grocery item ravens adore—they would not try to pull it up, apparently knowing that it was too heavy. Four ravens who were used to pulling up salami were presented with a pair of crossed strings, one of which led to salami and one of which led to a rock. To get the salami, a raven had to pull on the string above the rock. If it pulled on the string above the salami, it would get the rock. Matt, noting the path of the strings, pulled on the one that led to the salami. The other three, time after time, pulled on the string that led to the rock. Although they were not being rewarded for pulling that string, they pulled on it because it was above the salami, and so they *knew* it had to be the right string.

Heinrich also tied salami on strings to branches in the woods, but the wild ravens distrusted these and wouldn't go near them. As he later discovered, pulling up food dangling on strings is something that quite a few birds can do, including chickadees, Darwin's finches, and the plain titmouse. Some learn to do it, and some do it spontaneously.

Problem solving

Researchers in laboratories are always setting up bizarre and unnatural problems for animals to solve. In the wild animals sometimes encounter novel problems, but this is rare and we seldom happen to see it. Animals who live with or near people encounter weird problems more often.

Elizabeth Marshall Thomas's dog Sheilah alertly noticed when a parrot in the household dropped a piece of cheese onto the floor of his tall cage. She tried to reach into the cage with her paw to get the cheese, but couldn't. After surveying the situation, she took the edge of the paper lining the cage in her mouth and tugged it out, with the cheese riding on it. Clever, and like few situations wolves encounter.

The writer Philip Wylie once saw a squirrel trying to raid a bird feeder with a conical roof. It was hanging from a branch by a string too thin for the squirrel to climb down. Three times in a row the squirrel ran along the branch and leapt onto the roof of the feeder, and three times in a row the roof tipped and the squirrel slid off and fell to the ground. The squirrel then sat on the branch and stared at the feeder for a long time. Then it leaned forward and bit through the string. The feeder fell, and the squirrel ran down and ate birdseed.

The parts of smart

Biologist Rachel Smolker writes, "The more we have looked and conducted experiments and compared different species, the more we have been forced to rephrase the question 'How smart are they?' to ask 'How are they smart?'"

We can ask what attributes go into making a creature smart or able to learn. Memory is important, since otherwise there's no way to learn from the past. Sonja Yoerg would add that a "differentiated memory—for skills, for events, for facts—with varying expiration dates would keep things tidy." She'd also look for a "solid grip on cause and effect . . . hypothesis testing . . . serial order and pattern recognition, categorization, and concept formation. Some numerical skills might be a bonus ('Weren't there *three* pups here a minute ago?'), and being able to read the emotions of others might lead to a happier and longer life." Curiosity—the interest in gathering information—is another trait linked to learning, although not necessarily to intelligence.

Domains, modules, and systems

Some abilities are domain-specific. They only work in certain areas. You might have a great memory for spelling and a terrible memory for faces. As Yoerg writes in *Clever as a Fox*, "Birds that can remember their father's song for months can't remember the color of a triangle for twenty seconds. . . . Zebras are aces at discriminating one stripe width over another, but not better than horses at other visual discrimination tasks."

Birds who cache seeds for the winter have fantastic memories for hiding places. One pinyon jay might store 26,000 seeds in a season, and there wouldn't be much point if the bird could never find them again. Russell Balda and Alan Kamil looked at whether this great spatial memory capacity spills over into other areas. Sure enough, in two- and three-dimensional maze problems calling on spatial memory, seed-storing pinyon jays did better than scrub jays, who don't store food. On the other hand, they're not awfully good at remembering colors.

In the field of human as well as animal cognition, many theorists like the idea of modular abilities. "It is also essential to adopt a modular view of cognition as opposed to assuming some single entity such as learning ability or intelligence that all species possess to some degree," Sara Shettleworth writes. "Similarly, behavioral neuroscientists refer to memory systems, distinct areas of the brain that do distinct tasks or store distinct kinds of memories."

The innate, modular aspect of vervet monkey alarm calls produces a paradox. Vervets are born knowing how to make snake and leopard alarm calls, and with a partial idea of what these refer to. Through social learning, they refine their definition of the menace and learn how to react to such calls. But vervets who know that snakes are bad news do not notice much else about snakes. They will step casually along python tracks, real or faked, and scream in terror and shock if they then encounter a python. Cheney and Seyfarth call them "excellent primatologists but poor naturalists."

Experiments showed that it was far easier for hummingbirds to learn a task that rewarded a "win-shift" strategy than one that rewarded a "win-stay" strategy. Many laboratory tests are win-stay: an animal is supposed to find out which choice is rewarded and stay with that choice. Hummingbirds who had to pick between two fake flowers, only one of which contained syrup, took hundreds of trials to learn this. But if it was a win-shift test, in which they were supposed to go to the flower that was *not* rewarded last time, they learned quickly. In the wild, after all, flowers that hummingbirds drink from are depleted of nectar, and the hummingbird would be wasting its time to go back before the nectar is replenished. Win-shift is what works, and it's difficult and unnatural (though not impossible) to learn win-stay.

Sometimes it seems clear that performance in one domain improves ability in another. In recent years biologists have marveled at the dynamic hunting strategies of *Portia*, a spider-eating spider. *Portia* visits the webs of other spiders and experiments with different patterns of web twanging, attempting to pluck a rhythm that will delude the unsuspecting web owner to come out looking for a small helpless insect dinner or a spider of its own species seeking passion. *Portia* has a huge repertoire of these rhythms and varies them depending on the feedback it gets from its intended victim. This trial-and-error process is so scientific that scientists are filled with admiration. Sometimes *Portia* sneaks up on its victim by a circuitous route, and the part that impresses biologists is that *Portia* remembers where its victim is even when it's out of sight.

To see if *Portia*'s ability to employ trial and error applies only to hunting, researchers marooned spiders on tiny islands surrounded by an "atoll"—a ring of dry land, all within a larger tray of water. From the island, *Portia* could see both the atoll and the edge of the tray. (*Portia* has great eyesight, the better to hunt spiders.) There were two ways *Portia* could make a break for freedom: swimming to the atoll (and then the edge) or jumping to the atoll. The researchers randomly assigned each spider to either the swimming or the jumping group. If a spider tried the method they had assigned it, they allowed it to reach the atoll. If it tried the other method, a terrible storm came up and washed it back to the island (this storm took the shape of a researcher making waves with a little plastic scoop). Spiders who were washed back to the island switched methods. Spiders who made it to the atoll didn't switch, but used the same method to head for the edge of the tray. "In *Portia*, perhaps a predatory strategy that routinely demands fine control over the behaviour of dangerous prey has set the stage for the evolution of problem-solving abilities that, as a spin-off, can be readily applied to novel situations, including confinement problems," the proud researchers write. Although lauding *Portia*, biologists Eytan Avital and Eva Jablonka caution, "These are probably a small minority group among spiders." Maybe. But I'm keeping my eye on the whole bunch.

Object permanence

Marc Hauser uses the metaphor of an inborn mental toolkit, part of which is "a basic set of principles for recognizing objects and predicting their behavior." Many of these principles, such as object permanence and size constancy, were first identified by Jean Piaget, a pioneering developmental psychologist who described the intellectual stages human babies pass through.

Object permanence is the notion that things continue to exist even if you can't see them. If you take attractive food and show it to an animal and then, while it watches, put it under a cup, does the animal still think it's there under the cup, or has it ceased to exist?

In Irene Pepperberg's laboratory, they looked at object permanence with African grey parrots. By the time Griffin was 33 weeks old, his food preferences were clear. He was not wild about pellets of bird chow, but he loved cashews.* Under Griffin's watchful gray eyes, Pepperberg took a cashew, put it under a box, moved the box behind a screen, and then moved the box out again. Allowed to investigate, Griffin looked under the box, found no cashew, and accordingly checked behind the screen, where he found a cashew, which he ate. Pepperberg repeated this test, but this time secretly substituted a pellet of bird chow for the cashew. When Griffin looked behind the screen and found a pellet instead of a cashew, he stared at the pellet. He overturned every box in sight. He ran to Pepperberg in distress. The object wasn't permanent!

They tried this on Alex, who was older and more familiar with tests. "After upending the box and finding a pellet, Alex turned from the apparatus to the experimenters, narrowing his eyes to slits, a behavior we have come to interpret as 'anger.'" They gave him a trial in which the cashew stayed a cashew and then once again substituted bird chow. This time, in addition to giving Pepperberg the nasty slit-eyed

* I am the same way.

look, he banged on the table with his beak. I suggest that Alex has learned that if things are weird, it's Irene's fault.

I can see it, but I can't get my hands on it

Psychologist Adele Diamond worked with human babies to find out at what age they could inhibit one action in favor of another. She presented them with a clear plastic box containing an enchanting toy. The box was only open on one side. If the open side faced the babies, they could reach in and grab the toy, but if the open side was on the far side, they had to reach around to get it. Babies under nine months old could not figure this out and kept reaching straight ahead. "Like insects and birds banging into windows, they appeared to have in mind a simple yet rigidly fixed rule for reaching: if a desired object is in front of you, reach straight ahead," writes Marc Hauser.

Baby rhesus monkeys start out baffled, too, and can't comprehend reaching around until they're over four months old.* In Hauser's lab, they tried the clear box test on adult cotton-topped tamarins, who failed miserably, to the researchers' surprise. They tried again, with an opaque box. The tamarins knew that the food was there, even though they couldn't see it. But because they couldn't see it, the powerful impulse to reach straight ahead didn't kick in, and they could reach around to the side and get the food. In a clever follow-up, they gave the tamarins who had solved the opaque-box puzzle the same task with a clear box, and this time they passed. "Having learned about alternative reaching responses with the opaque box, the tamarins applied their knowledge to the transparent box. Moreover, they inhibited a seemingly potent tendency to reach straight ahead for food lying directly in front of them."

Psychologist Margaret Redshaw compared "4 hand-reared, lowland gorilla infants . . . resident in Jersey Zoo, and 2 mother-reared, male human infants, resident in London" for the first 18 months of their lives, looking at their achievement of the various Piagetian stages of the senso-

---

*Research with babies is good news for animals, because it is humane and noninvasive. People studying babies know they won't be giving them electric shocks or doing gratuitous brain surgery or having them raised by prairie voles, and they devise subtle ways of figuring out what's going on, which people working with animals may adopt later.

rimotor period. They passed through stages in the same order, but in general the gorilla babies started sooner and finished sooner. During the first year the little gorillas were 4 to 16 weeks ahead of the little humans. For example, being able to get a toy that is too far to reach by pulling on a cloth the toy is resting on is something that the gorillas could do at 18 to 30 weeks and people couldn't do until 34 weeks.

Before you go trading your kid for a gorilla, the gorillas later showed some ghastly deficits, such as (1) not making towers of two blocks and (2) not learning English. Also, "only one gorilla placed a wheeled toy on an incline, while the other ran away with the experimental apparatus." After the first year their behavior becomes more divergent. Redshaw writes, "The human infant learns to talk, play symbolically and to walk bipedally, while the gorilla knuckle-walks, chest-beats, wrestles, climbs and builds nests." The one thing on which gorilla and human babies were perfectly synchronized was the systematic dropping of objects, which enchanted both species from 34 weeks on.

Numbers and nature

In addition to checking animals for vocabulary and grammar abilities, it's also possible to look at their math apitude. Researchers have looked at whether apes, in particular, can display numerousness, ordination, counting, and subitization. Numerousness is the ability to judge quantities—a little, a lot, more. Ordination is putting things in a linear sequence. Some people think counting is the most sophisticated of these abilities, and others vote for subitization, the ability to rapidly label small quantities of things, usually fewer than six.

Sarah, Sheba, and the gumdrops

Sarah Boysen, looking at the numerical skills of chimpanzees, set up a task situation for two apes, Sarah and Sheba. Previous experiments had left Boysen marveling at their grasp of number concepts. She was now looking at their ability to make quantitative judgments. In such tasks animals usually are asked to point to the larger of two groups of items (here, gumdrops), which they then get as a reward. Instead, Boysen set out the task of pointing to the one of two groups of items which would be given to the other chimpanzee. Since Sarah and Sheba liked gumdrops and weren't particularly generous, they should have pointed to the dish with less candy. Then the other chimp would get the smaller amount and they would get the larger amount.

Neither chimpanzee could do it. Time after time the chimpanzee would point to the dish that held more gumdrops, and then, to their obvious frustration and distress, watch as it was given to the other chimpanzee. They would cry out in dismay and strike the apparatus. They simply could not learn to point to the smaller amount of candy. Like the tamarins, or babies under nine months, they couldn't inhibit the impulse to reach toward what they wanted.

Sheba had been taught Arabic numerals up to 6, and Boysen now tested to see if Sheba could indicate the smaller of two numerals, printed on cards, to decide how many gumdrops would be given to Sarah. Here Sheba had no difficulty. If the numbers were 2 and 6, she would point to the 2 and watch as two gumdrops were given to Sarah and then six gumdrops were given to her. But when they went back to pointing at actual candy, Sheba's performance fell apart. She understood the problem. She could do it with abstract numbers. But when she saw gumdrops, passion trumped reason.

Robert Shumaker repeated this experiment with Azy and Indah, orangutans at the National Zoo. Since they had not been taught Arabic numerals, and since zoos disapprove of gumdrops, he used grapes. At first they didn't know what Shumaker wanted. Azy just pointed to whichever bowl of grapes was on the right. This strategy resulted in Azy getting the larger amount of grapes only about 40 percent of the time. Then his performance suddenly jumped, and he was getting the larger amount almost 100 percent of the time. The jump happened overnight, Shumaker notes. "The only way I can explain it is that he was thinking about it overnight in his nest." Indah's performance went to 95 percent. The orangutans could do what the chimpanzees could not.

Shumaker does not argue that orangutans are smarter than chimpanzees, but that they are less impulsive. Chimps, who are intensely social, may have been selected for impulsive food grabbing. If they don't instantly snatch the biggest portion, another chimp will. Orangutans, who are far more solitary, can sit and ponder what to eat without worrying about another ape grabbing it. Thus social difference can lead to cognitive differences—or to apparent cognitive differences.

Another way to think of social-cognitive differences is in terms of personality. Primatologist Junichiro Itani was a pioneer in the study of animal personality (a subject only beginning to be considered respectable in English-language animal behavior research). Itani proposed that there is more variation in personality in primate groups as they become more complex. As P. J. Asquith summarizes, "In prosimians, the individual variability is very small. In Japanese macaques it becomes wider, in *Pan* [chimpanzee species] even wider, and in man the individuality is even larger." Curiously, recent studies show wide personality variation in octopuses.

Attempts to persuade a raccoon to reveal whether it could tell an array of three objects from arrays of one, two, four, or five objects were impeded by raccoon personality. Although at the beginning the raccoon was delighted to open a Plexiglas box and extract three grapes, Rocky soon became bored with grapes and would only play for raisins. Then he held out for chocolate-covered raisins, and then he demanded little metal bells. When he correctly opened the box with three bells in it, he was allowed to play with them, which involved batting and chasing them, and washing them in water. Rocky eventually paid enough attention often enough for the experimenter to conclude that Rocky was reliably selecting the number 3, whether through counting or subitizing. In the report the experimenter wonders why more people don't do research with raccoons, despite their reputation of being problem solvers "perhaps to a fault," and concludes that it has to do with lack of docility rather than lack of brainpower. "Rocky's attention span was less than ideal, and he became extremely difficult to manage during the early spring, when more salient motivational states than 'concern with number' were aroused."

Cognitive maps

Another area in cognitive research is the mental map. While the ability to form a cognitive map of the world around you may be innate, the actual details must be learned. Constellations, magnetic fields, the scent of water currents, the position of the sun, and other things some animals are predisposed to notice must be aligned with individual landmarks if the animal is to be able to find its way around when it grows up.

Frillfin gobies often live in tide pools. If, at low tide, you try to net, grab, or otherwise irk a frillfin goby that appears to be trapped in its small pool, the goby will simply leap out of the pool into the next pool and, as you follow, will by a series of up to six jumps bound, by a zigzag path if necessary, to open water. It seems that a goby learns the topography of the area at high tide, and maps it mentally, so at low tide it can calculate where to leap so that it lands in the next pool and not on the rocks or in your jaws.

Classic experiments with indigo buntings who grew up in a planetarium showed that while in the nest, the baby buntings learned to know the stars, and used them to orient themselves later when they migrated. They didn't need to see the whole sky, just a patch.* If biologist Steve Emlen showed them the wrong constellations in their youth, they would try to migrate in the wrong direction as adults. They learn this during a sensitive period. "Importantly . . . the way a bunting acquires song and the way it acquires spatial knowledge are different, involving different brain structures and developmental timetables," writes Marc Hauser. "If the organism depended on a domain-general learning system, one that was blind to the kind of experience or knowledge to be acquired, it would often learn the wrong things, or learn the right things too slowly, and thus decrease its chances of surviving."

The inquiring mind

While it is possible to be curious and stupid, curiosity is linked to knowledge, in that curiosity drives learning. An incurious animal won't learn much. Stephen Glickman and Richard Sroges did a classic study of curiosity by presenting objects to zoo animals and analyzing their reactions. The study was done in the 1960s, when enrichment was not a popular concept among zookeepers, and the introduction of blocks of wood, dowels, pieces of rubber tubing or chain, or a crumpled ball of paper was an earthshaking event for most zoo denizens. Some were too frightened to go near the objects. Some attacked the objects. A few found them erotic.

Lemurs sniffed the objects, rodents chewed on them, colobus monkeys surveyed them from a distance, baboons handled them extensively,

---

* Many of us grow up able to see the whole sky, yet neglect to learn to orient by the stars.

and the big cats caught and subdued them like prey. The baboons and the cats were particularly averse to having the objects removed, and in the case of the baboons, a fire hose had to be used. A wily chacma baboon was up to this challenge, and as soon as the hose was directed into the cage would rush to the front and divert the water with "a well-placed forepaw."

Reptiles did not care about the objects, except for a rather manic Orinoco crocodile who attacked them and a water monitor who administered a prolonged death grip to a block of wood. There was individual variation, so that one hedgehog ignored the objects and the other went wild for the blocks and tubing, chewing and toting them all over the place. Young animals displayed more curiosity than adults. In general, primates and predators showed the most curiosity.

Curious pigs

British scientists raised two groups of pigs in "modern intensive farming" conditions, with one difference. One group, the "substrate-impoverished" pigs, were kept on plain concrete floors. The "substrate-enriched" pigs had pens carpeted with straw and bark, each pen furnished with two tree branches.* The pigs were scored on their reactions to novel objects, which included a plastic toy tractor, a bicycle tire, a watering can, a bucket, a rubber boot, a three-legged stool, and a steering wheel. "These objects are of considerable interest to pigs," the authors assure us.

The impoverished pigs were apt to fear the novel objects. When interest conquered fear, they just chewed on the object relentlessly. In comparison, the enriched pigs spent more time scampering about, rooting the object with their noses, and if they were in a test pen rather than their home pen, examining its panels, bars, drain, and flooring. In other words, the plain flooring provided to the impoverished pigs had helped create unimaginative pigs.

Curiosity can benefit even a captive pig. At a commercial pig farm, sows wore collars with transponders. When the pig walked into the sow-feeder, the transponder told the feeder whether that pig had been fed that day. If not, the feeder dispensed a meal. One sow found a col-

---

* Wealth indeed.

lar that had fallen off another pig, and in the process of playing with it, discovered that if she carried it into the sow-feeder, she got fed again. Every day she'd use it as a meal ticket.

Curiosity carries risk. At a Kansas farm, 200 restless pigs lived in a field where nothing ever happened. They were thrilled one day when a teenager carrying an aluminum ladder entered their pen, and they were rapt at the spectacle of the teenager going to the concrete housing that held the irrigation pump, taking the heavy trapdoor off, lowering the ladder into the 20-foot-deep shaft, and descending into the shaft. The teenager was greasing the bearings of the pump when he was struck by a falling pig. Looking up with his lap full of pig, he beheld a ring of interested faces studying him. By the way the snouts jostled, he could tell that the pigs viewing him were being pushed by pigs behind them who also wanted a look, and that if he wanted to avoid being crushed by a pig avalanche, he needed to get out of the shaft. Clasping the pig in his arms, he slowly ascended the ladder, shouting to keep his fascinated porcine audience back.*

Memory

In 1964, the late psychologist Leslie Squier constructed a giant, rugged operant-conditioning test apparatus and tested some elephants at the zoo in Portland, Oregon, to see if their performance on such tests was like that of other animals. It was a simple enough task, and the elephants did well, pressing lit disks with their trunks, causing delightful sugar cubes to tumble into a steel hopper. The research was cut short after the elephants mastered the task. Eight years later, Squier and colleagues, wondering if it's true that elephants never forget, located the equipment on a scrap heap, buffed it up, and set out to repeat the tests on three elephants that had been tested in the 1960s: Rosy, Belle, and Tuy Hoa.

Tuy Hoa, a 20-year-old elephant from Saigon, did so well that it was clear she remembered all about it. (Hey! Sugar cubes are back in

---

*When the teenager later attended college he transformed this event into a philosophical fable warning that at all times we should consider ourselves surrounded by a horizon of snouts, of which it is essential that we be aware, and at which we must shout. His philosophy acquired a small following. The intellectual progress of the pig is unknown.

style!) It took her only six minutes to rattle off 20 correct choices in a row and scarf up 20 sugar cubes. Rosy and Belle did so badly that it seemed they must have terrible memories, but then it was discovered that they had been quietly going blind in the intervening eight years and had a hard time seeing the disks light up. Veterinary help was called in.

Shorter-term memory is also valuable. Early in the morning, wild hamadryas baboons in Ethiopia were walking along a dry riverbed in Stink Wadi. The other baboons took their usual route under the railway bridge, but the elderly female Narba went directly up on to the bridge and drank water from a shallow dent in an iron plate. She seemed certain that the water was there, even though water evaporates within a few hours in the desert. Apparently she remembered that it had rained the night before in this area. Narba hadn't been rained on, since it didn't rain at the sleeping cliffs where she was at the time, but she would have been able to see that it was raining over by the wadi.

You think you're so smart

Just what makes us so brilliant is a perennial topic of inquiry in the sciences. Often this leads to lengthy boasting about our giant brains. It's true that getting seconds on brains while the other animals were waiting for firsts on fur or feathers has been a good deal for us, but it's also true that our understanding of how the human brain differs from the brains of other animals undergoes constant revision, and the end of the revisions is nowhere in sight. Before you sneer at the puny cortex of birds, for example, you should keep in mind that some of the things we do with our cortex they do with their striatum. Dolphins have a big cortex, but they arrange it differently, in a thin layer, and they arrange the cortical neurons differently.

As for inquiries about how we *got* so smart, sometimes they refer to humans only, sometimes they are about the great apes, and sometimes they are about the entire order of primates. The previously mentioned (in chapter 2) arboreal clambering theory is about the great apes, for example.

Peel me a grape

Some scientists argue that most primates are omnivorous extractive foragers, and that takes brains. Extractive foraging means that you don't just wander around eating plants. You have to peel them or get past the thorns, the stinging hairs, the hard woody covering, or whatever means the plant has evolved to protect its tender nutritious parts.

Katharine Milton compared the foraging of spider monkeys and howler monkeys living in the same area. Howlers eat leaves, which are abundant, and spider monkeys eat fruit, which are harder to find. Perhaps spider monkey life involves more cognitive mapping, more analysis of fruit-ripening patterns, better color vision, more processing skills, and that explains why their brains are twice as big as those of howlers. (Others argue that eating leaves is not so simple, what with varying nutrients and toxins in leaves of the same species or even on the same tree.)

Machiavellian intelligence

As put forward by a group of scientists in the book *Machiavellian Intelligence*, edited by Richard Byrne and Andrew Whiten, the big selective advantage of intelligence might be the ability to navigate social situations. So the original plus to being smart might not have been about negotiating the physical environment or surviving its dangers, but about surviving the social environment of the group. Some suggest that primates (or some smart subset of the primates) may have a peculiarly social intelligence, geared toward understanding and manipulating relations with others. Other take the view that intelligence of a generalized kind is applied to social situations.

Nicholas Humphrey suggests that the technological challenges faced by Robinson Crusoe on the desert island are less daunting than the social difficulties presented by Man Friday. One of the demanding things about social life is that it fights back. While you figure out how to move from tree to tree, the trees just sit there, and while you figure out how to crack a nut, the nut waits patiently. But while you are figuring out how to deal with another monkey, it is figuring out how to deal with you, or a third monkey steps in to make the deal.

We are impressed with the technological sophistication of our lives, and we sometimes neglect to notice its interpersonal complexity. One study of situations that worry five-year-old humans found that 88 percent of their problems were social problems. Even some problems that seemed to be about objects were actually social problems, such as when another child was teasing them by withholding an object.

A corollary of the social intelligence idea is the grooming theory of language. Primatologist Robin Dunbar suggests that language evolved not to enable group hunting or tool use, but to substitute for and amplify the social bonding power of grooming. This is why we spend so much time—more than we think—gossiping. "Could it be that language evolved as a kind of vocal grooming to allow us to bond larger groups than was possible using the conventional primate mechanism of physical grooming?"* Language has the advantage that, while an ape can only groom one other ape at a time, you can talk to several people at once.

Byrne and Whiten hesitated before using the word "Machiavellian" to describe social intelligence. But the more they looked at the life of social primates, the more apt it seemed. "Co-operation is a notable feature of primate society, but its usual function is to out-compete other rivals for personal gain."

Often the gain is itself social. The bonobo Matata, Kanzi's mother, felt the need to reestablish her status in the colony after she had been away. In one incident, she jerked an infant's leg while its mother was elsewhere but another female was nearby. The infant screamed and the enraged mother burst on the scene to see Matata threatening the other female. The mother thrashed the other female, to whom Matata was now dominant again.

Matata was also known to grab things away from a trainer, scream, and use gestures to try to enlist researcher Sue Savage-Rumbaugh against the trainer, as if she had been attacked. No wonder Matata was not interested in learning to use lexigrams—they were not presented as a way to say the things she really wanted to express, such as, "If you're my friend, go beat her up."

When the young Kanzi was playing too roughly, his human friends would sometimes distract him by pretending they heard something in the woods and uttering alarm calls. As they made a show of

* Don't underestimate physical grooming, though. I want no ticks on me.

scanning the woods intently, Kanzi would stop misbehaving, come over, put his arm around someone, and join them in staring into the trees. When he was six, Kanzi began using the same fakery in awkward social situations. When there were a lot of people around him at once, Kanzi sometimes felt intimidated, and would look into the woods with a worried expression and make alarm calls in a strained tone. A hint that Kanzi was acting lay in the fact that his hair didn't stand on end, as it did when he was truly alarmed.

Wild olive baboons observed by Barbara Smuts live in bands in which young, strong adult males are dominant to older males. To researchers' surprise, the dominant males did not get to mate as frequently as their rank predicted. The reason was that the older males routinely formed coalitions and harassed the younger males when they were consorting with a female in estrus. Old males never harassed other old males, only young males, and they harassed them until the female ended up with another male.

In many social animals, some animals may take the roles of scroungers. They do not even try to learn some foraging skills, but instead learn how to get food other animals have obtained. Feral pigeons were presented with a task in which they had to peck a wooden stick inserted in a rubber stopper in an opaque upside-down test tube. When they did this, the stopper would come out and birdseed would fall on the ground. This is not obvious to pigeons, but they can learn to do it. If skilled birds were put in a small flock of pigeons, a few of them learned to do it too, and the rest waited until the seed scattered and rushed over to share. These scroungers followed the skilled birds, whom experimenters called producers, but did not try to produce food themselves.

If pigeons were in a cage next to a demonstrator, and there was a slanted tray so that some of the birdseed the producer got rolled into the observer's cage, they seldom learned. In some way, the ability to scrounge inhibited their urge to learn. Were these just stupid pigeons? No: when scroungers got demonstrations during which they could no longer scrounge, but could only watch, they learned to do it.

Kindly Doctor Machiavelli

Manipulating others is not always exploiting them. The young chimpanzee Loulis was obnoxiously and repeatedly attacking two adult females who were peacefully grooming each other. His friend Dar touched him on the arm as he charged them, but Loulis ignored the touch. On his next charge his mother, Washoe, reached out and touched his leg, and Dar signed "tickle" on Loulis's arm. Loulis turned and began to wrestle with Dar as Washoe tickled them both. Washoe and Dar had manipulated Loulis into behaving better.

At the Arnhem Zoo, two chimpanzee mothers watched as their small children played together. Next to the mothers, Jimmie and Tepel, slept Mama, the greatly respected matriarch of the colony. The children's wrestling turned into a screaming fight. Jimmie and Tepel were uneasy. They looked at each other, looked at the fighting kids, looked at each other, shifted restlessly. Apparently neither dared intervene for fear of antagonizing the other. Tepel poked Mama until she woke up, and pointed at the squabbling children. Mama rose to her feet, waved one arm, and gave a loud grunt-bark. The kids stopped fighting, and Mama went back to sleep.

"In the laboratory, when a normal primate has the choice of responding to a social cue or to objects, he turns first to the social cue," writes Alison Jolly. "During early insight tests, primates from lemur . . . to chimpanzee . . . would beg from the experimenter before attempting to solve a new problem (a quite accurate assessment of the real relationships of the situation)." Muni, a young gorilla, turned to humans to solve her problems. Confronted by an object suspended out of her reach, her first reaction was either to lead a person under the object and use them as a ladder or to lead them under the object and request that they get it for her. As for bonobos, it has been pointed out that "Kanzi's most complex symbol communications occur in social interactions involving three or more individuals."

The Case of the Unreliable Narrator

Vervet monkeys learn about other monkeys' reliability. They have three calls that indicate threat from a neighboring vervet group, the grunt, the wrr, and the chutter. If those tireless audio enthusiasts Robert Seyfarth and Dorothy Cheney play tapes over and over of a particular vervet's wrr when there is no other group around, the vervets who hear it eventually ignore that vervet's wrr. If they hear her chutter, they ignore that too, because they consider her unreliable. But if they instead hear the chutter of another vervet, one who hasn't worn them out with false alarms, they respond.

Learning your position in society

Greylag goose couples have ranks in the flock, and their goslings learn this. Goose families present united fronts against other families, with the goslings as well as the parents stretching out their necks and making threat displays. "It is from their participation in disputes involving the family that young geese come to recognize the rank their parents occupy in the goose flock. The youngsters automatically adopt the same rank, and it is amusing to see a half-grown goose cheekily approach a full-grown gander and, for example, drive him away from the food dish," Lorenz writes. If a gosling tries this when its family isn't nearby, the lower-ranking adult may give it a thrashing.

I want her to know her heritage

Japanese macaques inherit their mothers' rank. By six months they know their rank, largely from observing their mothers. One way high-ranking macaque mothers convey this information is by officious interference in children's play. A low-ranking mother will not intervene when her child is playing with another child unless her child utters a distress call. High-ranking mothers often swept down when their children were playing happily with others and "rescued" them from nonexistent peril, threatening those of lower rank. With such displays, "high-born females appeared to control and initiate the matrilineal transmission of rank, rather than act simply as models for the infants," write primatologists Bernard Chapais and Carole Gauthier.

Ringtailed lemurs do not inherit rank. Mothers often sit callously by and watch as young ringtails fight. The young lemurs establish dominance in play-fighting that turns to angry grappling. At the end the winner stays put and the loser runs to its mother for comfort. Learning whom you can beat takes place on a case-by-case basis, and it's not transitive. Just because lemur Alef is dominant to lemur Beth and lemur Beth is dominant to lemur Gimmel doesn't mean that Alef is also dominant to Gimmel. Gimmel can easily be dominant to Alef.

Zap! You're a sweetie

Not a recent experiment: there were four macaques in a monkey colony. Ali, the "boss," was strong, mean, and ill-tempered. Sometimes he expressed his anger by grimacing and biting his own hand, which doesn't sound like a happy monkey, but this was small comfort to Elsa, whom he bullied. He was friendly to the other female, Sarah, and either ignored or was hostile to Lou, the smaller male. The four monkeys were placed in a cage three feet by three feet by seven, with a lever at one end. Surgery was performed on Ali and Sarah to implant electrodes in their brains. The electrodes could be triggered by pushing the lever.

When the lever stimulated electrodes in Sarah's brain, no one cared. But when the lever stimulated Ali's caudate nucleus, he didn't feel like doing much of anything—he didn't eat, drink, walk, groom himself, or attack Elsa or Lou. Elsa figured this out the second day and spent a lot of time pushing the lever—while looking at Ali. Elsa was manipulating Ali, but who can blame her?

Theory of mind

If you're going to have a social life, it's very helpful to have a theory of mind, that is, to recognize that others have wishes, fears, and knowledge of their own. This is another of those hotly debated areas of study. It's not hard to find examples in which animals seem to have no conception whatsoever of the knowledge and feelings of others. The occasions when they do seem to have that insight are more interesting. (Small children may tell amusing lies because their theory of mind is not yet developed.)

Psychologists were astounded a few decades ago when it was found that babies often imitate people sticking out their tongues. Andrew Meltzoff made this scientific discovery (no doubt it had been the personal discovery of many people through the ages) in month-old babies and then arranged to be called to delivery rooms to see if newborns do it. Babies as young as 42 minutes stuck out their tongues in imitation. They also imitated lip protrusion and mouth opening. Somehow, without experience, babies make a connection between their own mouth or tongue, which they have never seen, and the face of the adult making faces at them. "Nature ingeniously gives us a jump start on the

Other Minds problem. We know, quite directly, that we are like other people and they are like us," write Meltzoff, Alison Gopnik, and Patricia Kuhl in *The Scientist in the Crib*. "Imitation is an innate mechanism for learning from adults, a culture instinct," they write. "In fact, recent research suggests that most other animals don't learn through imitation in this way."

Masako Myowa replicated these tests with an infant chimpanzee between 5 and 15 weeks old. Up until 12 weeks, the little ape opened her mouth when humans hovering over her opened their mouths, stuck out her tongue when they stuck out their tongues, and protruded her lips when they protruded theirs. After 12 weeks she gave it up.

Deception

Deception is often considered a sign that the deceiver may have a theory of mind. In an oceanarium in Hawaii, Ola, a baby *Pseudorca* whale, was friends with Keiki, a teenaged bottlenose dolphin. At night, Keiki would leap over a gate between tanks to be with Ola. (Ola was not as skilled a jumper.) The staff didn't like this, because it was a nuisance to separate the two animals for the morning show. They attached a wide, heavy plank over the partition to make it look too scary to jump over. A few days later, when the staff arrived in the morning, the plank was in the water and Keiki was in Ola's tank. Perhaps some softhearted employee had levered it off so the animals could be together. But it happened every night. Finally a trainer hid and watched Ola prop his tail against the floor of the tank and use that leverage to push the plank off the gate with his nose. The striking thing was that Ola knew not to do this when people were watching.

The plank was more firmly attached, but the staff instituted a daily playtime during which Keiki and Ola could be together in the main tank.

In the mirror, a monster! A beast! A vision of beauty!

A famous and controversial way of exploring animal mentality is the mirror test. People learn what a mirror image is, but most of us encountered mirrors so young that we don't remember learning it.

Most animals never figure out that a mirror image is of themselves. Kotar, a captive killer whale, was displeased with the killer whale he saw looking back at him from a big mirror his keepers rigged up. He touched the mirror, looked behind it, looked beneath it, and shook his head to threaten it. Lorenzo, a tame scrub jay, could never pass a mirror without delivering a threat or a peck. Dogs and cats learn that the mirror image is not another dog or cat and conclude that it is something they can ignore, but never show signs of grasping the connection between the image's movements and their own.

Captive chimpanzees presented with mirrors typically begin by treating the image as another chimpanzee, whom they usually try to menace. Then they look behind the mirror and then, often, they settle down for some serious primping. In 1970, Gordon Gallup Jr., wishing to test whether chimpanzees like this were really recognizing themselves, came up with the idea of making a mark on a chimpanzee's head while it was unconscious. The marking paint was odorless, so the only way the chimpanzee could know the mark was there would be if it looked in the mirror and realized that the painted ape it saw was itself.

The chimpanzees were startled at the sight. They ran their fingers over the marks on their heads—their real heads, not their reflected heads—and looked at their fingers to see if the paint had come off on them. They smelled their fingers.

The significance lies in the idea that a being who can recognize its image in this way must have a concept of self. Thus mirror self-recognition is thought to indicate self-awareness. "Without a sense of self, how would you know who you were seeing when confronted with your reflection in the mirror?" write scientists Gordon Gallup Jr., James Anderson, and Daniel Shillito. Perhaps recognizing self leads to inferring mental states to others—theory of mind. It is argued that this ability appears in children at the age they begin to recognize their mirror image.

In general, chimpanzees pass the mark-and-mirror test, bonobos pass, orangutans pass, and at least one gorilla passes. People pass after they're 18 to 24 months old. Outside the primate order, not much mirror skill has been found. Elephants in the National Zoo didn't pass the mark test. Neither did two young grey parrots.

Most researchers found no mirror self-recognition in macaques, but Maria Boccia argued that these investigations were flawed. They used infants, or macaques raised in isolation and hence socially abnormal, or they didn't give them enough mirror time. Although Boccia thought that mirror self-recognition was probably "at the upper limit of macaques' cognitive ability," she also thought that if you tested a lot of normal adults, and particularly if you gave them a chance to see other macaques looking in mirrors, you just might get something. More anecdotally, she had also noticed pigtail macaques flirting by looking at each other in mirrors. She let 15 adult female pigtail macaques and 7 teenaged macaques look in mirrors all day every day. These macaques had grown up around mirrors. Most of them showed behavior hinting at mirror self-recognition, and one female passed with flying colors, swiping at the mark on her face in an attempt to wipe the red dye off and, in a subsequent test, grabbing at the precise spot on her ear that had been marked.

Through the looking-glass

Barash, a young chimpanzee who had been exposed to a mirror outside his cage for six weeks and then hadn't seen it for a year, was given the mark test. He expressed curiosity about the marks, but what really got his interest was his teeth. The teeth that weren't there. Barash was four years old, and some of his baby teeth had fallen out since the last time he saw a mirror, and his adult teeth hadn't come in yet. He stared at the mirror with mouth wide open, stuck his finger in the gap, stuck his tongue in the gap, backed away, looked again, backed away, "vocalizing repeatedly" in apparent distress. My teeth!

The chimpanzees Sherman and Austin, at the Yerkes Regional Primate Center, have been known to use mirrors to apply marks—to put on makeup. Even better than mirrors are video monitors. The apes are often videotaped, and there are live monitors. Sherman and Austin have learned to make test motions to determine when they are seeing themselves live and when they are seeing an old tape. Sherman likes to dress up in fur shawls that make him look large and impressive, and to watch his enormous furry self bob and weave on-screen. Austin is more of a Food Channel viewer. He likes to watch himself eat on the monitors and likes to have the camera moved so that he can look down into his own throat, aided by a flashlight, which he shines into his mouth.

Also at Yerkes, the bonobo Matata, despite the fact that no one admits having said a word to her about her appearance, uses mirrors to pluck out her chin hairs. Her son Kanzi uses his image to assess his progress at blowing bubblegum bubbles and inflating balloons, and her daughter Panbanisha likes to gaze at her own face, particularly her large and fearsome canine teeth.

Anybody else?

When Chantek, the signing orangutan, was six, he was given a pair of sunglasses. He went into the bathroom, put them on, and examined the effect from various angles.* At the age of seven he used the bathroom mirror and his foster mother's eyelash curler in an attempt to curl his own eyelashes, simultaneously showing imitation, mirror self-recognition, and a hankering after chic.

Most gorillas are said to fail at mirror self-recognition. But Koko, the gorilla who was taught to sign by Penny Patterson, knows all about mirrors. A remotely operated video camera captured Koko enthusiastically brushing her teeth in front of a mirror. Afterward Koko inspected her gleaming teeth. But wait! What's this? Koko detected a black spot on one of her back teeth. This was a spot of black pigment, and perfectly normal but, owing to its remote location, Koko had never noticed it before. Leaning into the mirror, she touched it. She tried to brush it off with the brush end of her toothbrush. She poked at it with the handle end. It wouldn't come off. Patterson, knowing nothing of all this, entered the room. Koko rushed over, pried Patterson's mouth open, and carefully inspected her back teeth for a similar spot. Koko apparently has a theory of mind plus a theory of dentistry.

Not much of an ape

Gibbons, classified as "lesser" apes, did not pass mark tests. A group of researchers installed mirrors in the cages of gibbons in three European zoos, and got varying results. Buci, an aged white-handed gibbon in the zoo in Jászberény, Hungary, who had lived alone most of her life, seemed apprehensive about the mirror. She would not meet the gaze of the ape in the mirror and tended to sit sideways to it.

* He looked cool.

Todi, an adolescent wild-born red-cheeked gibbon in the Nyíregyháza Zoo in Hungary, spent a lot of time with the mirror. At first he carried food away from it (in case the gibbon in the mirror tried to take it?), but later he ate in front of it. He performed experiments such as raising one leg in the air, or doing a slow backward hanging somersault, peering in the mirror as he did so. One day he glanced in the mirror and instantly removed a speck of banana from his lip, a speck which had been there for a minute and half before he looked into the mirror.

Dodo, an adult wild-born white-cheeked gibbon in the Budapest Zoo, went wild for the mirror. At first he was startled by the sight of his own canines when he yawned, and he carried food away from it. After he relaxed, he spent a lot of time in experimental actions such as leg lifts, looking back and forth between parts of his body and their mirror reflection, and stepping back and forth in front of the mirror. He compared the mirror reflection with that in a puddle on the floor. He looked over his shoulder at his back.

Each gibbon got the mark test, and the researchers managed without anesthetizing them. An experimenter secretly put whipped cream on Buci's forehead while grooming her. Buci did not touch her face or try to remove the cream when she looked in the mirror. For Todi and Dodo, experimenters gave them a deep plastic cup with whipped cream at the bottom, which also had Day-Glo cosmetic cream on the rim. When the gibbons licked the whipped cream out of the cup, they got Day-Glo cream on their faces. Neither showed any reaction when they looked in the mirror, which is scored as failure. But they also got Day-Glo cream on their hands, and they didn't touch that or try to wipe it off.

The researchers arranged for Fadoro, a one-and-a-half-year-old siamang gibbon living with a family in Zürich, to be surreptitiously marked on the forehead while he was being brushed. Ten minutes later, he was offered a mirror. He glanced into it, paused, wiped his forehead (removing most of the mark), looked at his hand, looked back in the mirror, and went on his way.

Fadoro passed the test, perhaps because of his upbringing in a human family. But whether that says more about his theory of mind or his feeling that you don't want to have visible gunk on your fur is unknown. Dodo, the mirror addict, didn't pass, but the reason may have been his differing attitude toward personal grooming. Similarly

Todi may take the view that gunk on your face is of no concern, but a fleck of banana on your lip means that you may have missed some food—and that's *important*.

Who's the most streamlined of them all?

After all the to-do about how only the very most elite of apes have mirror self-recognition, it's refreshing to hear that bottlenose dolphins have it. Diana Reiss and Lori Marino worked with two dolphins at the New York Aquarium, 13-year-old Presley and 17-year-old Tab. They were housed in a pool with three semireflective glass walls, and later in a pool with an actual mirror on one wall. The dolphins were trained to be handled in a process that allowed Reiss and Marino to make marks on the dolphins. Sometimes they were marked (with a nontoxic temporary black marker), and sometimes they were sham-marked with a marker filled with water. Reiss and Marino drew triangles, circles, and cross-hatches on different parts of the dolphins.

First the researchers used the sham marker on Presley, with no effect. Then they made genuine marks on various parts of his body—over his eye, on his flipper, on his tail—which definitely got his attention. From spending around 8 seconds glancing at his reflection, he went to nearly 10 minutes at a time. If marked under his chin, he would go to the glass wall and stretch his neck upward repeatedly, and if marked behind his left pectoral fin, he would repeatedly present his left side to the glass and peer at his image. After he realized that sometimes the researchers were drawing on him, he'd rush right to his reflection and scan the parts of his body they had touched with the marker. Tab also used reflective surfaces to inspect marks drawn on his body.

Once, after the formal experimental sessions were over, one of the researchers made a mark on Presley's tongue, which they had never done before. Presley rocketed to the mirror and opened and closed his mouth in front of it, something he had never been seen to do.

How to get smarter

Is it possible to make animals smarter or to make animals learn better by providing them with an enriched environment? It's worth noting that most laboratory animals live in devastatingly impoverished, infantilizing conditions, from a physical and a social point of view.

Experimenters raised two groups of rats, one group in cages where they could see cardboard with circles and triangles on it, and one group that could only see plain cardboard. The rats with the wacky wallpaper did much better in later life at choosing between a circle and a triangle to get a rat chow reward. Okay, they did better on tests, but does that mean their lives were better?

Gorillas who were in stark conditions of captivity and did not reproduce began breeding when they were moved to an enriched enclosure. Jo Fritz, writing about taking in unwanted chimpanzees at the Primate Foundation, noted that of apes who came from laboratories, zoos, and circuses, the ones who had been performers were often the best adjusted. They did not rock back and forth, pluck themselves bald, eat their own dung, or mutilate themselves. "Dressing up and turning somersaults cannot be a prerequisite to chimpanzee social life, but the stimulation and occupation that is associated with training and learning (even circus-type tricks) may be one of the most important factors in understanding the differences seen among the singly reared, asocial chimpanzees."

Everybody loves Kermit

In a captive chimpanzee colony, there were initially two males, Darrell and Kermit. Occasionally they spent time with a female, Sheba. Darrell was large, aggressive, and dominant. When he saw Sheba he was very aggressive. Lowly, puny Kermit would protect Sheba with his body when Darrell tried to bite or strike her. Although chivalrous, Kermit was a poor student. The chimpanzees were being taught number concepts, and Kermit did badly. Researchers felt he had attention deficit disorder. After about six years of training that made little headway with Kermit (the others did better), all three were moved into more spacious, renovated quarters along with an older female, Sarah, and a new male, Bobby. There was a new touch-frame testing device.

All of a sudden Kermit could do number concepts. Maybe it was the new testing device, or maybe he'd matured, or maybe it was the fact that he had grown much larger, and in the new social environment, where the females liked and supported him, he was now the dominant male. That's enrichment.

Enrichment is a popular notion in zoos today, and that's good, but sometimes enrichment is overly focused on what looks enriching to visitors and not what's actually enriching for animals. An animal in a small cage with a rain forest scene painted on the walls is no better off than an animal in a small cage with plain walls. Zoos that have gone to great lengths to create "natural" enclosures for their animals, complete with uneven surfaces and real or fake trees and plants, are sometimes reluctant to bestow unnatural-looking "enrichment" on their animals. They are like doctrinaire counterculture parents—no plastic toys for you.

Who is smart?

This chapter has been dominated by primates, in whose intellect we take a nepotistic interest. Researchers invariably fall in love with the animals they study, no matter how repulsive others find them, so you can find scientists rhapsodizing about cockroaches and leeches. Often the paeans to unpopular animals consist of praise for how well adapted the creature is. But many researchers also make impressive cases for the intellectual capacities of nonprimates.

So who have we been wrongly underestimating? Even among the primates there are candidates. Robert Shumaker complains that the African ape model ignores the abilities of orangutans, and Anne Russon grumbles that "orangutans were sidelined once chimpanzees took center stage."

Gibbons have their grievances. Muriquis have been neglected. But let's look further.

Birds

Ornithologist Peter Marler, in a paper entitled "Are Primates Smarter than Birds?" doesn't answer directly but he matches up the feats of primates and the equally amazing feats of birds, throws in a thing or two birds do that nonhuman primates don't, and lets us make up our own minds. "Almost everyone has heard of Imo," grouses Marler, but he puts the titmice learning to open milk bottles up against Imo's macaque troop. Tool use, he points out, is as common in birds as in monkeys and apes. Cooperative hunting? Double-crested cormorants and white pelicans herd fish schools in groups, eagle and falcon pairs flush prey for each other, and ask any jackrabbit whether Harris's hawk families hunt cooperatively. Social learning? Many birds learn about enemies, foods, and foraging methods that way. Social intelligence? Get this: chickens can recognize up to 100 flockmates, and so, probably, can acorn woodpeckers. And vocal learning is "a social skill that, so far as we know, no non-human primate possesses." Bowerbird bowers—primates have nothing like bowers!

Cetaceans

As mentioned in chapter 4, the study of dolphin and whale cognition is still recovering from overblown claims made in the past for cetaceans as spiritual, altruistic, oceangoing geniuses. It's awkward to make the transition from brilliant angel to merely smart and nice. In addition, they live in the water and many employ ultrasonics in ways we only partially understand, so studying them involves more technology than studying primates. Following chimpanzees through the rain forest is easy compared to following dolphins through the sea. Nevertheless, we know that they are extremely social, are eminently capable of mimicry (including vocal learning), can hunt cooperatively, may teach their young, show tool use, and often display a twisted sense of humor.

Hyenas, octopuses, buffalo, bees

In 1984, Lawrence Frank startled attendees at an international conference on primatology with a paper called "Are Hyenas Primates?" in which he pointed out that hyena society has many of the socially complex and sophisticated aspects of, for example, macaque society.

Biologist Roland Anderson is in there boosting the octopus as "the most intelligent invertebrate in the world." Octopuses don't have some of the usual traits that correlate with intelligence. They don't live long, and the young don't spend time with their parents, either learning from them or being protected during a learning period. They appear to have no social lives. Yet they have versatile behavior, are capable of observational learning, and show individuality. Perhaps it has to do with the fact that cephalopods have unusually complicated bodies to run, what with tentacles and suckers and siphons and chromatophores, not to mention larval forms.

John Byers sticks up for the ungulates. Sort of. "I am going to argue that the ungulates are smarter than previously believed, but that their cognitive abilities are specialized, and most likely limited to just a few kinds of situation," he writes.

Writing of honeybees, citing their use of cognitive maps, discrimination learning, and dance language, James and Carol Gould extol their mental powers. To a point. "Despite a widespread unwillingness to take indications of insect intellect at face value, there is good evidence that even a few milligrams of highly specialized neural wiring can accomplish a limited set of individually impressive cognitive tasks essential to the natural history of the animal in question."

Fish brains

In a restrained 2002 paper, researchers Redouan Bshary, Wolfgang Wickler, and Hans Fricke note that theorists of cognition generally neglect to contemplate fish intellect, and are derelict in not doing so: "Most phenomena of interest for primatologists are found in fish as well. . . . We think that on [a] descriptive functional level, we are able to provide fish examples for almost all phenomena that are currently being discussed in the context of primate intelligence."

There are fish, they point out, who recognize other fish as individuals, fish who know each other's voices, fish who live in extended families, fish who appease higher-ranking fish in their group, fish who keep track of whether other fish are playing fair, fish who copy other fish in food choice or in mate choice, fish who learn traditional sites and routes from other fish, fish who can tell people apart, fish who apparently learn foraging techniques from other fish, fish who join other fish in ganging up on predators, and fish who remember landmarks they have not seen for at least six months. (These are not all the same fish.) Also, they note, fish as a group are far superior to primates at building structures such as nests. Not that anyone is keeping track.

Noting that "co-operative hunting has been cited as one of the hallmarks of hominid evolution," Bshary, Wickler, and Fricke describe cooperative hunts involving giant moray eels and groupers (lunartail groupers or red sea coral groupers). The groupers swim up to a cave in the coral where a moray is resting and shake their bodies in an exaggerated way. On about half the occasions that this was observed, the moray came out of the cave, and "the two predators would swim next to each other, searching for prey." Often they swam with their sides touching. When they found a prey fish hiding in the coral, the moray would prowl through holes in the coral, while the groupers waited to strike the fish the moray chased out.

Stylings by Mr. Fish

If Machiavellian intelligence is important in the development of primate brainpower, Bshary, Wickler, and Fricke suggest that cleaner fish are another place to look. Cleaner fish are fish of various species who nibble other fish clean of parasites, dead skin, and infected flesh. They typically wait at a known cleaning station* for fish to come by in search of their services. This is a source of food for them and a source of improved health and comfort for the "client" fish. Some fish clean other fish as a sideline and others are specialized to do nothing else. Cleaner wrasse may groom 2,300 fish a day, and they have been seen cleaning fish of more than 100 species. There is evidence that cleaner fish distinguish and discriminate between client fish that can't go to another cleaning station (because their territory or home range includes only that one) and clients who have other options. The clients with other options get cleaned first. Sometimes cleaner fish cheat and nibble off some perfectly healthy flesh, and if clients detect this, they can go into a frenzy of consumer outrage, chasing the cleaner. Cleaner fish are more obsequious to clients from predatory species, hovering above them and touching the client's dorsal fin with their own pelvic and pectoral fins—and they are more apt to do this if they get a second chance with a client who previously attacked them for cheating, a form of reconciliation. Cleaner fish behave better toward clients if other prospective clients are watching, especially if the waiting clients have the option of going elsewhere. Client fish use what they learn from watching to decide whether to keep their appointment or swim off and take their patronage elsewhere. In short, the cleaner-client relationship is complicated, with plenty of room for negotiation, and the better a fish is at keeping track of the behavior of other fish, the better off it will be. "Full-time cleaners will probably turn out to be the ultimate Machiavellian strategists among fish."

* Never referred to as a "salon."

It's bad enough that practically every time you say, "This is something that only the great apes can do," you are forced to add, "And Alex, the grey parrot. And dolphins." To have to add, "And certain species of octopus," "Also, hyenas," "Bats, too," "Plus, the cleaner wrasse" really detracts from the effect.

But if you believe the people who know them best, almost every animal is smarter than we knew. You learn something every day.

The relation between learning and intelligence is intimate. Both modular and generalized intelligence make it easier for animals to learn things, and make animals more likely to show insight. Yet learning is so vital to survival that even the dullest creatures learn a little. Every baby animal, from insect to ape, gathers information, perfects its skills, and goes forth into the world.

Conclusion:
Secrets of a Tiger's Success

BABY ANIMALS HAVE A LOT TO LEARN in order to survive and flourish. Even animals of comparatively feeble intellect, who rely most heavily on a suite of instincts, usually must learn the details of where they live and who their family is and whether there's anything left worth eating around here.

That tigers have serious learning to do is suggested by the fact that cubs stay with their mothers for more than two years. To begin with, the cub learns who its mother is, who its siblings are, and, when it is older, which other tigers its mother is friendly or unfriendly with. It learns to walk and climb trees. Playing with other cubs and with its mother, the young tiger practices the innate behavior patterns of stalking and pouncing, which give it joy, and learns how far it can jump, what are the most gratifying places to grab an opponent, and how hard it can bite before it makes another tiger angry. The cub also learns how tough it is and how big in comparison to the other cubs in the litter. In the time spent with its family, the tiger cub learns that it is a tiger, information that will be relevant when it wants to mate. The tiger Victoria, raised as the only child of the dog Rosemary, apparently missed this lesson when she was young enough to learn it.

In the world the cub encounters situations where it learns by conditioning and association, although if it is lucky it won't have the learning situation of being hit by a car, like the cattle-killing tiger of Trengganu. Like other cats, it will probably learn some actions by imitating its mother. Although tigers don't need to learn what to say, they probably learn the alarm calls of other species (who are mostly alarmed because there's a tiger in the neighborhood).

For a long time, the tiger cub works very hard at learning to catch prey. Even a zoo tiger learns that turkeys will fly over the fence, whereas ducklings won't. Some wild tigers learn about prey specialties such as horses, bears, or dogs. They learn that they don't really want to eat a civet cat or a crocodile. Like Tara, they may learn that otters are little, but they'll gang up on you.

Most tigers associate humans with danger. The tigers of Ranthambhore apparently learned too well that humans in jeeps are harmless types who just gape and take pictures, leaving them vulnerable to poachers. And some of the tigers of the Sundarbans have learned that humans are edible.

Predation is the area where tigers are at their best intellectually. Here is where we should look for tiger insight, forethought, and innovation. Clever tigers are not interested in the principle of the lever, they are interested (like Genghis, the tiger who hunted in water) in ingenious new ways to separate a deer from the herd. How to get close to a blesbok without being spotted and how to prevent a warthog from taking refuge in its burrow are the issues that preoccupy the tiger students Ron and Julie.

If you created an intelligence test for a tiger, it would be only fair to include sections devoted to stalking, pouncing, and killing, where the tiger would shine. It would probably not do as well on sections devoted to dragging prey. The tigers described by Locke, who kept trying to pull their prey in one direction without heeding the fact that its horns might be caught on a tree, were trying to solve the difficulty with brute force—pulling harder. But the solution to pigs taking refuge in a burrow is not to run faster, but to cut off the pigs' retreat. The second problem brings out more learning ability than the first.

We really don't know to what extent there may be a tiger culture, a social environment for learning. Although tigers are not profoundly social, mothers and cubs stay together for years, and after that may meet from time to time. Wild tigers spend more time associating with each other at kills than had been previously understood, which requires a certain amount of diplomacy. That male tigers, not famed for chivalry, shared kills with the hand-raised Tara, even when she was not in heat, was somewhat unexpected. Whether tigers learn from unrelated tigers, except in the important area of romance, is unknown. Tiger mating is a touchy situation for two touchy individuals, but they usually figure it out.

What I learned about learning

All animals, including us, evolve in response to change. But evolution is slow, takes place over generations, and is a response of the species. Learning is fast, and is the response of individuals. Culture, which creates an environment for learning, speeds it up.

We have evolved to learn from the world. Learning from the lecture, the blackboard, or the study materials is harder for us, and it's harder for other animals too. Animals do teach, but active teaching (as distinct from opportunity teaching, coaching, and apprenticeship) is rare because that's not the way most learning happens.

Looking at the stories of animals learning—and not learning, which can be even more illuminating—I was struck by the way learning interlocks with animal feelings and personalities. Rivalry, shyness, impatience, the desire for freedom and control can be as influential in the learning process as simple brainpower.

Examples include the observations that birds learn song better if they get to push the buttons. Parrots learn better if they can watch the competition, and apes learn more from watching someone else being taught than they do from being taught themselves. Animals prefer to try hard new things when no one is watching, whispering and mumbling the language skills they are mastering. You should learn language as young as possible; it may be more important to learn to interpret the communications of others (such as alarm calls) than to learn to make communications yourself; and most animals aren't nearly as interested in communicating with us as they are with each other. When they do want to communicate, it's usually not about the curriculum we had in mind.

A shocking revelation: being tested is boring, and boring things are harder to learn. (Is it true that the smartest kids get bored the quickest?)

Animals are more apt to imitate a creature they perceive as being like themselves than some stranger. Whether you are a cowbird or a chimpanzee, it messes you up to be raised in isolation. And raising kids is hard, but you get better at it.

Some skills or intelligences are only good for one thing and some carry over to many things. Learning to do something may or may not mean understanding what you're doing, as witness the prelearning dip seen in, for example, chickens, bees, and dogs. You can learn about how learning works and get better at it. As Christophe Boesch has pointed out, the richer the environment you give a captive animal, the smarter it turns out to be.

Learning comes with costs. There's the cost of running the brain, and there's also the cost of making mistakes. It's an adaptation of incredible power. Our extreme ability to learn has made us ridiculously successful as a species, and we share learning with other animals.

To learn is to change. Learning allows an animal child to finish the long, slow process of evolution by changing in its own lifetime. Tiger cubs, eaglets, or babies, nature brings us all into existence with the ability to learn, and the rest is up to us.

Notes

CHAPTER ONE: HOW TO DO OR KNOW SOMETHING NEW

Page
1 *birds are alike.* Morton 2002.
3 *open programs.* Mayr 1974.
4 *of birdsong.* Nicolai 1986.
4 *in the genome.* Immelmann 1975.
5 *or another drum."* Kummer cited in de Waal 2001.
5 *different habits.* Dilger 1962.
6 *correlation-learning device.* Cited in Byrne 1995.
6 *in the tank.* von Frisch 1938.
6 *isoamyl acetate.* Dukas 1998.
7 *problem tigers.* Locke 1954.
8 *Sonja Yoerg* Yoerg is an admirably clear writer who gets off some very funny lines in her book *Clever as a Fox*, which compels me to forgive her silliness in refuting what she says my erstwhile coauthor and I "would inevitably suggest" had we discussed her research.
8 *Skinner boxes.* Yoerg 2001.
9 *were universal."* Pepperberg 1999.
9 *performing animals.* Breland & Breland 1961.
10 *to do tricks.* Pryor 1975.
11 *people do this.* Keith-Lucas & Guttman 1975.
11 *latent learning.* Walker 1987; Gould & Gould 1999.
11 *suddenly understood.* Pryor 1999; the hula-ing chicken is in Pryor 1975.
11 *in honeybees.* James L. Gould, "Can Honey Bees Create Cognitive Maps?" pp. 41–45 in Bekoff et al. 2002.
12 *good at it.* Beck 1980.
12 *mussels on sand.* Burger 2001.
12 *trial and error."* de Waal 2001.

13 *looks like playing.* Ashmole & Tovar S. 1968.

13 *peck at it.* Hess 1956.

14 *chicken droppings.* Burton 1956.

15 *Kathleen Gibson.* Hilary O. Box & Kathleen R. Gibson, "New Perspectives in Studies of Social Learning: Editors' Comments," pp. 1–5 in Box & Gibson 1999.

15 *with their mothers.* Phyllis C. Lee & Cynthia J. Moss, "The Social Context for Learning and Behavioural Development among Wild African Elephants," pp. 102–125 in Box & Gibson 1999.

15 *her London home.* Kipps 1953.

16 *lever was pushed.)* Anthouard 1987.

16 *an Egyptian goose.* Krüger 2001.

17 *stimulus enhancement.* Sherry & Galef 1984, 1990.

17 *also took it up.* Summers-Smith 1963.

17 *writes Richard Byrne.* Byrne 1995.

17 *a culture instinct.'* Gopnik et al. 1999.

17 *monkeys and apes."* de Waal 2001.

18 *case of Okíchoro.* Bruce R Moore, "The Evolution of Imitative Learning," pp. 245–265 in Heyes & Galef 1996.

18 *in all the books.* Quoted in Zuk 2002.

18–19 *food) by observation.* John et al. 1968.

19 *good at imitating.* Pryor 1999.

19 *their own bill.* Dawson & Foss 1965.

19 *South American monkeys.* Voelkl & Huber 2000.

20 *in the same tank.* Tayler & Saayman 1973.

21 *trainers were present.* Morton 2002.

21 *breeding red pandas.* Greta McMillan, personal communication.

22 *for the first time.* Weigl & Hanson 1980.

23 *that actually worked.* Michael Tomasello, "Cultural Transmission in the Tool Use and Communicatory Signaling of Chimpanzees?" pp. 274–311 in Parker & Gibson 1990.

23 *started the car.* H. Lyn White Miles, "ME CHANTEK: The Development of Self-Awareness in a Signing Orangutan," pp. 254–272 in Parker et al. 1994; Temerlin 1975; Hayes 1951.

23 *as a shawl.* Prince-Hughes 2001; Dawn Prince-Hughes, personal communication.

23 *Camp Leakey.* Russon 2000.

24 *rod on the box.* Cited in Robert W. Shumaker and Karyl B. Swartz, "When Traditional Methodologies Fail: Cognitive Studies of Great Apes," pp. 335–343 in Bekoff et al. 2002.

24 *at the National Zoo.* Ibid.

25 *puppies were bred.* Slabbert & Rasa 1997.

26 *28 of them did.* Hosey et al. 1997.

26 *learned to do it.* Elisabetta Visalberghi and Dorothy Munkenbeck Fragaszy, "Do Monkeys Ape?," pp. 247–283 in Parker & Gibson 1990; Kummer 1995.

27 *to get it."* Pongrácz et al. 2001.

28 *a fascinating box.* Huber et al. 2001.

29 *many entertaining things.* Pryor et al. 1969; Pryor 1975.

29 *deal of English.* Savage-Rumbaugh et al. 1998.

CHAPTER TWO: LEARNING THE BASICS

Page

31 *chaparral below.* Snyder & Snyder 2000.

31 *coordinated fashion.* Wilson & Kleiman 1974.

32 *is coming from.* Knudsen & Knudsen 1985.

32 *water looks like.* Paul Rozin, "The Selection of Foods by Rats, Humans, and Other Animals," pp. 21–76 in Rosenblatt et al. 1976.

32 *her first snow.* Wayre 1976.

33 *learn about ice.* Henry 1986.

33 *a baby orangutan.* Laidler 1980.

33 *hand coordination.* Francesco Antinucci, "The Comparative Study of Cognitive Ontogeny in Four Primate Species," pp. 157–171 in Parker & Gibson 1990.

34 *shortly after birth.* Sarah McCarthy, personal communication.

35 *walk at the same age.* Gould & Gould 1999.

35 *they move off.* Rijt-Plooij & Plooij 1987.

36 *short distance away.* Maestripieri 1995.

36 *look for her.* Rijt-Plooij & Plooij 1987.

36 *about eight months old."* Temerlin 1975.

36 *he was eight months old.* Laidler 1980.

37 *told what to do.* Grunwald 1995.

37 *trees at 13 weeks.* Kilham & Gray 2002.

37 *ears, and eyes.* Baldwin & Baldwin 1978.

38 *assigning themselves "projects."* Fagen 1981.

38 *in a mimosa tree.* Temerlin 1975.

38 *a poor showing.* Carter 1981; Linden 1986.

39 *next tree over.* Galdikas cited in Fagen 1981.

39 *stepping off neatly."* Russon 2000.

39 *different distances.* Kim A. Bard, "'Social Tool Use' by Free-Ranging Orangutan: A Piagetian and Developmental Perspective on the Manipulation of an Animate Object," pp. 356–378 in Parker & Gibson 1990.

39 *requires self-conception.* Povinelli & Cant 1995.

40 *climb down again.* Pryor 1999.

41 *attempt to fly.* Corbo & Barras 1983.

41 *broke its neck.* Lorenz 1978.

42 *on the ground.* Miller 1894.

42 *his first flight.* Pittman 2003.

42 *plummy English accent.* Hancock 1977.

43 *raised by people.* Lorenz 1978.

44 *and tree smarts.* Durden 1972.

44 *at four weeks.* Wayre 1976, 1979.

45 *drag them in.* Liers 1951.

45 *edge of a pool.* Wayre 1976, 1979.

45 *a learned skill.* Blaisdell 1999.

45 *to the sea.* Miles 1984.

45 *coming to water.* Wayre 1976.

45 *and Janet Mann.* Hal Whitehead and Janet Mann, "Female Reproductive Strategies of Cetaceans: Life Histories and Calf Care," pp. 219–246 in Mann et al. 2000.

46 *the Letaba River.* Krüger 2001.

46 *on their own.* Dirk van Vuren, "Mammalian Dispersal and Reserve Design," pp. 369–393 in Caro 1998.

46 *with the area."* Karen Higginbottom and David B. Croft, "Social Learning in Marsupials," pp. 80–101 in Box & Gibson 1999.

46 *called* social play." Cited in Allen & Bekoff 1997.

46 *Fagen proposes.* Fagen 1981.

47 *characteristic of play.* Wilson & Kleiman 1974.

47 *making huge splashes.* Stirling 1988.

48 *floated down it.* Ficken 1977.

48 *on sunny days.* Gould & Gould 1999.

48 *in a tent.* Durrell 1977.

48 *hours at a time.* Ficken 1977.

49 *with a log.* Nishiwaki 1962.

49 *seen whales play.* Whitehead 1990.

49 *in New Zealand.* Diamond & Bond 1999.

50 *dolphins looked on.* Reiss 1991.

51 *make bubble rings.* Gewalt 1989.

51 *rings this way.* Pace 2000.

51 *with a ball.* Mather & Anderson 1999.

52 *peekaboo is beguiling.* Lyn Miles, "The Argument for Ape Personhood," presentation at Chimpanzoo 2002 conference.

52 *as an adult."* Wasser 1978.

52 *social ground rules.* Bekoff 2002.

52 *to learn discretion."* Diamond & Bond 1999.

52 *what they know.* Allen & Bekoff 1997.

53 *each other's tails.* Baldwin & Baldwin 1978.

53 *chow puppy, Nanuq.* Temerlin 1975.

54 *won't play with them.* Marc Bekoff and Colin Allen, "The Evolution of Social Play: Interdisciplinary Analyses of Cognitive Processes," pp. 429–435 in Bekoff et al. 2002.

54 *with other puppies.* Corbett 1995.

54 *were this virtuous."* Bekoff 2002.

54 *bravely with it.* Maxeen Biben, "Squirrel Monkey Playfighting: Making the Case for a Cognitive Training Function for Play," pp.161–182 in Bekoff & Byers 1998.

55 *most of the play.* David Powell, Lori Tarou, personal communication 2002.

55 *two muskox calves,* David R. Klein, "Comparative Social Learning among Arctic Herbivores: The Caribou, Muskox and Arctic Hare," pp. 126–140, Box & Gibson 1999.

55 *take it away.* Ficken 1977.

55 *keep-away with stones.* Ibid.

55 *play keep-away.* Fagen 1981.

55 *tail and "hula."* Pryor 1975.

55 *a streamside meadow* Park 1971.

56 *over the water.* Fagen 1981.

56 *the "carrying dance."* Kummer 1995.

56 *as a doll.* Hirata et al. 2001.

56 *a newborn chimp* Cited in de Waal 2001.

56 *play with dolls.* Savage-Rumbaugh et al. 1998.

57 *from a glass.* Lyn Miles, "The Argument for Ape Personhood," presentation at Chimpanzoo 2002 conference.

57 *came on television.)* Kathleen Rita Gibson, "Tool Use, Imitation, and Deception in a Captive Cebus Monkey," pp. 205–218 in Parker & Gibson 1990.

57 *Zoo in 1940.* Gordon M. Burghardt, "The Evolutionary Origins of Play Revisited: Lessons from Turtles," in Bekoff & Byers 1998.

CHAPTER THREE: LEARNING YOUR SPECIES

Page

59 *of various birds.* Oulié 1939.

60 *they never caught.* Thomas 2000.

60 *mother rejected him.* Laidler 1980.

60 *beware of them.* Blaisdell 1999.

62 *confidently after him."* Lorenz 1952, 1978.

63 *to the pens.* Laurance 2000.

63 *to eat it.* Mealy 2002.

64 *their foster siblings.* Penn & Potts 1998.

64 *hamster families.* Mateo & Johnston 2000.

64 *from the sheep.* Weisbord & Kachanoff 2000.

65 *they grew up.* Owens et al. 1999.

65 *for a year.)* Kendrick et al. 2001.

66 *studying cockatoos.* Rowley & Chapman 1986, 1991; Rowley 1991.

70 *were "double-imprinted."* Zann 1996.

70 *the black robin.* Butler & Merton 1992.

72 *Zoo in Vienna.* Lorenz 1952.

73 *into raising them.* Meine & Archibald 1996.

73 *and raised children.* Pittman 2003.

74 *and Helen Snyder.* Snyder & Snyder 2000.

74 *raised by puppets.* Graham 2000.

74 *flawless condor style.* Kay 2002.

74 *the London Zoo.* Perry 1969; David Powell, Lori Tarou, personal interview.

76 *dog, to raise.* Tilson et al. 1994; Steve Taylor, Gerry Brady, personal interviews.

76 *for mate material.* Patterson 1994.

77 *with other leopards.* Singh 1982.

78 *owls don't get rabies).* McKeever 1994.

78 *hoped to breed.* Berry 1972.

79 *from a nestling.* Zann 1996.

79 *laid an egg.* Lockley 1961.

79 *rock ptarmigans.* Hancock 1977.

80 *whales in sight.* Baraff & Asmutis 1995.

80 *nest parasitism.* Davies 2000.

81 *not my egg."* Heinrich 1999.

81 *breeding experience.")* Slater 1987.

83 *recognize each other.* Hauber et al. 2001.

83 *look like them.* Hauber et al. 2000.

83 *cowbirds standing by.* Meredith J. West, Andrew P. King, and Todd M. Freeberg, "Building a Social Agenda for the Study of Bird Song," pp. 41–56 in Snowdon & Hausberger 1997; ten Cate & Vos 1999.

85 *attempt was doomed.* Saba and Dudu Douglas-Hamilton, personal communication.

86 *one impala calf.* Nzioka 2003.

86 *dead fetus."* Thomas 2000.

87 *followed lions.* Schaller 1972.

87 *she brooded devotedly.* Durden 1972.

CHAPTER FOUR: HOW TO GET YOUR POINT ACROSS

Page

91 *chimpanzee sounds.* Temerlin 1975.

92 *same old tune.* Catchpole et al. 1986; Hasselquist et al. 1996.

92 *before giving birth.* Grunwald 1995

93 *answers its mother.* Charrier et al. 2001.

93 *with one sibling.* Nakagawa et al. 2001.

93 *flocks of up to 150.* Wanker et al. 1998.

94 *on the reefs.* Myrberg & Riggio 1985.

95 *at wild wolves.* Harrington & Mech 1979.

96 *writes Richard Zann.* Zann 1996.

96 *zebra finch song.* Adret 1993.

97 *with their family.* Zann 1996.

97 *trained musicians!"* Keyser 1894.

97 *to their babble.* Snowdon & Elowson 2001.

97 *than they show.* Nelson 1989.

98 *raised in isolation.* Meredith J. West, Andrew P. King, and Todd M. Freeberg, "Building a Social Agenda for the Study of Bird Song," pp. 41–56 in Snowdon & Hausberger 1997.

99 *matters differently.* Yamaguchi 2001.

99 *had been studied.* Martine Hausberger, "Social Influences on Song Acquisition and Sharing in the European Starling (*Sturnus vulgaris*)," pp. 128–156 in Snowdon & Hausberger 1997.

99 *nesting season begins.* Baptista et al. 1993.

100 *a Lincoln sparrow.* Baptista et al. 1981.

100 *for learning song.* Kiester & Kiester 2002; Nottebohm et al. 1981.

100 *in the winter.* Barnea & Nottebohm 1994.

101 *had heard before."* Nottebohm & Nottebohm 1969.

101 *North Carolina Zoological Park.* Marshall et al. 1999.

101 *your own folks.* Peter L. Tyack, "Functional Aspects of Cetacean Communication," pp. 270–307 in Mann et al. 2000.

102 *persecuted species."* Brown 1985.

102 *with other sparrows.* Kipps 1953.

103 *"a rudimentary song."* Summers-Smith 1963.

104 *found in the hummingbirds.* Baptista & Schuchmann 1990.

104 *called "baby talk."* Baptista 2001.

104 *are learners too.* Rusch 2001.

105 *also to communicate.* Dawson 1991.

105 *learn this very quickly* Scherrer & Wilkinson 1993.

105 *millions of pups.* Balcombe 1990.

105 *over to the loudspeaker.* Porter 1979.

105 *tropical rain forest.* Boughman 1997, 1998, telephone interview 2002.

106 *that resembled mimicry.* Cruickshank et al. 1993.

107 *by a wild raven.* Heinrich 1999.

107 *"See you soon, baboon."* Corbo & Barras 1983.

108 *ornate tail feathers.* Reilly 1988.

108 *of social virtue."* Pratt 1940.

109 *to him every morning.* Reilly 1988.

110 *are impressive mimics.* Oatley 1998.

110 *a use for mimicry.* Ibid.

110 *mimicry that goes unnoticed.* Goodwin 1986.

111 *Oatley in robin-chats.* Oatley 1998.

111 *separation of 20 minutes.* Ibid.

111 *the <u>Canadian Journal of Zoology</u>.* Ralls et al. 1985.

112 *in <u>The New Yorker</u>.* Hiss 1983.

112 *what people said to him.* Eaton 1979.

113 *didn't have to be learned.* Wickler 1980.

113 *with loud "tsreeee"s.* Oatley 1998.

113 *a new couple learned to duet.* Maples et al. 1989.

114 *an underwater canyon.* Pryor 1975.

115 *wolves, bears, and boars.* Bugnyar et al. 2001.

115 *ethologist Paul Leyhausen.* Leyhausen 1979.

116 *baboons, and unfamiliar humans.* Robert M. Seyfarth and Dorothy L. Cheney, "Some General Features of Vocal Development in Nonhuman Primates," pp. 249–273 in Snowdon & Hausberger 1997; Cheney & Seyfarth 1985.

116 *looking toward the sound.* Cheney & Seyfarth 1985; Hauser 1988.

117 *baby rabbits to adulthood.* Burton 1956.

117 *to humanity from the divine.* Amy Samuels and Peter Tyack, "Flukeprints: A History of Studying Cetacean Societies," pp. 9–44 in Mann et al. 2000; Collet 2000.

118 *regional dialectical differences.* McCowan & Reiss 2001.

118 *signature whistle hypothesis.* Peter L. Tyack, "Functional Aspects of Cetacean Communication," pp. 270–307 in Mann et al. 2000.

118 *which change their voices.* Ibid.

118 *dolphins—use signature whistles.* Janik & Slater 1995.

118 *six months to a year old."* Smolker 2001.

119 *seem to have signatures.* Randall 1995.

119 *were vocally normal.* Hammerschmidt et al. 2001.

120 *Ivory Coast in the 1980s.* Boesch 1991.

121 *decoded by the listening apes.* Arcadi 2000.

122 *bonobo proposes to take.* Ellen J. Ingmanson, "Tool-Using Behavior in Wild *Pan paniscus*: Social and Ecological Considerations," pp. 190–210 in Russon et al. 1996.

123 *one of several colored cloths.* Fernald 1984.

124 *talk just as children do.* Hayes 1951.

124 *intensive training for six years.* Viki lived to the age of six and a half.

124 *were similarly frustrating.* Laidler 1980.

125 *comprehension by Pepperberg.* Pepperberg 1999.

125 *and shape, for example.* Pepperberg 1983.

126 *water when he was thirsty.* Lorenz 1952.

128 *their communicative repertoire.* Fouts et al. 1984.

128 *gorilla could learn to sign.* Patterson & Linden 1981.

129 *her cross-fostered son.* H. Lyn White Miles, "The Cognitive Foundations for Reference in a Signing Orangutan," pp. 511–539 in Parker & Gibson 1990; H. Lyn White Miles, presentation at Chimpanzoo 2002 conference.

130 *for life in the wild.* Galdikas 1995.

131 *after linguist Noam Chomsky* Terrace 1979.

131 *whence they had come,* Linden 1986.

131 *drill on naming objects.* Chris O'Sullivan and Carey Page Yeager, "Communicative Context and Linguistic Competence: The Effects of Social Setting on a Chimpanzee's Conversational Skill," pp. 269–279 in Gardner et al. 1989.

132 *never be quizzed again.)* Amory 1997.

132 *to ask for things.* Sue Savage-Rumbaugh, Mary Ann Romski, William D. Hopkins, & Rose A. Sevcik, "Symbol Acquisition and Use by *Pan troglodytes, Pan paniscus, Homo sapiens,*" pp. 266–295 in Heltne & Marquardt 1989.

133 *meaning of 12 lexigrams.* Savage-Rumbaugh et al. 1998.

134 *University in Inuyama, Japan.* Tetsuro Matsuzawa, "Chimpanzee Ai and Her Son Ayumu: An Episode of Education by Master-Apprenticeship," pp. 189–195 in Bekoff et al. 2002.

135 *Rumbaugh and Roger Lewin.* Savage-Rumbaugh & Lewin 1994.

136 *yet to be discovered."* Boesch 1991.

136 *from birds to humans."* Snowdon & Hausberger 1997.

136 *and get a new male.)* Singapore Zoological Gardens Docents 2000.

136 *are safe from predators.* Kummer 1995.

137 *direction of the pointed finger.* Soproni et al. 2001, 2002.

138 *cling to their mothers.* Savage-Rumbaugh et al. 1998.

138 *his group had not noticed.* Veà & Sabater-Pi 1998.

138 *learned this from watching.* Savage-Rumbaugh et al. 1998.

138 *at the Yerkes Primate Center.* Savage-Rumbaugh, E. Sue, Beverly J. Wilkerson, and Roger Bakeman, "Spontaneous Gestural Communication among Conspecifics in the Pygmy Chimpanzee (*Pan paniscus*)," pp. 97–116 in Bourne 1977.

139 *via an underwater keyboard.* Xitco et al. 2001.

140 *Wolves do terribly.* Hare et al. 2002.

140 *loved riding in the truck.* Krüger 2001.

CHAPTER FIVE: HOW TO MAKE A LIVING

Page

141 *tiger expert Ron Tilson.* Ron Tilson, personal communication.

142 *walking in the forest.* Singh 1982.

143 *closed the door.* Russon 2000.

143 *for him to perch in.* Kipps 1953.

143 *generation has been born.* Boyd 2002.

143 *the middle of pine forests.* Referenced in Stanley A. Temple, "Manipulating Behavioral Patterns of Endangered Birds: A Potential Management Technique," pp. 435–443 in Temple 1977.

143 *species of sea anemone.* Reebs 2001.

144 *great danger of predation.* Scott R. Derrickson and Noel F. R. Snyder, "Potentials and Limits of Captive Breeding in Parrot Conservation," pp. 133–163 in Beissinger & Snyder 1992.

144 *the birds and eaten them.* James W. Wiley, Noel F. R. Snyder, and Rosemarie S. Gnam, "Reintroduction as a Conservation Strategy for Parrots," pp. 165–200 in Beissinger & Snyder 1992.

144 *perched on top of the pen.* DeBlieu 1993.

145 *the researchers noted.* Devra G. Kleiman, Benjamin B. Beck, James M. Dietz, Lou Ann Dietz, Jonathan D. Ballou, and Adelmar F. Coimbra-Filho, "Conservation Program for the Golden Lion Tamarin," pp. 959–979 in Benirschke 1986.

145 *was better survival.* Brooke 1989.

145 *sleep on the ground.* Carter 1981.

145 *stay off the street.* DeBlieu 1993.

145 *around on the surface.* Miller et al. 1996.

145 *toward artificial structures.* Snyder & Snyder 2000.

146 *landed 15 at a time. San Francisco Chronicle* September 2, 1996; Whitaker 1999.

146 *off parked trucks. The Economist* January 25, 2003.

146 *slunk out in single file."* Milner 1999/2000.

146 *campground one day.* Eaton 2003.

146 *warehouse in Oxford.* Macdonald 1987.

147 *"potentially difficult prey."* Mueller & Berger 1970.

147 *open to new menu items.* Zwank et al. 1988.

147 *object" they come across.* J. M. Whitehead, "Development of Feeding Selectivity in Mantled Howling Monkeys, *Alouatta palliata*," pp. 105–117 in Else & Lee 1986.

148 *found he'd been eating.* Kilham & Gray 2002.

148 *an island in the Gambia.* Carter 1981; Linden 1986.

148 *than to hunting and gathering.* Temerlin 1975.

148 *about wild foods.* Russon 2000.

149 *seeking what they might devour.* Macdonald 1987.

150 *meet in the wild.* Devra G. Kleiman, Benjamin B. Beck, James M. Dietz, Lou Ann Dietz, Jonathan D. Ballou, and Adelmar F. Coimbra-Filho, "Conservation Program for the Golden Lion Tamarin," pp. 959–979 in Benirschke 1986.

151 *endangered least terns.* Avery et al. 1995.

152 *tapeworms and nodule worms.* Huffman & Caton 2001.

153 *chew, and spit them out.* Toshisada Nishida, "Individuality and Flexibility of Cultural Behavior Pattens in Chimpanzees," pp. 392–413 in de Waal & Tyack 2003.

153 *rub themselves with millipedes.* Valderrama et al. 2000.

153 *given an inch of cigar daily.* Kathleen Rita Gibson, "Tool Use, Imitation, and Deception in a Captive Cebus Monkey," pp. 205–218 in Parker & Gibson 1990.

153 *as her daughters watched.* Hauser 2000.

154 *what's on the menu.* Diamond & Bond 1999.

155 *prey the species eats.* Riedman 1996.

155 *clutching a different can.* McCleneghan & Ames 1976.

156 *dwindling prey, elk.* Matthiessen 2000.

156 *harmed by the tiger.* Locke 1954

156 *magpie geese to wombats.* Corbett 1995.

157 *snapping up caterpillars.* de Ruiter 1952.

157 *stems, honey, and funguses."* Galdikas 1995.

158 *once they've fledged.* Snyder & Snyder 2000.

158 *how to rob nests.* Carter 1981.

158 *inept foragers" among birds.* Hauser 2000.

159 *62 percent for the adults.* Quinney & Smith 1980.

159 *teenaged terns* Dunn 1972.

159 *lights to attract insects.* French & Whitaker 2002.

160 *rocks in shallow water.* Yoerg 1994, 1998.

160 *fledglings in an Oxford garden.* Davies 1976.

160 *to achieve independence.* Heinsohn 1991.

161 *food items with babies.* Boinski & Fragaszy 1989.

162 *prowess gradually improves.* Stirling 1988.

163 *a bunch of fish at once.* Smolker 2001.

163 *tumbled to the bottom.* Bshary et al. 2002.

164 *genuinely dangerous horns.* F. C. Eloff, "Ecology and Behavior of the Kalahari Lion," pp. 90–126 in Eaton 1973.

165 *without human protection.* "Living with Tigers," a Discovery Channel Quest show, viewed at http://dsc.discovery.com/convergence/tigers/tigers.html.

166 *not quite two years old.* Shah & Shah 1996.

166 *motor pattern is innate.* Leyhausen 1979.

167 *a Kenyan national park.* Eaton 1970.

167 *South African nature reserves.* Pettifer 1980.

168 *they're a year old.* Shah & Shah 1996.

168 *cubs to the bait.* Schaller 1967.

169 *on the hindquarters.* Thapar 1986.

169 *prey they had caught.* Sunquist & Sunquist 1988.

169 *sprang on him.* Hedren & Taylor 1985.

170 *partridges, squirrels, and mice.* Thapar 1989.

170 *bring himself to eat it.* Shah & Shah 1996.

170 *than most great cats.* Schaller 1972.

170 *from watching adults.* Stander 1992.

170 *she spotted a chital.* Shah & Shah 1996.

171 *three yearling cubs.* Ibid.

171 *hunting in water.* Thapar 1986, 1989; Shah & Shah 1996.

172 *in one of 20 hunts.)* Sunquist & Sunquist 1988.

173 *preserve in northern India.* Singh 1982.

173 *prowess of a great cat.* Krüger 2001.

175 *if they're really hungry.* I owe this analogy to John McCarthy.

175 *certain point in the hunt.* Adamson 1980.

175 *'What next?'"* *Queen of Shaba*, the book describing Penny's life and release, was published after Joy Adamson was killed. Penny had by this time mated and had cubs. She was an excellent mother, but it is not clear whether she was self-supporting, since the Adamsons were still bringing her food and water.

175 *several litters in the wild.* Adamson 1972.

176 *Twycross Zoo in the UK.* Singh 1981, 1984.

177 *no difficulty finding food.* Byrne & Byrne 1993.

178 *where to find pinecones.* Snyder et al. 1994.

178 *of Jerusalem pine.* Joseph Terkel, "Cultural Transmission of Feeding Behavior in the Black Rat (*Rattus rattus*)," pp. 17–47 in Heyes & Galef 1996.

180 *they eat the first one.* Riedman 1996.

181 *a goat for them.* Sunquist & Sunquist 1988.

181 *place to begin eating.* Locke 1954.

181 *had never known shortage.* Macdonald 1987.

CHAPTER SIX: HOW NOT TO BE EATEN

Page

183 *them to her presence.* Strier 1999.

184 *not view as threats.* Nice & ter Pelkwyk 1941.

185 *and hand-raised chicks.* Thaler 1986.

186 *release into the wild.* Snyder et al. 1994.

186 *thick-billed parrot alarm call.* James W. Wiley, Noel F. R. Snyder, and Rosemarie S. Gnam, "Reintroduction as a Conservation Strategy for Parrots," pp.165–200 in Beissinger & Snyder 1992.

186 *alarm-called like crazy.* Coss & Ramakrishnan 2000.

187 *will fear snakes too.* Mineka, Susan, and Michael Cook, "Social Learning and the Acquisition of Snake Fear in Monkeys," pp. 51–73, in Zentall & Galef 1988.

187 *they couldn't do it.* Ibid.

187 *had never seen a snake.* Hayes 1951.

188 *great horned owl, Bubo.* Heinrich 1987.

188 *with hostile intent.* Krüger 2001.

189 *back toward the snake.* Sheldrick 1966.

189 *in a cottonwood grove.* Miller 1894.

189 *mobbing some object.* E. Curio, "Cultural Transmission of Enemy Recognition by Birds," pp. 75–97 in Zentall & Galef 1988.

190 *stoats are dangerous.* Maloney & McLean 1995.

190 *hand-reared takahe chicks.* E. Curio, "Cultural Transmission of Enemy Recognition by Birds," pp. 75–97 in Zentall & Galef 1988.

191 *whether it's hungry.* Hamerstrom, 1957.

191 *and bands baby crows.* Heinrich 1999.

191 *hawk on the wing.* Buitron 1983.

191–192 *as a fox or a cat.* Lorenz 1952.

192 *took off for the open sea.* Richard C. Connor, "Group Living in Whales and Dolphins," pp. 199–218 in Mann et al. 2000.

192 *harassment on their mind.* Reebs 2001.

192 *in the Serengeti.* Rood 1983.

192 *gazelles inspecting cheetahs.* FitzGibbon 1994.

193 *killed by lions.* Schaller 1972.

193 *at this inexplicable combination.* Krüger 2001.

193 *at first huddle together.* Pitcher et al. 1986; Griffin 1992.

194 *pike to the other fish.* Dugatkin & Godin 1992.

194 *no predatory mammals.* Griffin et al. 2000.

195 *with the greatest alarm.* McLean et al. 2000.

195 *predators, particularly red foxes.* van Heezik & Maloney 1997; van Heezik et al. 1999.

195 *reduce bustard mortality.* van Heezik & Seddon 2001.

196 *rabbits dive for a hole.* Corbett 1995.

196 *their noses above ground.* Miller et al. 1996.

197 *wolves and grizzly bears.* Joel Berger, "Future Prey: Some Consequences of the Loss and Restoration of Large Carnivores," pp. 80–100 in Caro 1998; Bekoff 2002.

198 *mostly eats algae.* Losey 1982.

199 *before help came.* Todd 1971.

199 *friendliness to newcomers.* Macdonald 1987.

199 *vehicle containing humans.* Stirling 1988.

199 *George and Joy Adamson.* Adamson 1961.

200 *the coast of France.* Collet 2000.

200 *where they can hear.* Lord 1998.

200 *targets for poachers.* Shah & Shah 1996.

200 *cage birds were used.* Collins et al. 1998.

201 *a bottle of liquor.* Borner 1985; Matsumoto-Oda 2002.

201 *socializing with humans.* Santos 1995.

201 *the wild beavers were.* Żurowski 1979.

201 *water with open mouth.* Smolker 2001.

202 *up to 300 times a day.* Goodey & Liley 1986.

202 *them in the laboratory.* Tulley & Huntingford 1987.

203 *white hair on their haunches.* Allen & Bekoff 1997.

204 *where it had been nosing about.* Sullivan 1979.

204 *flapping and quacking.* Cited in Armstrong 1947.

204 *in a Norwegian forest.* Sonerud 1988.

204 *who didn't get distracted.* Marcström 1986.

205 *get near their nests.* Ristau 1991.

205 *for the sake of the bait.* Cypher et al. 2000.

205 *trapped every night.* Mary Lynn Fischer, personal communication.

206 *easy to trap red foxes.* Macdonald 1987.

206 *local small mammals.* Laurance 2000.

207 *ingenuity of all concerned.* Diamond & Bond 1999.

CHAPTER SEVEN:
INVENTION, INNOVATION, AND TOOLS

Page
209 *for dealing with this.* Jones & Kamil 1973.

210 *a case of extreme old age.* Corbett 1946.

211 *getting over porcupine fights.* Montgomery 1995.

211 *a painful dental abscess.* Patterson & Neiburger 2001.

211 *with innate technologies."* Kummer & Goodall 1985.

212 *a troop of vervets.* Cambefort 1981.

212 *they were innovators.* Laland & Reader 1999.

213 *small island of Koshima.* Bennett G. Galef Jr., "Social Transmission of
    Acquired Behavior: A Discussion of Traditions and Social Learning in
    Vertebrates," pp. 77–100 in Rosenblatt et al. 1976; Michael A. Huffman,
    "Acquisition of Innovative Cultural Behaviors in Nonhuman Primates: A
    Case Study of Stone Handling, a Socially Transmitted Behavior in Japanese
    Macaques," pp. 267–289 in Heyes & Galef 1996; de Waal 2001.

214 *an Einstein among macaques.* Avital & Jablonka 2000.

215 *putting it in the water.* Lovell 1958.

215 *been spotted doing this.* Higuchi 1987, 1988.

215 *fish to their doom.* Sisson 1974.

215 *south Florida, and western Africa.* Baptista 1994.

216 *do it, without success.* Gould & Gould 1999.

216 *bits of wood, and anvils.* Henty 1986.

216 *whale-watching cruises.* Weinrich et al. 1992.

217 *Special about Using Tools?"* Hansell 1987.

218 *of cognitive skills.* Robert Shumaker, "Numerical Competence in Orangutans
    (*Pongo pygmaeus*)," presentation at Chimpanzoo 2002 conference.

218 *were closely meshed.* Gibson & Ingold 1993.

218 *physical causality."* Cited in Elisabetta Visalberghi and Luca Limongelli,
    "Acting and Understanding: Tool Use Revisited through the Minds of
    Capuchin Monkeys," pp. 57–79 in Russon et al. 1996.

218 *in their environment.* Richard W. Byrne, "The Misunderstood Ape: Cognitive
    Skills of the Gorilla," pp. 111–130 in Russon et al. 1996.

219 *high for her to reach.* Prince-Hughes 2001, personal communication.

219 *trees in their enclosure.* Nakamichi 1999.

219 *they make rain hats.* Ellen J. Ingmanson, "Tool-Using Behavior in Wild *Pan
    paniscus*: Social and Ecological Considerations," pp. 190–210 in Russon et al.
    1996.

220 *peat swamp forests.* van Schaik & Fox 1996.

220 *easygoing and gregarious.* de Waal 2001

221 *macaques of Cayo Santiago.* Cited by Elisabetta Visalberghi and Dorothy
    Munkenbeck Fragaszy, "Do Monkeys Ape?" pp. 247–273 in Parker & Gibson
    1990.

221 *head as an umbrella.* Galdikas 1995.

221 *her cross-fostered son.* H. Lyn Miles, "The Argument for Ape Personhood," presentation at Chimpanzoo 2002 conference.

222 *the Temerlin family.* Temerlin 1975.

222 *project was in peril.* Jane Goodall, speech at Chimpanzoo 2002 conference.

223 *a competent termite dipper.* Cited in Michael Tomasello, "Cultural Transmission in the Tool Use and Communicatory Signaling of Chimpanzees?" pp. 274–311 in Parker & Gibson 1990.

223 *try to fish for termites.* Beck 1980.

224 *for nothing at all.* Jane Goodall, panel discussion, Chimpanzoo 2002 conference.

224 *wield them ineptly.* Beck 1980.

224 *the ant hole, and stir.* Cited in Michael Tomasello, "Cultural Transmission in the Tool Use and Communicatory Signaling of Chimpanzees?" pp. 274–311 in Parker & Gibson 1990.

224 *preserve in the Ivory Coast.* Andrew Whiten, "From the Field to the Laboratory and Back Again: Culture and 'Social Mind' in Primates," pp. 385–392 in Bekoff et al. 2002.

224 *found that both work.* Christophe Boesch, "Three Approaches for Assessing Chimpanzee Culture," pp. 404–429 in Russon et al. 1996.

224 *ants, usually in trees.* Nishida & Hiraiwa 1982.

225 *fruit of the umbrella tree.* Yamakoshi 1998. The umbrella tree is *Musanga cecropoides*.

225 *attracted chimpanzees.* Tetsuro Matsuzawa, "Field Experiments on Use of Stone Tools by Chimpanzees in the Wild," pp. 351–370 in Wrangham et al. 1994.

225 *to level his anvil stone.* Tetsuro Matsuzawa, "Chimpanzee Intelligence in Nature and Captivity: Isomorphism of Symbol Use and Tool Use," pp. 196–209 in McGrew et al. 1996.

225 *pestle pounding.* Tetsuro Matsuzawa and Gen Yamakoshi, "Comparison of Chimpanzee Material Culture between Bossou and Nimba, West Africa," pp. 211–232 in Russon et al. 1996.

226 *algae from ponds.* Tetsuro Matsuzawa, "Chimpanzee Ai and Her Son Ayumu: An Episode of Education by Master-Apprenticeship," pp. 189–195 in Bekoff et al. 2002.

226 *sticks to get honey.* Bermejo & Illera 1999.

226 *flowers of kapok trees.* Alp 1997.

226 *captive chimpanzees.* Beck 1980; McGrew et al. 1975.

227 *in unexpected ways.* Savage-Rumbaugh & Lewin 1994.

229 *Tietê River in Brazil.* Ottoni & Mannu 2001.

229 *with a boiled potato.* Elisabetta Visalberghi, "Insight from Capuchin Monkey Studies: Ingredients of, Recipes for, and Flaws in Capuchins' Success," pp. 405–411 in Bekoff et al. 2002.

230 *unfastening his leash."* Kathleen Rita Gibson, "Tool Use, Imitation, and Deception in a Captive Cebus Monkey," pp. 205–218 in Bekoff et al. 2002.

230 *apparently simple tasks."* Elisabetta Visalberghi and Luca Limongelli, "Acting and Understanding: Tool Use Revisited through the Minds of Capuchin Monkeys," pp. 57–79 in Russon et al. 1996.

230 *maple-flavored syrup.* Westergaard & Fragaszy 1987.

231 *to death with a branch.* Boinski 1988.

231 *understand about tool use.* Visalberghi & Limongelli 1994.

232 *a different response.* Tokida et al. 1994.

232 *trunks with their feet.* Wood 1988; Baptista 1994. Note that the species is *Probosciger aterrimus.*

233 *to crack open clams.* Hancock 1977.

233 *no one knows how.* Smolker 2001.

233 *out of a crevice.* Brown & Norris 1956.

234 *babies ought to be.* Keenleyside & Prince 1976.

234 *had broken off.* Hart et al. 2001.

235 *Automobiles as Nutcrackers?"* Maple 1974.

235 *Automobiles as Nutcrackers."* Grobecker & Pietsch 1978.

235 *Anecdote to the Test."* Cristol et al. 1997.

236 *wheels of cars.* Pollack 1992; Davies n.d.; Nikon Web Magazine 2000.

236 *to younger crows.* Nikon Web Magazine 2000.

236 *nonhuman tool wielders.* Schmid 2002.

237 *snap off nearby trees.* Hunt 2000.

237 *use hooked instruments."* Hunt & Gray 2002.

237 *for ravens' nests.* Janes 1976.

238 *it wasn't pretty.* Cited in Beck 1980.

238 *who know how.* Thouless et al. 1989.

239 *the yellowhead jawfish?* Colin 1973.

239 *sweet little nests.)* Charles-Dominique 1977.

240 *chimpanzees and bonobos.* Barbara Fruth and Gottfried Hohmann, "Comparative Analyses of Nest Building Behavior in Bonobos and Chimpanzees," pp. 109–128 in Wrangham et al. 1994.

240 *tool we see today."* Barbara Fruth and Gottfried Hohmann, "Nest Building Behavior in the Great Apes: The Great Leap Forward?" pp. 225–240 in McGrew et al. 1996.

240 *make ground nests.* Anne E. Russon, "Exploiting the Expertise of Others," pp. 174–206 in Whiten & Byrne 1997.

241 *in the village weaver.* Collias & Collias 1984.

243 *water from Otter Creek.* Hasler 1966.

243 *remarkable old age.* Kipps 1953.

244 *crippled by polio.)* Kummer & Goodall 1985.

CHAPTER EIGHT: HOW TO GET CULTURED

Page

245 *fish cooperatively.* Pryor et al. 1990; Simões-Lopes et al. 1998.

247 *learning from others."* de Waal 2001.

247 *no culture without language.* Cited in Whiten et al. 1999.

247 *nonsubsistence, and naturalness.* Christophe Boesch, "Three Approaches for Assessing Chimpanzee Culture," pp. 404–429 in Russon et al. 1996.

247 *considers cultural behavior.* Michael Tomasello, "Cultural Transmission in the Tool Use and Communicatory Signaling of Chimpanzees?" pp. 274–311 in Parker & Gibson 1990.

247 *and Kinji Imanishi.* de Waal 2001; Sue Taylor Parker and Anne E. Russon, "On the Wild Side of Culture and Cognition in the Great Apes," pp. 430–450 in Russon et al. 1996.

248 *to dig for edible roots.* Kummer 1995.

248 *assets to elephant herds.* McComb et al. 2001.

248 *breed after 40.* James R. Boran and Sara L. Heimlich, "Social Learning in Cetaceans: Hunting, Hearing and Hierarchies," pp. 282–307 in Box & Gibson 1999.

248 *his 85-year-old mother.* Diamond 2001.

248 *trait of female dispersal.* Sue Taylor Parker and Anne E. Russon, "On the Wild Side of Culture and Cognition in the Great Apes," pp. 430–450 in Russon et al. 1996.

249 *struggled to master.* Duane M. Rumbaugh, E. Sue Savage-Rumbaugh, and Rose A. Sevcik, "Biobehavioral Roots of Language: A Comparative Perspective of Chimpanzee, Child, and Culture," pp. 319–334 in Wrangham et al. 1994.

249 *chart their genealogies.* Price & Wiley 2000.

250 *sophistication and grandeur.* Collias & Collias 1984; Borgia 1986.

251 *into the bower itself.* Vellenga, cited in Diamond 1982.

251 *distinctly different bowers.* Diamond 1986, 1988.

252 *slight genetic differences.* Uy 2002.

252 *case of stone handling.* Michael A. Huffman, "Acquisition of Innovative Cultural Behaviors in Nonhuman Primates: A Case Study of Stone Handling, a Socially Transmitted Behavior in Japanese Macaques," pp. 267–289 in Heyes & Galef 1996; de Waal 2001.

254 *went wild over headstands.* Morton 2002.

254 *dolphins one summer.* Pryor 1975.

254 *hours of humpback song.* Noad et al. 2000; Davidson 2000.

255 *relaxing in trees.* Schaller 1972.

255 *to their own devices.* Katharine Milton, "Diet and Social Organization of a Free-Ranging Spider Monkey Population: The Development of Species-Typical Behavior in the Absence of Adults," pp. 173–181 in Pereira & Fairbanks 1993.

256 *two species of macaques.* de Waal 1996.

257 *colonies of chimpanzees.* Richard W. Wrangham, Frans B. M. de Waal, and W. C. McGrew, "The Challenge of Behavioral Diversity," pp. 1–18 in Wrangham et al. 1994.

257 *5 meters by 10 meters.* Kathleen Morgan, Fabienne Mondesir, Katherine Buell, Phyllis Guy, Sonia Pizarro, Victoria Carmella, Kate Hunt, Nicki Anderson, and Sarah Leahy, "Changes in Animal Behavior, Visitor Behavior, and Visitor

Attitude with a Change in Exhibit," a presentation at Chimpanzoo 2002 conference; "Southwick's Wild Animal Farm, a Gem in Central Massachusetts," viewed at http://www.epinions.com/content_22753480324.

258 *females were giving birth.* Itani 1959.

259 *partly reared in captivity.* Galdikas 1995.

259 *large, gentle silverback.* Shumaker 1993.

260 *in the Kalahari Desert.* Thomas 1994.

261 *male-female relationships.* Kummer 1995.

262 *on the coral reef.* Helfman & Schultz 1984.

262 *in the late afternoon.* Prins 1996.

264 *stormy days in fall.* Lorenz 1978.

264 *with a different route.* International Crane Foundation, "Whooping Crane Migration."

265 *same routes each year.* David R. Klein, "Comparative Social Learning among Arctic Herbivores: The Caribou, Muskox and Arctic Hare," pp. 126–140 in Box & Gibson 1999; Espmark 1970.

266 *groups across Africa.* Whiten et al. 1999.

266 *the Sassandra River.* Boesch et al. 1994; Christophe Boesch, "Three Approaches for Assessing Chimpanzee Culture," pp. 404–429 in Russon et al. 1996.

267 *off the coast of Liberia.* Agoramoorthy & Hsu 1999.

269 *together in large cages.* The Economist, January 25, 2003.

270 *in the subsequent fall.* Ross 2002.

270 *getting food from people.* Russon 2000.

271 *Chantek her foster son.* H. Lyn White Miles, "The Argument for Ape Personhood," presentation at Chimpanzoo 2002 conference.

271 *to imitate a demonstrator.* Tomasello et al. 1993; Michael Tomasello, "The Question of Chimpanzee Culture," pp. 301–317 in Wrangham et al. 1994.

272 *takes a different view.* de Waal 2001.

272 *and took it up herself.* Sumita et al. 1985.

272 *are to imitate strangers.* Cited in Christophe Boesch, "Three Approaches for Assessing Chimpanzee Culture," pp. 404–429 in Russon et al. 1996.

272 *animals seem to get smarter.* Ibid.

CHAPTER NINE: PARENTING AND TEACHING

Page

273 *gorilla, Dolly.* Joines 1977.

275 *review some strategies.* Choudhury & Black 1993.

276 *not a robot or dummy.* Mason 1945.

277 *mating-for-life camp.* Yamamoto et al. 1989.

277 *an inexperienced male.* Butler & Merton 1992.

278 *and they're decisive.* Dilger 1962.

278 *not be so simple for beginners.* Miller et al. 1996.

279 *expert Ron Tilson.* Ron Tilson, personal communication.

279 *normal sexual behavior.* Rogers & Davenport 1969.

280 *had not worked this out.* Snyder & Snyder 2000.

281 *cold snaps in the weather.* Marzluff 1988.

282 *rather late in the game.* Stanley A. Temple, "Manipulating Behavioral Patterns of Endangered Birds: A Potential Management Technique," pp. 435–443 in Temple 1977.

282 *raise the resultant ducklings.* Robertson 1998.

283 *cubs who aren't their own.* Holden 2001.

283 *to adopt orphaned chicks* White & Millham 1993.

283 *brought to Simon Thomsett.* Thomsett 1983.

284 *giving birth to nine pups.* Askins 2002.

284 *will try to get adopted.* Davies 2000.

285 *foaling for the first time.* Sarah McCarthy, personal communication.

286 *the calf up, and caress it.* Grunwald 1995.

286 *is an endangered species.* Kunz et al. 1994.

287 *had never had a baby before.* Savage-Rumbaugh et al. 1998.

287 *at the Seattle Zoo.* Keiter et al. 1983.

288 *every time it approached.* Morton 2002.

288 *Maud was weaning him.* Galdikas 1995.

288 *mother when weaning.* Toshisada Nishida, "Deceptive Behavior in Young Chimpanzees: An Essay," pp. 285–292 in Nishida 1990.

289 *mother Fifi had another baby.* Frans B. M. de Waal, "Conflict as Negotiation," pp. 159–172 in McGrew et al. 1996.

289 *the trials of her son Kanzi.* Savage-Rumbaugh et al. 1998.

289 *flick it away from them.* Kummer & Goodall 1985.

290 *moved PN's hand away.* Mariko Huraiwa-Hasegawa, "A Note on the Ontogeny of Feeding," pp. 277–283 in Nishida 1990.

290 *successfully hatched two chicks.* Pittman 2003.

290 *their care more successfully.* Cameron et al. 2000.

290 *which are long-lived.* Eileen C. Rees, Pia Lievesley, Richard A. Pettifor, and Christopher Perrins, "Mate Fidelity in Swans: An Interspecific Comparison," pp. 118–137 in Black 1996.

291 *they are together.* Jeffrey M. Black, Sharmila Choudhury, and Myrfen Owen, "Do Barnacle Geese Benefit from Lifelong Monogamy?" pp. 91–117 in Black 1996.

291 *newlyweds.* Susan Hannon & Kathy Martin, "Mate Fidelity and Divorce in Ptarmigan: Polygyny Avoidance on the Tundra," pp. 192–210 in Black 1996.

291 *excel at keeping chicks alive.* William J. Sydeman, Peer Pyle, Steven D. Emslie, and Elizabeth D. McLaren, "Causes and Consequences of Long-Term Partnerships in Cassin's Auklets," pp. 211–222 in Black 1996.

291 *earlier in the season.* James A. Mills, John Yarrall, and Deborah A. Mills, "Causes and Consequences of Mate Fidelity in Red-Billed Gulls," pp. 286–304 in Black 1996.

291 *at the breeding ground?* Tony D. Williams, "Mate Fidelity in Penguins," pp. 269–285 in Black 1996.

291 *attracted to actual dotards.* Avital & Jablonka 2000.

291 *with an older bird.* I. Newton and I. Wylie, "Monogamy in the Sparrowhawk," pp. 249–267 in Black 1996.

291 *and tattered ears.* Goodall 1986.

291 *approached the older female.* Cited in Goodall 1986.

292 *started their own families.* Komdeur 1996.

292 *some they do not.* Marzluff & Balda 1992.

292 *leave all child care to the mother.* Wang & Insel 1996.

293 *coaching, and active teaching.* Caro & Hauser 1992.

295 *the infant's mental state.* Maestripieri 1995.

295 *infants to crawl and walk.* Ibid.

295 *put a lot of time into it.* Katharine Milton, "Foraging Behaviour and the Evolution of Primate Intelligence," pp. 285–305 in Byrne & Whiten 1988.

295 *orphaned pallid bats one year.* Patricia Winters, personal communication.

296 *followed by their cubs.* Barrie K. Gilbert, "Opportunities for Social Learning in Bears," pp. 225–235 in Box & Gibson 1999; Bailey & Faust 1984.

296 *in the dusky wood rat.* Burton 1956.

297 *may nip them on the nose.* Liers 1951.

297 *reason for this caution.* Reiss 1991.

297 *this dangerous activity.* Goodall 1986; Charles H. Janson and Carel P. van Schaik, "Ecological Risk Aversion in Juvenile Primates: Slow and Steady Wins the Race," pp. 57–74 in Pereira & Fairbanks 1993.

297 *parents wherever they go.* Lorenz 1978.

297 *very little direct instruction.* Thomas Wynn, "Layers of Thinking in Tool Behavior," pp. 389–406 in Gibson & Ingold 1993.

298 *facilitation, and active teaching.* Christophe Boesch, "Aspects of Transmission of Tool-Use in Wild Chimpanzees," pp. 171–183 in Gibson & Ingold 1993.

299 *both animals and man."* Avital & Jablonka 2000.

299 *in the Maine woods.* Heinrich 1999.

300 *raised a family of chicks.* Avital & Jablonka 2000.

300 *father stays with them.* Jan A. J. Nel, "Social Learning in Canids: An Ecological Perspective," pp. 259–278 in Box & Gibson 1999.

300 *earthworms are a big food item.* Macdonald 1987.

301 *elephant seal pups.* Guinet 1991; Guinet & Bouvier 1995.

302 *the coast of Patagonia.* Lopez & Lopez 1985.

302 *giving them fish.* Ashmole & Tovar S. 1968.

302 *of osprey school.* Meinertzhagen 1954.

303 *more skilled parrot Alex.* Irene Maxine Pepperberg, "Social Influences on the Acquisition of Human-Based Codes in Parrots and Nonhuman Primates," pp. 157–177, in Snowdon & Hausberger 1997.

304 *Often she'll walk away."* Patterson & Linden 1981.

304 *and contraband shellfish.* Examples of most of these are given in Weisbord & Kachanoff 2000

304 *stored drug samples in.* Derr 2002.

304 *herd trout into a net.* O'Neill 2001.

305 *sharks on command.* Irvine et al. 1973.

306 *catch it* and *kill it."* McKeever 1994.

307 *using conditioning techniques.* Reiss 1991; Linden 2002.

308 *tricks he did on request.* Kipps 1953.

308 *she handled him perfectly.* Shumaker 1993.

309 *Mouila, became pregnant.* Mager 1981.

309 *in the Kasekala troop.* Goodall 1986.

CHAPTER TEN: WHAT LEARNING
TELLS US ABOUT INTELLIGENCE

Page

311 *cracks and dead wood.* Hunt 1996.

311 *right tool for the job.* Weir et al. 2002; Schmid 2002.

312 *human average is 100).* Patterson & Linden 1981.

312 *where the toy had fallen.* Penny Patterson, panel discussion, Chimpanzoo 2002 conference.

312 *on intelligence tests,* James L. Gould, "Can Honey Bees Create Cognitive Maps?" pp. 41–45 in Bekoff et al. 2002.

312 *were equally accurate.* Hollard & Delius 1982.

313 *that has gotten stuck.* Locke 1954.

313 *they are safe from foxes.* Lorenz 1978.

314 *getting a fish reward.* Pryor 1975.

314 *we see as 'intelligent.')"* Byrne 1995.

314 *as well as species-specific.* Sue Taylor Parker and Patricia Poti, "The Role of Innate Motor Patterns in Ontogenetic and Experiential Development of Intelligent Use of Sticks in Cebus Monkeys," pp. 219–243 in Parker & Gibson 1990.

315 *in order to survive."* Byrne 1995.

315 *flecks of suet to eat.* Heinrich 1999.

316 *seen string in their lives.* Heinrich 1995.

316 *and the plain titmouse.* Millikan & Bowman 1967.

317 *floor of his tall cage.* Thomas 2000.

317 *with a conical roof.* Pryor 1975.

317 *'How are they smart?'"* Smolker 2001.

317 *keep things tidy."* Yoerg 2001.

318 *never find them again.* Russell P. Balda and Alan C. Kamil, "Spatial and Social Cognition in Corvids: An Evolutionary Approach," pp. 128–134 in Bekoff et al. 2002.

318 *Sara Shettleworth writes.* Sara J. Shettleworth, "Spatial Behavior, Food Storing, and the Modular Mind," pp. 123–128 in Bekoff et al. 2002.

318 *much else about snakes.* Cheney & Seyfarth 1985.

318 *a "win-stay" strategy.* Alan C. Kamil, "Adaptation and Cognition: Knowing What Comes Naturally," pp. 533–544 in Roitblat et al. 1984.

319 *a spider-eating spider.* Jackson & Pollard 2001.

319 *a larger tray of water.* Jackson et al. 2001.

320 *minority group among spiders.*" Avital & Jablonka 2000.

320 *predicting their behavior.*" Hauser 2000.

320 *African grey parrots.* Pepperberg 1999.

321 *in favor of another.* Cited in Hauser 2000.

321–322 *the sensorimotor period.* Redshaw 1978.

322 *counting, and subitization.* Robert Shumaker, "Numerical Competence in Orangutans (*Pongo pygmaeus*)," presentation at Chimpanzoo 2002 conference.

322 *two apes, Sarah and Sheba.* Sarah T. Boysen, "'More Is Less': The Elicitation of Rule-Governed Resource Distribution in Chimpanzees," pp. 177–189 in Russon et al. 1996.

323 *at the National Zoo.* Robert Shumaker, "Numerical Competence in Orangutans (*Pongo pygmaeus*)," presentation at Chimpanzoo 2002 conference.

324 *animal behavior research).* P. J. Asquith, "Anthropomorphism and the Japanese and Western Traditions in Primatology," pp. 61–71 in Else & Lee 1986.

324 *variation in octopuses.* Sinn et al. 2001.

324 *the raccoon's personality.* Davis 1984.

325 *to open water.* Reebs 2001; Aronson 1971.

325 *later when they migrated.* Hauser 2000.

325 *analyzing their reactions.* Glickman & Sroges 1966.

326 *with one difference.* Wemelsfelder et al. 2000.

326 *collars with transponders.* Donald M. Broom, "Social Transfer of Information in Domestic Animals," pp. 158–168 in Box & Gibson 1999.

327 *where nothing ever happened.* Bradburn Young, personal communication.

327 *like that of other animals.* Markowitz et al. 1975.

328 *riverbed in Stink Wadi.* Sigg 1980.

328 *they do with their striatum.* Pepperberg 1999.

328 *cortical neurons differently.* Smolker 2001.

329 *and that takes brains.* K. R. Gibson, "Cognition, Brain Size and the Extraction of Embedded Food Resources," pp. 93–103 in Else & Lee 1986.

329 *living in the same area.* Katharine Milton, "Foraging Behaviour and the Evolution of Primate Intelligence," pp. 285–305 in Byrne & Whiten 1988.

330 *problems were social problems.* Cited in Alain Schmitt and Karl Grammer, "Social Intelligence and Success: Don't Be Too Clever in Order to Be Smart," pp. 86–111 in Whiten & Byrne 1997.

330 *bonding power of grooming.* Dunbar 1996.

330 *and uttering alarm calls.* Sue Savage-Rumbaugh and Kelly McDonald, "Deception and Social Manipulation in Symbol-Using Apes," pp. 224–237 in Byrne & Whiten 1988.

331 *dominant to older males.* Barbara Smuts, "Gestural Communication in Olive Baboons and Domestic Dogs," pp. 301–306 in Bekoff et al. 2002.

331 *opaque upside-down test tube.* Giraldeau & Lefebvre 1987.

332 *small children played together.* Cited in Christopher Boehm, "Pacifying Interventions at Arnhem Zoo and Gombe," pp. 211–226 in Wrangham et al. 1994.

332 *writes Alison Jolly.* Alison Jolly, "Lemur Social Behaviour and Primate Intelligence" pp. 27–33 in Byrne & Whiten 1988.

332 *humans to solve her problems.* Juan Carlos Gómez, "The Emergence of Intentional Communication as a Problem-Solving Strategy in the Gorilla," pp. 333–355 in Parker & Gibson 1990.

332 *three or more individuals."* Sue Savage-Rumbaugh, Mary Ann Romski, William D. Hopkins, and Rose A. Sevcik, "Symbol Acquisition and Use by *Pan troglodytes, Pan paniscus, Homo sapiens,*" pp. 266–295 in Heltne & Marquardt 1989.

332 *ignore that vervet's wrr.* Robert M. Seyfarth & Dorothy L. Cheney, "Do Monkeys Understand Their Relations?" pp. 69–84 in Ristau 1991.

333 *their goslings learn this.* Lorenz 1978.

333 *inherit their mothers' rank.* Bernard Chapais and Carole Gauthier, "Early Agonistic Experience and the Onset of Matrilineal Rank Acquisition in Japanese Macaques," pp. 245–258 in Pereira et al. 1993.

333 *lemurs do not inherit rank.* Michael E. Pereira, "Agonistic Interaction, Dominance Relation, and Ontogenetic Trajectories in Ringtailed Lemurs," pp. 285–305 in Pereira et al. 1993.

334 *macaques in a monkey colony.* Delgado 1963.

334 *sticking out their tongues.* Gopnik et al. 1999.

335 *and 15 weeks old.* Myowa 1996.

335 *teenaged bottlenose dolphin.* Pryor 1975.

335 *is the mirror test.* Gordon G. Gallup Jr., James R. Anderson, and Daniel J. Shillito, "The Mirror Test," pp. 325–333, in Bekoff et al. 2002.

336 *his keepers rigged up.* Linden 2002.

336 *a threat or a peck.* Leslie 1985.

336 *your reflection in the mirror?"* Gordon G. Gallup Jr., James R. Anderson, and Daniel J. Shillito, "The Mirror Test," pp. 325–333, in Bekoff et al. 2002.

336 *investigations were flawed.* Maria Boccia, "Mirror Behavior in Macaques," pp. 350–360 in Parker et al. 1994.

337 *was given the mark test.* Calhoun & Thompson 1988.

337 *to put on makeup.* Savage-Rumbaugh et al. 1998.

338 *pluck out her chin hairs.* Savage-Rumbaugh & Lewin 1994.

338 *a hankering after chic* H. Lyn White Miles, "ME CHANTEK: The Development of Self-Awareness in a Signing Orangutan," pp. 254–272 in Parker et al. 1994.

338 *knows all about mirrors.* Penny Patterson, Q & A session at Chimpanzoo 2002 conference.

338 *and got varying results.* Ujhelyi et al. 2000.

340 *bottlenose dolphins have it.* Reiss & Marino 2001.

341 *only see plain cardboard.* Gibson & Walk 1956.

341 *to an enriched enclosure.* Nadja Wielebnowski, "Contributions of Behavioral Studies to Captive Management and Breeding of Rare and Endangered Mammals," pp. 130–162 in Caro 1998.

341 *often the best adjusted.* Jo Fritz, "Resocialization of Asocial Chimpanzees," pp. 351–359 in Benirschke 1986.

341 *time with a female, Sheba.* Sarah T. Boysen, "Individual Differences in the Cognitive Abilities of Chimpanzees," pp. 335–350 in Wrangham et al. 1994.

342 *"enrichment" on their animals.* Prince-Hughes 2001.

342 *chimpanzees took center stage."* Russon 2000.

342 *make up our own minds.* Peter Marler, "Social Cognition: Are Primates Smarter than Birds?" in Nolan & Ketterson 1996.

343 *for example, macaque society.* Thelma Rowell, "The Myth of Peculiar Primates," pp. 6–16 in Box & Gibson 1999.

343 *invertebrate in the world."* Angier 1998.

343 *for the ungulates.* John A. Byers, "The Ungulate Mind," pp. 35–39 in Bekoff et al. 2002.

343 *their mental powers.* Gould & Gould 1999.

343 *of primate intelligence."* Bshary et al. 2002.

Bibliography

Adamson, Joy. 1961. *Living Free: The Story of Elsa and Her Cubs*. New York: Harcourt, Brace & World.

———. 1972. *Pippa's Challenge*. New York: Harcourt Brace Jovanovich.

———. 1980. *Queen of Shaba*. New York: Harcourt Brace Jovanovich.

Adret, Patrice. 1993. "Operant Conditioning, Song Learning and Imprinting to Taped Song in the Zebra Finch." *Animal Behaviour* 46: 149–159.

Agoramoorthy, Govindasamy, and Minna J. Hsu. 1999. "Rehabilitation and Release of Chimpanzees on a Natural Island." *Journal of Wildlife Rehabilitation* 22 (1): 3–7.

Allen, Colin, and Marc Bekoff. 1997. *Species of Mind: The Philosophy and Biology of Cognitive Ethology*. Cambridge, Massachusetts: MIT Press.

Alp, Rosalind. 1997. "'Stepping-Sticks' and 'Seat-Sticks': New Types of Tools Used by Wild Chimpanzees (*Pan troglodytes*) in Sierra Leone." *American Journal of Primatology* 41: 45–52.

Amory, Cleveland. 1997. *Ranch of Dreams: The Heartwarming Story of America's Most Unusual Animal Sanctuary*. New York: Viking.

Angier, Natalie. August 11, 1998. "At Love and Play under the Sea in Octopus's Garden." *New York Times*, p. B7.

Anthouard, Michel. 1987. "A Study of Social Transmission in Juvenile Dicentrarchus Labras (Pisces, Serranidae), in an Operant Conditioning Situation." *Behaviour* 103: 266–275.

Arcadi, Adam Clark. 2000. "Vocal Responsiveness in Male Wild Chimpanzees: Implications for the Evolution of Language." *Journal of Human Evolution* 39: 205–223.

Armstrong, Edward A. 1947. *Bird Display and Behaviour: An Introduction to the Study of Bird Psychology*. Revised edition. New York: Oxford University Press.

Aronson, Lester R. 1971. "Further Studies on Orientation and Jumping Behavior in the Gobiid Fish, *Bathygobius soporator.*" *Annals of the New York Academy of Sciences* 188: 378–392.

Ashmole, N. Philip, and Humberto Tovar S. 1968. "Prolonged Parental Care in Royal Tern and Other Birds." *The Auk* 85: 90–100.

Askins, Renée. 2002. *Shadow Mountain: A Memoir of Wolves, a Woman, and the Wild.* New York: Doubleday.

Avery, Michael L., Mark A. Pavelka, David L. Bergman, David G. Decker, C. Edward Knittle, and George M. Linz. 1995. "Aversive Conditioning to Reduce Raven Predation on California Least Tern Eggs." *Colonial Waterbirds* 18: 131–138.

Avital, Eytan, and Eva Jablonka. 2000. *Animal Traditions: Behavioural Inheritance in Evolution.* Cambridge: Cambridge University Press.

Bailey, Edgar P., and Nina Faust. 1984. "Distribution and Abundance of Marine Birds Breeding between Amber and Kamishak Bays, Alaska, with Notes on Interactions with Bears." *Western Birds* 15: 161–174.

Balcombe, Jonathan P. 1990. "Vocal Recognition of Pups by Mother Mexican Free-tailed Bats, *Tadarida brasiliensis mexicana.*" *Animal Behaviour* 39: 960–966.

Baldwin, J. D., and J. I. Baldwin. 1978. "Exploration and Play in Howler Monkeys (*Alouatta palliata*)." *Primates* 19 (3): 411–422.

Baptista, Luis F. 1994. "Instinct and Imitation: Cultural Tradition in Birds." *Pacific Discovery* 47 (4): 8–15.

———. 2001. "The Song and Dance of Hummingbirds." *California Wild* 54: 14–20.

Baptista, Luis F., and Karl-Ludwig Schuchmann. 1990. "Song Learning in the Anna Hummingbird (*Calypte anna*)." *Ethology* 84: 15–26.

Baptista, Luis F., Martin L. Morton, and Maria E. Pereyra. 1981. "Interspecific Song Mimesis by a Lincoln Sparrow." *Wilson Bulletin* 93: 265–267.

Baptista, Luis F., Pepper Trail, Barbara B. DeWolfe, and Martin L. Morton. 1993. "Singing and Its Functions in Female White-crowned Sparrows." *Animal Behaviour* 46: 511–524.

Baraff, L., and R. Asmutis. 1995. "Long-Term Association of an Individual Long-finned Pilot Whale with Atlantic White-sided Dolphins." *Abstracts, Eleventh Biennial Conference on the Biology of Marine Mammals, 14–18 December, Orlando, Florida,* p. 7.

Barnea, Anat, and Fernando Nottebohm. 1994. "Seasonal Recruitment of Hippocampal Neurons in Adult Free-Ranging Black-capped Chickadees." *Proceedings of the National Academy of Sciences* 91: 11217–11221.

Barry, Dave. November 21, 1989. "Giving Thanks for Technology." *Atlanta Journal-Constitution,* p. B4.

Beck, Benjamin B. 1980. *Animal Tool Behavior: The Use and Manufacture of Tools by Animals.* New York: Garland STPM Press.

Beissinger, Steven R., and Noel F. R. Snyder. 1992. *New World Parrots in Crisis: Solutions from Conservation Biology.* Washington: Smithsonian Institution Press.

Bekoff, Marc. 2002. *Minding Animals: Awareness, Emotions, and Heart.* Oxford: Oxford University Press.

Bekoff, Marc, and John A. Byers, eds. 1998. *Animal Play: Evolutionary, Comparative, and Ecological Perspectives.* Cambridge: Cambridge University Press.

Bekoff, Marc, Colin Allen, and Gordon M. Burghardt. 2002. *The Cognitive Animal: Empirical and Theoretical Perspectives on Animal Cognition.* Cambridge, Massachusetts: MIT Press.

Benirschke, Kurt, ed. 1986. *Primates: The Road to Self-Sustaining Populations.* New York: Springer-Verlag.

Bermejo, Magdalena, and German Illera. 1999. "Tool-Set for Termite-Fishing and Honey Extraction by Wild Chimpanzees in the Lossi Forest, Congo." *Primates* 40 (4): 619–627.

Berry, Robert B. 1972. "Reproduction by Artificial Insemination in Captive American Goshawks." *Journal of Wildlife Management* 36: 1283–1288.

Black, Jeffrey M., ed. 1996. *Partnerships in Birds: The Study of Monogamy.* Oxford: Oxford University Press.

Blaisdell, Frank. 1999. "Rehabilitation of River Otters." *National Wildlife Rehabilitators Association Quarterly* 17 (2).

Boesch, Christophe. 1991a. "Symbolic Communication in Wild Chimpanzees?" *Human Evolution* 6 (1): 81–90.

———. 1991b. "Teaching among Wild Chimpanzees." *Animal Behaviour* 41: 530–532.

Boesch, Christophe, Paul Marchesi, Nathalie Marchesi, Barbra Fruth, and Frédéric Joulian. 1994. "Is Nutcracking in Wild Chimpanzees a Cultural Behaviour?" *Journal of Human Evolution* 26: 325–338.

Boinski, S. 1988. "Use of a Club by a Wild White-faced Capuchin (*Cebus capucinus*) to Attack a Venomous Snake (*Bothrops asper*)." *American Journal of Primatology* 14: 177–179.

Boinski, S., and D. M. Fragaszy. 1989. "The Ontogeny of Foraging in Squirrel Monkeys, *Saimiri oerstedi*." *Animal Behaviour* 37: 415–428.

Borgia, Gerald. 1986. "Sexual Selection in Bowerbirds." *Scientific American* 254 (6): 92–100.

Borner, Monica. 1985. "The Rehabilitated Chimpanzees of Rubondo Island." *Oryx* 19: 151–154.

Boughman, Janette Wenrick. 1997. "Greater Spear-nosed Bats Give Group-Distinctive Calls." *Behavioral Ecology and Sociobiology* 40: 61–70

———. 1998. "Vocal Learning by Greater Spear-nosed Bats." *Proceedings of the Royal Society of London B* 265: 227–233.

Bourne, Geoffrey H., ed. 1977. *Progress in Ape Research.* New York: Academic Press.

Box, Hilary O., and Kathleen R. Gibson, eds. 1999. *Mammalian Social Learning: Comparative and Ecological Perspectives.* Symposia of the Zoological Society of London 72. Cambridge: Cambridge University Press.

Boyd, Lee. 2002. "Reborn Free." *Natural History* 111 (6): 56–61.

Breland, Keller, and Marian Breland. 1961. "The Misbehavior of Organisms." *American Psychologist* 16: 681–684.

Brooke, James. October 17, 1989. "Gold Monkeys Learn How to Live in Wild in Brazilian Preserve." *New York Times,* p. C4.

Brown, David H., and Kenneth S. Norris. 1956. "Observations of Captive and Wild Cetaceans." *Journal of Mammalogy* 37: 311–326.

Brown, Eleanor D. 1985. "The Role of Song and Vocal Imitation among Common Crows (*Corvus brachyrhynchos*)." *Zeitschrift für Tierpsychologie* 68: 115–136.

Bshary, Redouan, Wolfgang Wickler, and Hans Fricke. 2002. "Fish Cognition: A Primate's Eye View." *Animal Cognition* 5: 1–13.

Bugnyar, Thomas, Maartje Kijne, and Kurt Kotrschal. 2001. "Food Calling in Ravens: Are Yells Referential Signals?" *Animal Behaviour* 61: 949–958.

Buitron, Deborah. 1983. "Variability in the Responses of Black-billed Magpies to Natural Predators." *Behaviour* 87: 209–236.

Burger, Joanna. 2001. *The Parrot Who Owns Me: The Story of a Relationship.* New York: Villard.

Burton, Maurice. 1956. *Infancy in Animals.* New York: Roy Publishers.

Butler, David, and Don Merton. 1992. *The Black Robin: Saving the World's Most Endangered Bird.* Auckland: Oxford University Press.

Byrne, Richard W. 1995. *The Thinking Ape: Evolutionary Origins of Intelligence.* Oxford: Oxford University Press.

Byrne, Richard W., and Jennifer M. E. Byrne. 1993. "Complex Leaf-Gathering Skills of Mountain Gorillas (*Gorilla g. beringei*): Variability and Standardization." *American Journal of Primatology* 31: 241–261.

Byrne, Richard W., and Andrew Whiten, eds. 1988. *Machiavellian Intelligence: Social Expertise and the Evolution of Intellect in Monkeys, Apes, and Humans.* Oxford: Clarendon Press.

Calhoun, Suzanne, and Robert L. Thompson. 1988. "Long-Term Retention of Self-Recognition by Chimpanzee." *American Journal of Primatology* 15: 361–365.

Cambefort, J. P. 1981. "A Comparative Study of Culturally Transmitted Patterns of Feeding Habits in the Chacma Baboon *Papio ursinus* and the Vervet Monkey *Cercopithecus aethiops*." *Folia primatologica* 36: 243–263.

Cameron, Elissa Z., Wayne L. Linklater, Kevin J. Stafford, and Edward O. Minot. 2000. "Aging and Reproductive Success in Horses: Declining Residual Reproductive Value or Just Older and Wiser?" *Behavioral Ecology and Sociobiology* 47: 243–429.

Caro, T. M., ed. 1998. *Behavioral Ecology and Conservation Biology*. New York: Oxford University Press.

Caro, T.M., and M. D. Hauser, 1992. "Is There Teaching in Nonhuman Animals?" *The Quarterly Review of Biology* 67 (2):151–174.

Carter, Janis. 1981. "A Journey to Freedom." *Smithsonian* 12 (1): 90–101.

Catchpole, Clive, Bernd Leisler, and John Dittami. 1986. "Sexual Differences in the Responses of Captive Great Reed Warblers (*Acrocephalus arundinaceus*) to Variation in Song Structure and Repertoire Size." *Ethology* 73: 69–77.

Charles-Dominique, Pierre. 1977. *Ecology and Behaviour of Nocturnal Primates: Prosimians of Equatorial West Africa*. Translated by R. D. Martin. New York: Columbia University Press.

Charrier, Isabelle, Nicolas Mathevon, and Pierre Jouventin. 2001. "Mother's Voice Recognition by Seal Pups." *Nature* 412: 873.

Cheney, D. L., and R. M. Seyfarth. 1985. "Social and Non-Social Knowledge in Vervet Monkeys." *Philosophical Transactions of the Royal Society of London, Series B* 308: 187–201.

Choudhury, Sharmila, and Jeffrey M. Black. 1993. "Mate-Selection Behaviour and Sampling Strategies in Geese." *Animal Behaviour* 46: 747–757.

Colin, Patrick L. 1973. "Burrowing Behavior of the Yellowhead Jawfish, *Opistognathus aurifrons*." *Copeia* 1: 84–90.

Collet, Anne. 2000. *Swimming with Giants: My Encounters with Whales, Dolphins, and Seals*. Translated by Gayle Wurst. Minneapolis: Milkweed Editions.

Collias, Nicholas E., and Elsie C. Collias. 1984. *Nest Building and Bird Behavior*. Princeton, N.J.: Princeton University Press.

Collins, Mark S., Thomas B. Smith, Robert E. Seibels, and I Made Wedana Adi Putra. 1998. "Approaches to the Reintroduction of the Bali Mynah." *Zoo Biology* 17: 267–284.

Corbett, Jim. 1946. *Man-Eaters of Kumaon*. New York: Oxford University Press.

Corbett, Laurie. 1995. *The Dingo in Australia and Asia*. Sydney: University of New South Wales Press.

Corbo, Margarete Sigl, and Diane Marie Barras. 1983. *Arnie, the Darling Starling*. New York: Fawcett Crest.

Coss, Richard A., and Uma Ramakrishnan. 2000. "Perceptual Aspects of Leopard Recognition by Wild Bonnet Macaques (*Macaca radiata*)." *Behaviour* 137: 315–336.

Cristol, Daniel A., Paul V. Switzer, Kara L. Johnson, and Leah S. Walke. 1997. "Crows Do Not Use Automobiles as Nutcrackers: Putting an Anecdote to the Test." *The Auk* 114 (2): 296–298.

Cruickshank, Alick, Jean-Pierre Gautier, and Claude Chappuis. 1993. "Vocal Mimicry in Wild African Grey Parrots, *Psittacus erithacus*." *Ibis* 135: 293–299.

Cypher, Brian L., Gregory D. Warrick, Mark R. M. Otten, Thomas P. O'Farrell, William H. Berry, Charles E. Harris, Thomas T. Kato, Patrick M. McCue, Jerry H. Scrivner, and Bruce W. Zoellick. 2000. "Population Dynamics of San Joaquin Kit Foxes at the Naval Petroleum Reserves in California." *Wildlife Monographs* 145: 1–43.

Davidson, Keay. November 10, 2000. "When It Comes to Tunes, Whales Can Be Teenyboppers." *San Francisco Chronicle*, pp. A4–A5.

Davies, Gareth Huw. No date. "Bird Brains." www.pbs.org/lifeofbirds/brain/index.html

Davies, N. B. 1976. "Parental Care and the Transition to Independent Feeding in the Young Spotted Flycatcher (*Muscicapa striata*)." *Behaviour* 59: 280–295.

———. 2000. *Cuckoos, Cowbirds and Other Cheats*. London: T. & A. D. Poyser.

Davis, Hank. 1984. "Discrimination of the Number Three by a Raccoon (*Procyon lotor*)." *Animal Learning & Behavior* 12: 409–413.

Dawson, Betty V., and B. M. Foss. 1965. "*Observational Learning in Budgerigars*." *Animal Behaviour* 13 (4): 470–474.

Dawson, Stephen M. 1991. "Clicks and Communication: The Behavioural and Social Contexts of Hector's Dolphin Vocalizations." *Ethology* 88: 265–276.

DeBlieu, Jan. 1993. *Meant to Be Wild: The Struggle to Save Endangered Species through Captive Breeding*. Golden, Colo.: Fulcrum Publishing.

Delgado, José M. R. 1963. "Cerebral Heterostimulation in a Monkey Colony." *Science* 141: 161–163.

Derr, Mark. December 24, 2002. "With Dog Detectives, Mistakes Can Happen." *New York Times,* pp. D1, D4.

de Ruiter, L. 1952. "Some Experiments on the Camouflage of Stick Caterpillars." *Behaviour* 4: 222–232.

de Waal, Frans. 1996. *Good Natured: The Origins of Right and Wrong in Humans and Other Animals*. Cambridge, Mass.: Harvard University Press.

———. 2001. *The Ape and the Sushi Master: Cultural Reflections by a Primatologist*. New York: Basic Books.

de Waal, Frans, and Peter L. Tyack. 2003. *Animal Social Complexity: Intelligence, Culture, and Individualized Societies*. Cambridge, Mass.: Harvard University Press.

Diamond, Jared. 1982. "Evolution of Bowerbirds' Bowers: Animal Origins of the Aesthetic Sense." *Nature* 297: 99–102.

———. 1986. "Animal Art: Variation in Bower Decorating Style among Male Bowerbirds, *Amblyornis inornatus*." *Proceedings of the National Academy of Sciences* 83: 3042–3046.

———. 1988. "Experimental Study of Bower Decoration by the Bowerbird, *Amblyornis inornatus*, Using Colored Poker Chips." *The American Naturalist* 131 (5): 631–653.

———. 2001. "Unwritten Knowledge." *Nature* 410: 521.

Diamond, Judy, and Alan B. Bond. 1999. *Kea, Bird of Paradox: The Evolution and Behavior of a New Zealand Parrot.* Berkeley: University of California Press.

Dilger, William C. 1962. "The Behavior of Lovebirds." *Scientific American* 206 (1): 88–98.

Dugatkin, Lee Alan, and Jean-Guy J. Godin. 1992. "Prey Approaching Predators: A Cost-Benefit Perspective." *Annales Zoologici Fennici* 29: 233–252.

Dukas, Reuven. 1998. "Ecological Relevance of Associative Learning in Fruit Fly Larvae." *Behavioral Ecology and Sociobiology* 19: 195–200.

Dunbar, Robin. 1996. *Grooming, Gossip, and the Evolution of Language.* Cambridge, Mass.: Harvard University Press.

Dunn, Euan K. 1972. "Effect of Age on the Fishing Ability of Sandwich Terns *Sterna sandvicensis.*" *Ibis* 114: 360–366.

Durden, Kent. 1972. *Gifts of Eagle.* New York: Simon & Schuster.

Durrell, Gerald. 1977. *Golden Bats and Pink Pigeons.* New York: Simon & Schuster.

Eaton, Joe. 2003. "Gymnogyps Rising?" *Faultline.* http://www.faultline.org/place/2003/04/gymnogypsrising1.html

Eaton, Randall L. 1970. "The Predatory Sequence, with Emphasis on Killing Behavior and Its Ontogeny, in the Cheetah (*Acinonyx jubatus* Schreber)." *Zeitschrift für Tierpsychologie* 27: 492–504.

———. 1979. "A Beluga Whale Imitates Human Speech." *Carnivore* 2 (3): 22–23.

———, ed. 1973. *The World's Cats: Volume 1: Ecology and Conservation.* Winston, Ore.: World Wildlife Safari.

The Economist. January 25, 2003. "Condor Breeding: Spreading Wings." p. 77.

Else, James G., and Phyllis C. Lee, eds. 1986. *Primate Ontogeny, Cognition and Social Behaviour. Selected Proceedings of the Tenth Congress of the International Primatological Society, Volume 3.* Cambridge: Cambridge University Press.

Epinions, "Southwick's Wild Animal Farm, a Gem in Central Massachusetts," a review. Viewed at http://www.epinions.com/content_22753480324 October 4, 2003.

Espmark, Yngve. 1970. "Abnormal Migratory Behaviour in Swedish Reindeer." *Arctic* 23 (3): 199–200.

Fagen, Robert. 1981. *Animal Play Behavior.* New York and Oxford: Oxford University Press.

Fernald, Dodge. 1984. *The Hans Legacy: A Story of Science.* Hillsdale, N.J.: Lawrence Erlbaum Associates.

Ficken, Millicent. 1977. "Avian Play." *The Auk* 94: 573–582.

FitzGibbon, Clare D. 1994. "The Costs and Benefits of Predator Inspection Behaviour in Thomson's Gazelles." *Behavioral Ecology and Sociobiology* 34: 139–148.

Fouts, Roger S., Deborah H. Fouts, and Donna Schoenfeld. 1984. "Sign Language Conversational Interactions between Chimpanzees." *Sign Language Studies* 42: 1–12.

French, Barbara, and John O. Whitaker Jr. 2002. "Helping Orphans Survive." *Bats* 20 (4): 1–3.

Galdikas, Biruté M. F. 1995. *Reflections of Eden: My Years with the Orangutans of Borneo.* Boston: Little, Brown.

Gardner, R. Allen, Beatrix T. Gardner, and Thomas E. Van Cantfort. 1989. *Teaching Sign Language to Chimpanzees.* Albany, N.Y.: SUNY Press.

Gewalt, Wolfgang. 1989. "Orinoco Freshwater Dolphins (*Inia geoffrensis*) Using Self-Produced Air Bubble 'Rings' as Toys." *Aquatic Mammals* 15 (2): 73–79.

Gibson, Eleanor J., and Richard D. Walk. 1956. "The Effect of Prolonged Exposure to Visually Presented Patterns on Learning to Discriminate Them." *Journal of Comparative and Physiological Psychology* 49: 239–242.

Gibson, Kathleen R., and Tim Ingold. 1993. *Tools, Language and Cognition in Human Evolution.* Cambridge: Cambridge University Press.

Giraldeau, Luc-Alain, and Louis Lefebvre. 1987. "Scrounging Prevents Cultural Transmission of Food-Finding Behaviour in Pigeons." *Animal Behaviour* 35: 387–394.

Glickman, Stephen E., and Richard W. Sroges. 1966. "Curiosity in Zoo Animals." *Behaviour* 26: 151–188.

Goodall, Jane. 1986. *The Chimpanzees of Gombe: Patterns of Behavior.* Cambridge, Mass.: Harvard University Press.

Goodey, Wayne, and N. R. Liley. 1986. "The Influence of Early Experience on Escape Behavior in the Guppy (*Poecilia reticulata*)." *Canadian Journal of Zoology* 64: 885–888.

Goodwin, Derek. 1986. *Crows of the World,* 2nd ed. Seattle: University of Washington Press.

Gopnik, Alison, Andrew N. Meltzoff, and Patricia K. Kuhl. 1999. *The Scientist in the Crib: What Early Learning Tells Us About the Mind.* New York: Perennial/HarperCollins.

Gould, James L., and Carol Grant Gould. 1999. *The Animal Mind.* New York: Scientific American Library/HPHLP.

Graham, Frank, Jr. January/February 2000. "The Day of the Condor." *Audubon* 46–53.

Griffin, Andrea S., Daniel T. Blumstein, and Christopher S. Evans. 2000. "Training Captive-Bred or Translocated Animals to Avoid Predators." *Conservation Biology* 14 (5): 1317–1326.

Griffin, Andrea S., Christopher S. Evans, and Daniel T. Blumstein. 2001. "Learning Specificity in Acquired Predator Recognition." *Animal Behaviour* 62: 577–589.

Griffin, Donald R. 1992. *Animal Minds.* Chicago: University of Chicago Press.

Grobecker, David B., and Theodore W. Pietsch. 1978. "Crows Use Automobiles as Nutcrackers." *The Auk* 95: 760–761.

Grunwald, Lisa. May 1995. "Animal Babies." *Life* 18 (5): 60–76.

Guinet, Christophe. 1991. "Intentional Stranding Apprenticeship and Social Play in Killer Whales (*Orcinus orca*)." *Canadian Journal of Zoology* 69: 2712–2716.

Guinet, Christophe, and Jérome Bouvier. 1995. "Development of Intentional Stranding Hunting Techniques in Killer Whale (*Orcinus orca*) Calves at Crozet Archipelago." *Canadian Journal of Zoology* 73: 27–33.

Hamerstrom, Frances. 1957. "The Influence of a Hawk's Appetite on Mobbing." *The Condor* 59 (3): 192–194.

Hammerschmidt, Kurt, Tamara Freudenstein, and Uwe Jürgens. 2001. "Vocal Development in Squirrel Monkeys." *Behaviour* 138: 1179–1204.

Hancock, Lyn. 1977. *There's a Raccoon in My Parka*. New York: Doubleday.

Hansell, Michael. 1987. "What's So Special about Using Tools?" *New Scientist* 113 (January 8): 54–56.

Hare, Brian, Michelle Brown, Christina Williamson, and Michael Tomasello. 2002. "The Domestication of Social Cognition in Dogs." *Science* 298: 1634–1636.

Harrington, Fred H., and L. David Mech. 1979. "Wolf Howling and Its Role in Territory Maintenance." *Behaviour* 68: 207–249.

Hart, Benjamin L., Lynette A. Hart, Michael McCoy, and C. R. Sarath. 2001. "Cognitive Behaviour in Asian Elephants: Use and Modification of Branches for Fly Switching." *Animal Behaviour* 62: 839–847.

Hasler, Arthur D. 1966. *Underwater Guideposts: Homing of Salmon*. Madison: University of Wisconsin Press.

Hasselquist, Dennis, Staffan Bensch, and Torbjörn von Schantz. 1996. "Correlation between Male Song Repertoire, Extra-Pair Paternity and Offspring Survival in the Great Reed Warbler." *Nature* 381: 229–232.

Hauber, Mark E., Stefani A. Russo, and Paul W. Sherman. 2001. "A Password for Species Recognition in a Brood-Parasitic Bird." *Proceedings of the Royal Society of London B* 268: 1041–1048.

Hauber, Mark E., Paul W. Sherman, and Dóra Paprika. 2000. "Self-Referent Phenotype Matching in a Brood Parasite: The Armpit Effect in Brown-headed Cowbirds (*Molothrus ater*)." *Animal Cognition* 3: 113–117.

Hauser, Marc D. 1988. "How Infant Vervet Monkeys Learn to Recognize Starling Alarm Calls: The Role of Experience." *Behaviour* 105: 187–201.

Hauser, Marc D. 2000. *Wild Minds: What Animals Really Think*. New York: Henry Holt.

Hayes, Cathy. 1951. *The Ape in Our House*. New York: Harper & Brothers.

Hedren, Tippi, and Theodore Taylor. 1985. *The Cats of Shambala*. New York: Simon & Schuster.

Heinrich, Bernd. 1987. *One Man's Owl*. Princeton, N.J.: Princeton University Press.

———. 1995. "An Experimental Investigation of Insight in Common Ravens (*Corvus corax*)." *The Auk* 112 (4): 994–1003.

———. 1999. *Mind of the Raven: Investigations and Adventures with Wolf-Birds*. New York: Cliff Street Books/HarperCollins.

Heinsohn, Robert G. 1991. "Slow Learning of Foraging Skills and Extended Parental Care in Cooperatively Breeding White-winged Choughs." *The American Naturalist* 137(6): 864–881.

Helfman, Gene S., and Eric T. Schultz. 1984. "Social Transmission of Behavioural Traditions in a Coral Reef Fish." *Animal Behaviour* 32: 379–384.

Heltne, Paul G., and Linda A. Marquardt, eds. 1989. *Understanding Chimpanzees*. Cambridge, Mass.: Harvard University Press.

Henry, J. David. 1986. *Red Fox: The Catlike Canine*. Washington, D.C.: Smithsonian Institution Press.

Henty, C. J. 1986. "Development of Snail-Smashing by Song Thrushes." *British Birds* 79: 277–281.

Hess, Eckhard H. 1956. "Space Perception in the Chick." *Scientific American* 195 (1): 71–80.

Heyes, Cecilia M., and Bennett G. Galef Jr. 1996. *Social Learning in Animals: The Roots of Culture*. San Diego: Academic Press.

Higuchi, Hiroyoshi. 1987. "Cast Master." *Natural History* 96 (8): 40–43.

———. 1988. "Individual Differences in Bait-Fishing by the Green-backed Heron *Ardeola striata* Associated with Territory Quality." *Ibis* 130: 39–44.

Hirata, Satoshi, Gen Yamakoshi, Shiho Fujita, Gaku Ohashi, and Tetsuro Matsuzawa. 2001. "Capturing and Toying with Hyraxes (*Dendrohyrax dorsalis*) by Wild Chimpanzees (*Pan troglodytes*) at Bossou, Guinea." *American Journal of Primatology* 53: 93–97.

Hiss, A. 1983. "Hoover." *The New Yorker* 58 (January 3): 25–27.

Holden, Karina. 2001. "Polar Bears Swap Cubs." *Nature Australia* 27 (1): 10.

Hollard, Valerie D., and Juan D. Delius. 1982. "Rotational Invariance in Visual Pattern Recognition by Pigeons and Humans." *Science* 218: 804–806.

Hosey, Geoffrey R., Marie Jacques, and Angela Pitts. 1997. "Drinking from Tails: Social Learning of a Novel Behaviour in a Group of Ring-tailed Lemurs (*Lemur catta*)." *Primates* 38 (4): 415–422.

Huber, Ludwig, Sabine Rechberger, and Michael Taborsky. 2001. "Social Learning Affects Object Exploration and Manipulation in Keas, *Nestor notabilis*." *Animal Behaviour* 62: 945–954.

Huffman, M. A., and J. M. Caton. 2001. "Self-Induced Increase of Gut Motility and the Control of Parasitic Infections in Wild Chimpanzees." *International Journal of Primatology* 22 (3): 329–346.

Hunt, Gavin R. 1996. "Manufacture and Use of Hook-Tools by New Caledonian Crows." *Nature* 379: 249–251.

———. 2000. "Tool Use by the New Caledonian Crow *Corvus moneduloides* to Obtain Cerambycidae from Dead Wood." *Emu* 100: 109–114.

Hunt, Gavin R., and Russell D. Gray. 2002. "Species-Wide Manufacture of Stick-Type Tools by New Caledonian Crows." *Emu* 102: 349–353.

Immelmann, Klaus. 1975. "Ecological Significance of Imprinting and Early Learning." *Annual Review of Ecology and Systematics* 6: 15–37.

International Crane Foundation. "Whooping Crane Migration." Viewed at http://www.savingcranes.org/whatsnew/WhoopingMigration2002.asp, April 10, 2003.

Irvine, Blair, Randall S. Wells, and Perry W. Gilbert. 1973. "Conditioning an Atlantic Bottle-nosed Dolphin, *Tursiops truncatus*, to Repel Various Species of Sharks." *Journal of Mammalogy* 54: 503–505.

Itani, Junichiro. 1959. "Paternal Care in the Wild Japanese Monkey, *Macaca fuscata fuscata*." *Primates* 2 (1): 61–93.

Jackson, Robert R., and Simon D. Pollard. 2001. "How to Stalk a Spitting Spider." *Natural History* 110 (October): 16–18.

Jackson, Robert R., Chris M. Carter, and Michael S. Tarsitano. 2001. "Trial-and-Error Solving of a Confinement Problem by a Jumping Spider, *Portia fimbriata*." *Behaviour* 138: 1215–1234.

Janes, Stewart. 1976. "The Apparent Use of Rocks by a Raven in Nest Defense." *The Condor* 78: 409.

Janik, Vincent M., and Peter J. B. Slater. 1995. "Whistle Matching in Wild Bottlenose Dolphins." *Abstracts, Eleventh Biennial Conference on the Biology of Marine Mammals, 14–18 December, Orlando, Florida*, p. 57.

John, E. Roy, Phyllis Chesler, Frank Bartlett, and Ira Victor. 1968. "Observation Learning in Cats." *Science* 159 (3822): 1489–1491.

Joines, Steven. 1977. "A Training Programme Designed to Induce Maternal Behaviour in a Multiparous Female Lowland Gorilla *Gorilla g. gorilla* at the San Diego Wild Animal Park." *International Zoo Yearbook* 17: 185–188.

Jones, Thony B., and Alan C. Kamil. 1973. "Tool-Making and Tool-Using in the Northern Blue Jay." *Behaviour* 180: 1076–1078.

Kay, Jane. April 16, 2002. "Ball of Fluff Is Giant Step in Condor Comeback." *San Francisco Chronicle*.

Keenleyside, Miles H. A., and Cameron E. Prince. 1976. "Spawning-Site Selection in Relation to Parental Care of Eggs in *Aequidens paraguayensis* (Pisces: Ciclidae)." *Canadian Journal of Zoology* 54: 2135–2139.

Keiter, Mary D., Timothy Reichard, and Jack Simmons. 1983. "Removal, Early Hand Rearing, and Successful Reintroduction of an Orangutan (*Pongo pygmaeus pygmaeus abelii*) to Her Mother." *Zoo Biology* 2: 55–59.

Keith-Lucas, Timothy, and Norman Guttman. 1975. "Robust Single Trial

Delayed Backward Conditioning." *Journal of Comparative Psychology* 88 (1): 468–476.

Kendrick, Keith M., Ana P. da Costa, Andrea E. Leigh, Michael R. Hinton, and Jon W. Peirce. 2001. "Sheep Don't Forget a Face." *Nature* 414: 165–166.

Keyser, Leander S. 1894. *In Bird Land*. Chicago: A. C. McClurg.

Kiester, Edwin, Jr., and William Kiester. 2002. "Birdbrain Breakthrough." *Smithsonian* 33 (3): 36–38.

Kilham, Benjamin, and Ed Gray. 2002. *Among the Bears: Raising Orphan Cubs in the Wild*. New York: Henry Holt.

Kipps, Clare. 1953. *Sold for a Farthing*. London: Frederick Muller.

Knudsen, Eric I., and Phyllis F. Knudsen. 1985. "Vision Guides the Adjustment of Auditory Localization in Young Barn Owls." *Science* 230: 545–548.

Komdeur, Jan. 1996. "Influence of Helping and Breeding on Reproductive Performance in the Seychelles Warbler: A Translocation Experiment." *Behavioral Ecology* 7 (3): 326–333.

Krüger, Kobie. 2001. *The Wilderness Family: At Home with Africa's Wildlife*. New York: Ballantine Books.

Kummer, Hans. 1995. *In Quest of the Sacred Baboon*. Translated by M. Ann Biederman-Thorson. Princeton, N.J.: Princeton University Press.

Kummer, Hans, and Jane Goodall. 1985. "Conditions of Innovative Behaviour in Primates." *Philosophical Transactions of the Royal Society of London, Series B* 308: 203–214.

Kunz, T. H., A. L. Allgaier, J. Seyjagat, and R. Caligiuri. 1994. "Allomaternal Care: Helper-Assisted Birth in the Rodrigues Fruit Bat, *Pteropus rodricensis* (Chiroptera: Pteropodidae)." *Journal of the Zoological Society of London* 232: 691–700.

Laidler, Keith. 1980. *The Talking Ape*. New York: Stein & Day.

Laland, Kevin N., and Simon M. Reader. 1999. "Foraging Innovation in the Guppy." *Animal Behaviour* 57: 331–340.

Laurance, William. 2000. *Stinging Trees and Wait-a-Whiles: Confessions of a Rainforest Biologist*. Chicago: University of Chicago Press.

Leslie, Robert Franklin. 1985. *Lorenzo the Magnificent: The Story of an Orphaned Blue Jay*. New York: W. W. Norton and Company.

Leyhausen, Paul. 1979. *Cat Behavior: The Predatory and Social Behavior of Domestic and Wild Cats*. New York: Garland STPM Press.

Liers, Emil E. 1951. Notes on the River Otter (*Lutra canadensis*)." *Journal of Mammalogy* 32 (1): 1–9.

Linden, Eugene. 1986. *Silent Partners: The Legacy of the Ape Language Experiments*. New York: Times Books/Random House.

———. 2002. *The Octopus and the Orangutan: More True Tales of Animal Intrigue, Intelligence, and Ingenuity*. New York: Dutton.

"Living with Tigers," a Discovery Channel Quest show. Viewed at http://dsc.discovery.com/convergence/tigers/tigers.html

Locke, A. 1954. *The Tigers of Trengganu*. New York: Scribner's.

Lockley, R.M. 1961. *Shearwaters*. Garden City, N.Y.: Doubleday.

Lopez, Juan Carlos, and Diana Lopez. 1985. "Killer Whales (*Orcinus orca*) of Patagonia, and Their Behavior of Intentional Stranding While Hunting Near Shore." *Journal of Mammalogy* 66 (1): 181–183.

Lord, Jeanne. 1998. "The Red Fox in Rehabilitation—an Overview." *National Wildlife Rehabilitators Association Quarterly* 16 (4).

Lorenz, Konrad Z. 1952. *King Solomon's Ring*. Translated by Robert Martin. New York: Thomas Y. Crowell.

———. 1978. *The Year of the Greylag Goose*. New York: Harcourt Brace Jovanovich.

Losey, George S., Jr. 1982. "Ecological Cues and Experience Modify Interspecific Aggression by the Damselfish, *Stegastes fasciolatus*." *Behaviour* 81: 14–37.

Lovell, Harvey B. 1958. "Baiting of Fish by a Green Heron." *Wilson Bulletin* 70: 280–281.

McCleneghan, Kim, and Jack A. Ames. 1976. "A Unique Method of Prey Capture by a Sea Otter, *Enhydra lutris*." *Journal of Mammalogy* 57 (2): 410–412.

McComb, Karen, Cynthia Moss, Sarah M. Durant, Lucy Baker, and Soila Sayialel. 2001. "Matriarchs as Repositories of Social Knowledge in African Elephants." *Science* 292: 491–494.

McCowan, Brenda, and Diana Reiss. 2001. "The Fallacy of 'Signature Whistles' in Bottlenose Dolphins: A Comparative Perspective of 'Signature Information' in Animal Vocalizations." *Animal Behaviour* 62: 1151–1162.

Macdonald, David. 1987. *Running with the Fox*. London: Unwin Hyman.

McGrew, W. C., Linda F. Marchant, and Toshisada Nishida, eds. 1996. *Great Ape Societies*. Cambridge: Cambridge University Press.

McGrew, W. C., C. E. G. Tutin, and P. S. Midgett Jr. 1975. "Tool Use in a Group of Captive Chimpanzees." *Zeitschrift für Tierpsychologie* 37: 145–162.

McKeever, Kay. 1994. "Foundling Owls—at Risk?" *National Wildlife Rehabilitators Association Quarterly* 12 (1 & 2).

McLean, Ian G., Natalie T. Schmitt, Peter J. Jarman, Colleen Duncan, and C. D. L. Wynne. 2000. "Learning for Life: Training Marsupials to Recognise Introduced Predators." *Behaviour* 137: 1361–1376.

Maestripieri, Dario. 1995. "Maternal Encouragement in Nonhuman Primates and the Question of Animal Teaching." *Human Nature* 6 (4): 361–378.

Mager, Wim. 1981. "Stimulating Maternal Behaviour in the Lowland Gorilla *Gorilla g. gorilla* at Apeldoorn." *International Zoo Yearbook* 21: 138–143.

Maloney, Richard F., and Ian G. McLean. 1995. "Historical and Experimental Learned Predator Recognition in Free-Living New Zealand Robins." *Animal Behaviour* 50: 1193–1201.

Mann, Janet, Richard C. Connor, Peter L. Tyack, and Hal Whitehead. 2000. *Cetacean Societies: Field Studies of Dolphins and Whales.* Chicago: University of Chicago Press.

Maple, Terry. 1974. "Do Crows Use Automobiles as Nutcrackers?" *Western Birds* 5: 97–98.

Maples, E. G., M. M. Haraway, and C. W. Hutto. 1989. "Development of Coordinated Singing in a Newly Formed Siamang Pair (*Hylobates syndactylus*)." *Zoo Biology* 8: 367–378.

Marcström, V. 1986. "Märkningavripkycklingar—Enutmaning för Hunder." *Svensk Jakt* 124: 508–510.

Markowitz, Hal, Michael Schmidt, Leonie Nadal, and Leslie Squier. 1975. "Do Elephants Ever Forget?" *Journal of Applied Behavior Analysis* 8: 333–335.

Marshall, Andrew J., Richard W. Wrangham, and Adam Clark Arcadi. 1999. "Does Learning Affect the Structure of Vocalizations in Chimpanzees?" *Animal Behaviour* 58: 825–830.

Marzluff, John. 1988 "Do Pinyon Jays Alter Nest Placement Based on Prior Experience?" *Animal Behaviour* 36: 1–10.

Marzluff, John, and Russell P. Balda. 1992. *The Pinyon Jay: Behavioral Ecology of a Colonial and Cooperative Corvid.* London: T & AD Poyser.

Mason, A. G. 1945. "The Display of the Corn-Crake." *British Birds* 38: 350–352.

Mateo, J. M., and R. E. Johnston. 2000. "Kin Recognition and the 'Armpit Effect': Evidence of Self-Referent Matching." *Proceedings of the Royal Society of London, Series B* 267: 695–700.

Mather, Jennifer A., and Roland C. Anderson. 1999. "Exploration, Play, and Habituation in Octopuses (*Octopus dofleini*)." *Journal of Comparative Psychology* 113 (3): 333–338.

Matsumoto-Oda, Akiko. 2002. "Chimpanzees in the Rubondo Island National Park, Tanzania." http://mahale.web.infoseek.co.jp/PAN/7_2/7(2)-02.html

Matthiessen, Peter. 2000. *Tigers in the Snow.* New York: Farrar, Straus & Giroux.

Mayr, Ernst. 1974. "Behavior Programs and Evolutionary Strategies." *American Scientist* 62: 650–659.

Mealy, Nora Steiner. 2002. "The Mark of Cain; Sibling Rivalry to the Death." *California Wild* 55 Winter: 30–34.

Meine, Curt D., and George W. Archibald, eds. 1996. *The Cranes: Status Survey and Conservation Action Plan.* IUCN, Gland, Switzerland, and Cambridge, U.K.: Northern Prairie Wildlife Research Center Home Page. Viewed at http://www.npwrc.usgs.gov/resource/distr/birds/cranes/cranes.htm (Version 02MAR98)/ and http://www.npsc.nbs.gov/resource/distr/birds/cranes/crane.htm October 3, 2003.

Meinertzhagen, R. 1954. "The Education of Young Ospreys." *The Ibis* 96: 153–155.

Miles, Hugh. 1984. *The Track of the Wild Otter*. New York: St. Martin's Press.

Miller, Brian, Richard P. Reading, and Steve Forrest. 1996. *Prairie Night: Black-footed Ferrets and the Recovery of Endangered Species*. Washington, D.C.: Smithsonian Institution Press.

Miller, Olive Thorne. 1894. *A Bird-Lover in the West*. Boston and New York: Houghton Mifflin.

Millikan, George C., and Robert I. Bowman. 1967. "Observations in Galápagos Tool-Using Finches in Captivity." *Living Bird* 6: 23–41.

Milner, Richard. December 1999/January 2000. "Cultureless Vultures." *Natural History* 108:92:

Montgomery, Sy. 1995. *Spell of the Tiger: The Man-Eaters of the Sundarbans*. Boston: Houghton Mifflin.

Morton, Alexandra. 2002. *Listening to Whales: What the Orcas Have Taught Us*. New York: Ballantine Books.

Mueller, Helmut C., and Daniel D. Berger. 1970. "Prey Preferences in the Sharp-shinned Hawk: The Roles of Sex, Experience, and Motivation." *The Auk* 87: 452–457.

Myowa, Masako. 1996. "Imitation of Facial Gestures by an Infant Chimpanzee." *Primates* 37 (2): 207–213.

Myrberg, Arthur A., Jr., and Robert J. Riggio. 1985. "Acoustically Mediated Individual Recognition by a Coral Reef Fish (*Pomacentrus partitus*)." *Animal Behaviour* 33: 411–416.

Nakagawa, Shinichi, Joseph R. Waas, and Masamine Miyazaki. 2001. "Heart Rate Changes Reveal That Little Blue Penguin Chicks (*Eudyptula minor*) Can Use Vocal Signatures to Discriminate Familiar from Unfamiliar Chicks." *Behavioral Ecology and Sociobiology* 50: 180–188.

Nakamichi, Masayuki. 1999. "Spontaneous Use of Sticks as Tools by Captive Gorillas (*Gorilla gorilla gorilla*)." *Primates* 40 (3): 487–498.

Nelson, Katherine, ed. 1989. *Narratives from the Crib*. Cambridge, Mass.: Harvard University Press.

Nice, Margaret M., and Joost ter Pelkwyk. 1941. "Enemy Recognition by the Song Sparrow." *The Auk* 58: 195–214.

Nicolai, Jürgen. 1986. "The Effect of Age on Song-Learning Ability in Two Passerines." *Acta XIX Congressus Internationalis Ornithologici*: 1098–1105.

Nikon Web Magazine. 2000. "Japan—A Land of Birds: Familiar Wild Birds of Japan." Viewed at
http://www.nikon.co.jp/main/eng/society/birds99__00/200005/

Nishida, Toshisada, ed. 1990. *The Chimpanzees of the Mahale Mountains: Sexual and Life History Strategies*. Japan: University of Tokyo Press.

Nishida, Toshisada, and Mariko Hiraiwa. 1982. "Natural History of a Tool-Using Behavior by Wild Chimpanzees in Feeding upon Wood-Boring Ants." *Journal of Human Evolution* 11: 73–99.

Nishiwaki, Masaharu. 1962. "Aerial Photographs Show Sperm Whales' Interesting Habits." *Norsk Hvalfangst-Tidende* 10: 395–398.

Noad, Michael J., Douglas H. Cato, M. M. Bryden, Micheline-N. Jenner, and K. Curt S. Jenner. 2000. "Cultural Revolution in Whale Songs." *Nature* 408: 537.

Nolan, V., Jr., and E. D. Ketterson, eds. 1996. *Current Ornithology*, vol. 13. New York: Plenum Press.

Nottebohm, Fernando, and Marta Nottebohm. 1969. "The Parrots of Bush Bush." *Animal Kingdom* 72: 19–23.

Nottebohm, Fernando, Susan Kasparian, and Constantine Pandazis. 1981. "Brain Space for a Learned Task." *Brain Research* 213: 99–109.

Nzioka, Mary. February 2, 2003. "Samburu Lioness Now Goes for Baby Impala." *East African Standard*. Viewed at http://www.eastandard.net/ archives/February/sun02022003/national/nat02022003003.htm

Oatley, Terry. 1998. *Robins of Africa*. Randburg, South Africa: Acorn Books.

O'Neill, Helen. January 1, 2001. "The Dogs Don't Hunt—They Herd Fish." *San Francisco Chronicle*, p. F1.

Ottoni, Eduardo, and Massimo Mannu. 2001. "Semifree-Ranging Tufted Capuchins (*Cebus apella*) Spontaneously Use Tools to Crack Open Nuts." *International Journal of Primatology* 22 (3): 347–358.

Oulié, Marthe. 1939. *Charcot of the Antarctic*. New York: E. P. Dutton.

Owens, Ian P. F., Candy Rowe, and Adrian L. R. Thomas. 1999. "Sexual Selection, Speciation and Imprinting: Separating the Sheep from the Goats." *Trends in Ecology and Evolution* 14: 131–132.

Pace, Daniela S. 2000. "Fluke-Made Bubble Rings As Toys in Bottlenose Dolphin Calves (*Tursiops truncatus*)." *Aquatic Mammals* 26 (1): 57–64

Park, Ed. 1971. *The World of the Otter*. Philadelphia: J. B. Lippincott.

Parker, Sue Taylor, and Kathleen Rita Gibson, eds. 1990. *"Language" and Intelligence in Monkeys and Apes*. Cambridge: Cambridge University Press.

Parker, Sue Taylor, Robert W. Mitchell, and Maria Boccia, eds. 1994. *Self-Awareness in Animals and Humans: Developmental Perspectives*. Cambridge: Cambridge University Press.

Patterson, Bruce D., and Ellis J. Neiburger. 2001. "Lion with a Sore Tooth." *Nature Australia* 26 (12): 12.

Patterson, Francine, and Eugene Linden. 1981. *The Education of Koko*. New York: Holt, Rinehart & Winston.

Patterson, Gareth. 1994. *Last of the Free*. New York: St. Martin's Press.

Penn, Dustin, and Wayne Potts. 1998. "MHC-Disassortative Mating Preferences Reversed by Cross-Fostering." *Proceedings of the Royal Society of London B* 265: 1299–1306.

Pepperberg, Irene Maxine. 1983. "Cognition in the African Grey Parrot: Preliminary Evidence for Auditory/Vocal Comprehension of the Class Concept." *Animal Learning & Behavior* 11 (2): 179–185.

———. 1999. *The Alex Studies: Cognitive and Communicative Abilities of Grey Parrots*. Cambridge, Mass.: Harvard University Press.

Pereira, Michael E., and Lynn A. Fairbanks, eds. 1993. *Juvenile Primates: Life History, Development, and Behavior*. New York: Oxford University Press.

Perry, Richard. 1969. *The World of the Giant Panda*. New York: Bantam.

Pettifer, Howard L. 1980. "The Experimental Release of Captive-Bred Cheetah (*Acinonyx jubatus*) into the Natural Environment." In *Worldwide Furbearer Conference Proceedings, August 3–11, 1980*, vol. 2. Edited by Joseph A. Chapman and Duane Pursley. Falls Church, Va.: R. R. Donnelly.

Pitcher, T. J., D. A. Green, and A. E. Magurran. 1986. "Dicing with Death: Predator Inspection Behaviour in Minnow Shoals." *Journal of Fish Biology* 28: 439–448.

Pittman, Craig. January 2003. "Making Whoopee." *Smithsonian* 33 (10): 92–95.

Pollack, Andrew. December 5, 1992. "Quoth Japan's Crows: Evermore and Everywhere." *New York Times*, p. 2.

Pongrácz, Péter, Ádám Miklósi, Enikö Kubinyi, Kata Gurobi, József Topál, and Vilmos Csányi. 2001. "Social Learning in Dogs: The Effect of a Human Demonstrator on the Performance of Dogs in a Detour Task." *Animal Behaviour* 62: 1109–1117.

Porter, Fran L. 1979. "Social Behavior in the Leaf-nosed Bat, *Carollia perspicillata*." *Zeitschrift für Tierpsychologie* 50: 1–8.

Povinelli, Daniel J., and John G. H. Cant. 1995. "Arboreal Clambering and the Evolution of Self-Conception." *The Quarterly Review of Biology* 70 (4): 393–421.

Pratt, Ambrose. 1940. *The Lore of the Lyrebird*. Melbourne: Robertson & Mullens.

Price, Jordan, and R. Haven Wiley. 2000. "Duets and Drawls." *Natural History* 109 (March): 50–53.

Prince-Hughes, Dawn. 2001. *Gorillas among Us: A Primate Ethnographer's Book of Days*. Tucson, Ariz.: University of Arizona Press.

Prins, H. H. T. 1996. Ecology and Behavior of the African Buffalo: Social Inequality and Decision Making. London: Chapman & Hall.

Pryor, Karen. 1975. *Lads Before the Wind*. New York: Harper & Row.

———. 1999. *Don't Shoot the Dog! The New Art of Teaching and Training*. Rev. ed. New York: Bantam Books.

Pryor, Karen, Richard Haag, and Joseph O'Reilly. 1969. "The Creative Porpoise: Training for Novel Behavior." *Journal of the Experimental Analysis of Behavior* 12: 653–661.

Pryor, Karen, Jon Lindbergh, Scott Lindbergh, and Raquel Milano. 1990. "A Dolphin-Human Fishing Cooperative in Brazil." *Marine Mammal Science* 6 (1): 77–82.

Quinney, T. E., and P. C. Smith. 1980. Comparative Foraging Behaviour and Efficiency of Adult and Juvenile Great Blue Herons. *Canadian Journal of Zoology* 58 (1): 1168–1173.

Ralls, Katherine, Patricia Fiorelli, and Sheri Gish. 1985. "Vocalizations and Vocal Mimicry in Captive Harbor Seals, *Phoca vitulina*." *Canadian Journal of Zoology* 63:1050–1056.

Randall, Jan A. 1995. "Modification of Footdrumming Signatures by Kangaroo Rats: Changing Territories and Gaining New Neighbours." *Animal Behaviour* 49: 1227–1237.

Redshaw, Margaret. 1978. "Cognitive Development in Human and Gorilla Infants." *Journal of Human Evolution* 7: 133–141.

Reebs, Stéphan. 2001. *Fish Behavior in the Aquarium and in the Wild*. Ithaca, N.Y.: Cornell University Press.

Reilly, Pauline. 1988. *The Lyrebird: A Natural History.* Kensington, New South Wales: New South Wales University Press.

Reiss, Diana. 1991. "The Secrets of the Dolphins." New York: Avon Camelot.

Reiss, Diana, and Lori Marino. 2001. "Mirror Self-Recognition in the Bottlenose Dolphin: A Case of Cognitive Convergence." *Proceedings of the National Academy of Sciences* 98 (16): 5937–5942.

Riedman, Marianne. 1996. "A Smorgasbord for Sea Otters: Cuisine and Table Manners Learned from Mom." *Pacific Discovery* 49 (4): 16–23.

Riedman, M. L., M. M. Staedler, J. A. Estes, B. Hrabrich, and D. R. Carlson. 1995. "Dietary Specialization and Kleptoparasitism in California Sea Otters." *Abstracts, Eleventh Biennial Conference on the Biology of Marine Mammals, 14–18 December, Orlando, Florida*, p. 97.

Rijt-Plooij, Hedwig H. C. van de, and Frans X. Plooij. 1987. "Growing Independence, Conflict and Learning in Mother-Infant Relations in Free-Ranging Chimpanzees." *Behaviour* 101 (1–3): 1–86.

Ristau, Carolyn A., ed. 1991. *Cognitive Ethology: The Minds of Other Animals: Essays in Honor of Donald R. Griffin*. Hillsdale, N.J.: Lawrence Erlbaum Associates.

Robertson, Gregory J. 1998. "Egg Adoption Can Explain Joint Egg-Laying in Common Eiders." *Behavioral Ecology and Sociobiology* 43: 289–296.

Rogers, Charles M., and Richard K. Davenport. 1969. "Effects of Restricted Rearing on Sexual Behavior of Chimpanzees." *Developmental Psychology* 1(3): 200–204.

Roitblat, H. L., T. G. Bever, and H. S. Terrace, eds. 1984. *Animal Cognition*. Hillsdale, N.J.: Lawrence Erlbaum Associates.

Rood, Jon P. 1983. "Banded Mongoose Rescues Pack Member from Eagle." *Animal Behaviour* 31: 1261–1262.

Rosenblatt, Jay S., Robert A. Hinde, Evelyn Shaw, and Colin Beer, eds. 1976. *Advances in the Study of Behavior,* vol. 6. New York: Academic Press.

Ross, John F. 2002 "No Exit." *Smithsonian* 33 (6): 62–69.

Rowley, Ian. 1991. "A Confusion of Cockatoos." *Natural History* 100 (11): 44–51.

Rowley, Ian, and Graeme Chapman. 1986. "Cross-Fostering, Imprinting and Learning in Two Sympatric Species of Cockatoo." *Behaviour* 96: 1–16.

———. 1991. "The Breeding Biology, Food, Social Organisation, and Conservation of the Major Mitchell or Pink Cockatoo, *Cacatua leadbeateri*, on the Margin of the Western Australian Wheatbelt." *Australian Journal of Zoology* 39: 211–61.

Rusch, Kathryn. 2001. "Whispers in the Canyon." *Natural History* 110: (July–August): 33–36.

Russon, Anne E. 2000. *Orangutans: Wizards of the Rain Forest*. Buffalo: Firefly Books.

Russon, Anne E., Kim A. Bard, and Sue Taylor Parker, eds. 1996. *Reaching into Thought: The Minds of the Great Apes*. Cambridge: Cambridge University Press.

San Francisco Chronicle. September 2, 1996. "Condors Just Don't Learn Hate: Endangered Birds Are Flying into Lots of Trouble." p. A8.

Santos, M. C. de O. 1995. "Behavior of a Lone, Wild, Sociable Bottlenose Dolphin, *Tursiops truncatus*, and a Case of Human Fatality in Brazil." *Abstracts, Eleventh Biennial Conference on the Biology of Marine Mammals, 14–18 December, Orlando, Florida*, p. 101.

Savage-Rumbaugh, Sue, and Roger Lewin. 1994. *Kanzi: The Ape at the Brink of the Human Mind*. New York: John Wiley & Sons.

Savage-Rumbaugh, Sue, Stuart G. Shanker, and Talbot J. Taylor. 1998. *Apes, Language, and the Human Mind*. New York: Oxford University Press.

Schaller, George. 1967. *The Deer and the Tiger: A Study of Wildlife in India*. Chicago: University of Chicago Press.

———. 1972. *The Serengeti Lion: A Study of Predator-Prey Relations*. Chicago: University of Chicago Press.

Scherrer, J. Andrew, and Gerald S. Wilkinson. 1993. "Evening Bat Isolation Calls Provide Evidence for Heritable Signatures." *Animal Behaviour* 46: 847–860.

Schmid, Randolph E. 2002. "Tool-Making Crow Impresses Scientists." Associated Press Syndicate. *San Francisco Chronicle*, August 9, 2002. p. A1.

Shah, Anup, and Manoj Shah. 1996. *A Tiger's Tale: The Indian Tiger's Struggle for Survival in the Wild*. Kingston-upon-Thames, Surrey: Fountain Press.

Sheldrick, Daphne. 1966. *The Orphans of Tsavo*. New York: David McKay.

Sherry, David F., and B. G. Galef Jr. 1984. "Cultural Transmission without Imitation: Milk Bottle Opening by Birds." *Animal Behaviour* 32: 937–938.

———. 1990. "Social Learning without Imitation: More About Milk Bottle Opening by Birds." *Animal Behaviour* 40: 987–989.

Shumaker, Robert W. 1993. "How Four Strangers Became a Family and Mothers Swapped Babies." *Gorilla* 16 (2): 2–4.

Shumaker, Robert W., Ann M. Palkovich, Benjamin B. Beck, Gregory A. Guagnano, and Harold Morowitz. 2001. "Spontaneous Use of Magnitude

Discrimination and Ordination by the Orangutan (*Pongo pygmaeus*)." *Journal of Comparative Psychology* 115 (4): 385–391.

Sigg, Hans. 1980. "Differentiation of Female Positions in Hamadryas One-Male Units." *Zeitschrift für Tierpsychologie* 53: 265–302.

Simões-Lopes, Paulo C., Marta E. Fabián, and João O. Menegheti. 1998. "Dolphin Interactions with the Mullet Artisanal Fishing on Southern Brazil: A Qualitative and Quantitative Approach." *Revista Brasileira de Zoologia* 15 (3): 709–726.

Singapore Zoological Gardens Docents. 2000. "Patas Monkey (*Erythrocebus patas*)." Viewed at http://www.szgdocent.org/pp/p-patas.htm

Singh, Arjan. 1981. *Tara—A Tigress*. London: Quartet Books.

———. 1982. *Prince of Cats*. London: Jonathan Cape.

———. 1984. *Tiger! Tiger!* London: Jonathan Cape.

Sinn, David L., Nancy A. Perrin, Jennifer A. Mather, and Roland C. Anderson. 2001. "Early Temperamental Traits in an Octopus (*Octopus bimaculoides*). *Journal of Comparative Psychology* 115 (4): 351–364.

Sisson, Robert F. 1974. "Aha! It Really Works!" *National Geographic* 145: 142–147.

Slabbert, J. M., and O. Anne E. Rasa. 1997. "Observational Learning of an Acquired Maternal Behaviour Pattern by Working Dog Pups: An Alternative Training Method?" *Applied Animal Behaviour Science* 53: 309–316.

Slater, Peter J. B., ed. 1987. *Encyclopedia of Animal Behavior*. New York: Facts on File.

Smolker, Rachel. 2001. *To Touch a Wild Dolphin: A Journey of Discovery with the Sea's Most Intelligent Creatures*. New York: Doubleday.

Snowdon, Charles T., and A. Margaret Elowson. 2001. "'Babbling' in Pygmy Marmosets: Development after Infancy." *Behaviour* 138: 1235–1248.

Snowdon, Charles T., and Martine Hausberger. 1997. *Social Influences on Vocal Development*. Cambridge: Cambridge University Press.

Snyder, Noel, and Helen Snyder. 2000. *The California Condor: A Saga of Natural History and Conservation*. London and San Diego: Academic Press.

Snyder, Noel, Susan E. Koenig, James Koschmann, Helen Snyder, and Terry B. Johnson. 1994. "Thick-billed Parrot Releases in Arizona." *The Condor* 96 (4): 845–862.

Sonerud, Geir A. 1988. "To Distract Display or Not: Grouse Hens and Foxes." *Oikos* 51: 233–237.

Soproni, Krisztina, Ádám Miklósi, József Topál, and Vilmos Csányi. 2001. "Comprehension of Communicative Signs in Pet Dogs (*Canis familiaris*)." *Journal of Comparative Psychology* 115 (2): 122–126.

———. 2002. "Dogs' (*Canis familiaris*) Responsiveness to Human Pointing Gestures." *Journal of Comparative Psychology* 116 (1): 27–34.

Stander, P. E. 1992. "Cooperative Hunting in Lions: The Role of the Individual." *Behavioral Ecology and Sociobiology* 29: 445–454.

Stirling, Ian. 1988. *Polar Bears.* Ann Arbor: University of Michigan Press.

Strier, Karen B. 1999. *Faces in the Forest: The Endangered Muriqui Monkeys of Brazil.* Reprint with new preface. Cambridge, Mass.: Harvard University Press.

Sullivan, M. G. 1979. "Blue Grouse Hen–Black Bear Confrontation." *Canadian Field–Naturalist* 93: 200.

Sumita, Kyoko, Jean Kitahara-Frisch, and Kohshi Norikoshi. 1985. "The Acquisition of Stone-Tool Use in Captive Chimpanzees." *Primates* 26 (2): 168–181.

Summers-Smith, D. 1963. *The House Sparrow.* London: Collins.

Sunquist, Fiona, and Mel Sunquist. 1988. *Tiger Moon.* Chicago: University of Chicago Press.

Tayler, C. K., and G. S. Saayman. 1973. "Imitative Behaviour by Indian Ocean Bottlenose Dolphins (*Tursiops aduncus*) in Captivity." *Behaviour* 44: 286–298.

Temerlin, Maurice K. 1975. *Lucy: Growing Up Human: A Chimpanzee Daughter in a Psychotherapist's Family.* Palo Alto, Calif.: Science and Behavior Books.

Temple, Stanley A., ed. 1977. *Endangered Birds: Management Techniques for Preserving Threatened Species.* Madison: University of Wisconsin Press.

ten Cate, Carel and Dave R. Vos. 1999. "Sexual Imprinting and Evolutionary Processes in Birds: A Reassessment." *Advances in the Study of Behavior* 28: 1–31. New York: Academic Press.

Terrace, Herbert S. 1979. *Nim.* New York: Alfred A. Knopf.

Thaler, Ellen. 1986. "Studies on the Behaviour of Some Phasianidae Chicks at the Alpenzoo–Innsbruck." *Proceedings of the Third International Symposium on Pheasants in Asia, Chiang Mai, Thailand.* Fordingbridge, Hampshire: World Pheasant Association.

Thapar, Valmik. 1986. *Tiger: Portrait of a Predator.* London: Collins.

———. 1989. *The Secret Life of Tigers.* Oxford: Oxford University Press.

Thomas, Elizabeth Marshall. 1994. *The Tribe of Tiger: Cats and Their Culture.* New York: Simon & Schuster.

———. 2000. *The Social Lives of Dogs: The Grace of Canine Company.* New York: Simon & Schuster.

Thomsett, Simon. 1983. "The Successful Rehabilitation of a Crowned Eagle." *International Zoo Yearbook* 23: 62–4.

Thouless, C. R., J. H. Fanshawe, and B. C. R. Bertram. 1989. "Egyptian Vultures *Neophron percnopterus* and Ostrich *Struthio camelus* Eggs: The Origins of Stone-Throwing Behaviour." *Ibis* 131: 9–15.

Tilson, Ronald, Gerald Brady, Kathy Traylor-Holzer, and Douglas Armstrong, eds. 1994. *Management and Conservation of Captive Tigers,* 2nd ed. Apple Valley, Minn.: Minnesota Zoo.

Todd, John H. 1971. "The Chemical Languages of Fishes." *Scientific American* 224 (5): 98–108.

Tokida, Eishi, Ichirou Tanaka, Haruo Takefushi, and Toshio Hagiwara. 1994. "Tool-Using in Japanese Macaques: Use of Stones to Obtain Fruit from a Pipe." *Animal Behaviour* 47: 1023–1030.

Tomasello, Michael, Sue Savage-Rumbaugh, and Ann Cale Kruger. 1993. "Imitative Learning of Actions on Objects by Children, Chimpanzees, and Enculturated Chimpanzees." *Child Development* 64:1688–1705.

Tulley, J. J., and F. A. Huntingford. 1987. "Paternal Care and the Development of Adaptive Variation in Anti-Predator Responses in Sticklebacks." *Animal Behaviour* 35 (5): 1570–1572.

Ujhelyi, Mária, Björn Merker, Pál Buk, and Thomas Geissmann. 2000. "Observations on the Behavior of Gibbons (*Hylobates leucogenys, H. gabriellae*, and *H. lar*) in the Presence of Mirrors." *Journal of Comparative Psychology* 114 (3): 253–262.

Uy, J. Albert C. 2002. "Say It with Bowers." *Natural History* 111 (2): 76–83.

Valderrama, Ximena, John G. Robinson, Athula B. Attygalle, and Thomas Eisner. 2000. "Seasonal Anointment with Millipedes in a Wild Primate: A Chemical Defense against Insects?" *Journal of Chemical Ecology* 26 (12): 2781–2790.

van Heezik, Yolanda, and Richard Maloney. 1997. "Update on the Houbara Bustard Reintroduction Programme in Saudi Arabia." *Reintroduction News* 13: 3–4.

van Heezik, Yolanda, and Philip J. Seddon. 2001. "Born to Be Tame." *Natural History* 110 (June): 58–63.

van Heezik, Yolanda, Philip J. Seddon, and Richard F. Maloney. 1999. "Helping Reintroduced Houbara Bustards Avoid Predation: Effective Anti-Predator Training and the Predictive Value of Pre-Release Behaviour." *Animal Conservation* 2: 155–163.

van Schaik, C. P., and E. A. Fox. 1996. "Manufacture and Use of Tools in Wild Sumatran Orangutans." *Naturwissenschaften* 83: 186–188.

Veà, Joaquim J., and Jordi Sabater-Pi. 1998. "Spontaneous Pointing Behaviour in the Wild Pygmy Chimpanzee (*Pan paniscus*)." *Folia Primatologica* 69: 289–290.

Visalberghi, Elisabetta, and Luca Limongelli. 1994. "Lack of Comprehension of Cause-Effect Relations in Tool-Using Capuchin Monkeys (*Cebus apella*)." *Journal of Comparative Psychology* 108 (1):18–22.

Voelkl, Bernhard, and Ludwig Huber. 2000. "True Imitation in Marmosets." *Animal Behaviour* 60: 195–2002.

von Frisch, Karl. 1938. "The Sense of Hearing in Fish." *Nature* 141: 8–11.

Walker, Stephen. 1987. *Animal Learning: An Introduction.* New York: Routledge & Kegan Paul.

Wang, Zuoxin, and Thomas R. Insel. 1996. "Parental Behavior in Voles." *Advances in the Study of Behavior* 25: 361–384.

Wanker, Ralf, Jasmin Apcin, Bert Jennerjahn, and Birte Waibel. 1998. "Discrimination of Different Social Companions in Spectacled Parrotlets (*Forpus conspicillatus*): Evidence for Individual Vocal Recognition." *Behavioral Ecology and Sociobiology* 43: 197–202.

Wasser, Samuel K. 1978. "Structure and Function of Play in the Tiger." *Carnivore* 1 (3): 27–40.

Wayre, Philip. 1976. *The River People*. New York: Taplinger Publishing.

———. 1979. *The Private Life of the Otter*. London: B. T. Batsford.

Weidensaul, Scott. 2002. *The Ghost with Trembling Wings: Science, Wishful Thinking, and the Search for Lost Spaces*. New York: North Point Press.

Weigl, Peter D., and Elinor V. Hanson. 1980. "Observational Learning and the Feeding Behavior of the Red Squirrel (*Tamiasciurus hudsonicus*)" The Ontogeny of Optimization." *Ecology* 61(2): 213–218.

Weinrich, Mason T., Mark R. Schilling, and Cynthia R. Belt. 1992. "Evidence for Acquisition of a Novel Feeding Behaviour: Lobtail Feeding in Humpback Whales, *Megaptera novaeangliae*." *Animal Behaviour* 44: 1059–1072.

Weir, Alex A. S., Jackie Chappell, and Alex Kacelnik. 2002. "Shaping of Hooks in New Caledonian Crows." *Science* 287: 981.

Weisbord, Merrily, and Kim Kachanoff. 2000. *Dogs with Jobs: Working Dogs around the World*. New York: Pocket Books.

Wemelsfelder, Françoise, Marie Haskell, Michael T. Mendl, Sheena Calvert, and Alistair B. Lawrence. 2000. "Diversity of Behavior during Novel Object Tests Is Reduced in Pigs Housed in Substrate-Impoverished Conditions." *Animal Behaviour* 60: 385–394.

Westergaard, Gregory Charles, and Dorothy Munkenbeck Fragaszy. 1987. "The Manufacture and Use of Tools By Capuchin Monkeys (*Cebus apella*)." *Journal of Comparative Psychology* 101 (2): 159–168.

Whitaker, Barbara. November 2, 1999. "Released to the Wild, Condors Choose a Nice Peopled Retreat." *New York Times*, p. A8.

White, Jan, and Cheryl Millham. 1993. "Rehabilitation Notes: Common Merganser (*Mergus merganser*)." *Journal of Wildlife Rehabilitation* 16 (4): 8–12.

Whitehead, Hal. 1990. *Voyage to the Whales*. Post Mills, Vt.: Chelsea Green.

Whiten, Andrew, and Richard W. Byrne, eds. 1997. *Machiavellian Intelligence II: Extensions and Evaluations*. Cambridge: Cambridge University Press.

Whiten, Andrew, J. Goodall, W. C. McGrew, T. Nishida, V. Reynolds, Y. Sugiyama, C. E. G. Tutin, R. W. Wrangham, and C. Boesch. 1999. "Cultures in Chimpanzees." *Nature* 399: 682–685.

Wickler, Wolfgang. 1980. "Vocal Dueting and the Pair Bond." *Zeitschrift für Tierpsychologie* 52: 201–209.

Wilson, Susan C., and Devra G. Kleiman. 1974. "Eliciting Play: A Comparative Study." *American Zoologist* 14: 341–370.

Wood, G. A. 1988. "Further Field Observations of the Palm Cockatoo *Probosciger aterrimus* in the Cape York Peninsula, Queensland." *Corella* 12 (2): 48–52.

Wrangham, Richard W., W. C. McGrew, Frans B. M. de Waal, and Paul G. Heltne, with assistance from Linda A. Marquardt, eds. 1994. *Chimpanzee Cultures*. Cambridge, Mass.: Harvard University Press.

Xitco, Mark J., Jr., John D. Gory, and Stan A. Kuzcaj II. 2001. "Spontaneous Pointing by Bottlenose Dolphins (*Tursiops truncatus*)." *Animal Cognition* 4: 115–123.

Yamaguchi, Ayako. 2001. "Sex Differences in Vocal Learning in Birds." *Nature* 411: 257–258.

Yamakoshi, Gen. 1998. "Dietary Responses to Fruit Scarcity of Wild Chimpanzees at Bossou, Guinea: Possible Implications for Ecological Importance of Tool Use." *American Journal of Physical Anthropology* 106: 283–295.

Yamamoto, J. T., K. M. Shields, J. R. Millam, T. E. Roudybush, and C. R. Grau. 1989. "Reproductive Activity of Force-Paired Cockatiels (*Nymphicus hollandicus*)." *The Auk* 106: 86–93.

Yoerg, Sonja I. 1994. "Development of Foraging Behaviour in the Eurasian Dipper, *Cinclus cinclus*, from Fledging until Dispersal." *Animal Behaviour* 47: 577–588.

———. 1998. "Foraging Behavior Predicts Age at Independence in Juvenile Eurasian Dippers (*Cinclus cinclus*)." *Behavioral Ecology* 9 (5): 471–477.

———. 2001. *Clever as a Fox: Animal Intelligence and What It Can Teach Us about Ourselves*. New York: Bloomsbury.

Zann, Richard A. 1996. *The Zebra Finch: A Synthesis of Field and Laboratory Studies*. Oxford: Oxford University Press.

Zentall, Thomas R., and Bennett G. Galef Jr, eds. 1988. *Social Learning: Psychological and Biological Perspectives*. Hillsdale, N.J.: Lawrence Erlbaum Associates.

Zuk, Marlene. 2002. *Sexual Selections: What We Can and Can't Learn about Sex from Animals*. Berkeley: University of California Press.

Żurowski, Wirgiliusz. 1979. "Preliminary Results of European Beaver Reintroduction in the Tributary Streams of the Vistula River." *Acta Theriologica* 24 (7): 85–91.

Zwank, Phillip J., James P. Geaghan, and Donna A. Dewhurst. 1988. "Foraging Differences between Native and Released Mississippi Sandhill Cranes: Implications for Conservation." *Conservation Biology* 2: 386–390.

Acknowledgments

I'VE ENCOUNTERED so many interesting and helpful people that I know I am leaving some out.

For their generous willingness to give me information and tell me stories, thanks to Terri Block, Alan Bond and Judy Diamond, Janette Wenrick Boughman, Gerry Brady, Richard G. Coss, Martha White Coyote, Michael Dee, Saba and Dudu Douglas-Hamilton, Mike Dulaney, Barbara French, Stephen Glickman, Sarah McCarthy, Greta McMillan, Jan Mosterd, David G. Myers, Irene Pepperberg, David Powell, Bill Rainey, Pat Savage of the World Pheasant Association, Lori Tarou, Steve Taylor, Janet Tilson, Ron Tilson, Beth Weise (who translated from the Swedish), Patricia Winters, and Bradburn Young.

For their assistance, kindness, and encouragement on this project, thanks to Paulina Borsook, Linda Dyer, Mary Lynn Fischer, Jim Fisher, Thaisa Frank, Andrew Gunther, Barbara Gunther, Sumana Harihareswara, Cynthia Heimel, Janet Jones, Karen Kienitz, Jack King, Bruce Koball, Mary Susan Kuhn, Jeffrey Moussaieff Masson, John McCarthy, Teresa Moore, and Judith Newman. As promised, I grovel before the kindness and skills of Brady Lea, David Gallagher, and Su. Suttle Taggart. Abjectly. The libraries of San Francisco State University, the Mechanics' Institute, the California Academy of Sciences, the University of California at San Francisco, and the University of California at Berkeley were tremendously helpful, but I owe the greatest debt to the interlibrary loan staff at the San Francisco Public Library, who hunted out any number of obscure journal articles and books with never a peep of complaint or incredulity.

Thanks also to the crowd at the water cooler, also known as The Well (well.com). Get back to work. Also Salon.com, with special thanks to Andrew Leonard.

My patient, smart, and cheerful agent, Stuart Krichevsky, made it all happen.

Thanks to my wonderful editor at HarperCollins, Jennifer Brehl, who took on a twice-orphaned book with the enthusiasm and understanding of someone who'd been there from the beginning, and who generously sent me Terry Pratchett books when she must have known I'd drop work to read them.

Thanks most of all to my family, who heard all these animal stories and many more, and continued to encourage me: Joseph Gunther, Kitty McCarthy, and Daniel Gunther. I'm a lucky person.

Index

active teaching, 293, 294, 298–99, 349
Adachi, Yasuhiro, 236
Adamson, George, 76, 77, 175, 199
Adamson, Joy, 76, 175, 199–200
adoption, 282–85, 310. *See also* cross-fostering
Adret, Patrice, 96
Agoramoorthy, Govindasamy, 267, 268
Aisner, Ran, 178
alarm calls, 116–17, 136, 185, 186, 190, 196, 294, 318
  mimicry in birds and, 111
  modular aspect of, 318
  unreliable, 332–33
"alarm" stage, 184
albatrosses, 81
Alex (grey parrot), 125–27, 140, 303–4, 320–21
Allen, Colin, 53
Alp, Rosalind, 226
American Sign Language, 91, 127–30, 131–32, 148, 222, 304
Anderson, James, 336
Anderson, Roland, 343
anemonefish, 143–44
antelopes, 203
Antinucci, Francesco, 33
Ape and the Sushi Master, The (de Waal), 247
apprenticeship, 297–98
arboreal clambering theory, 39–40, 328
Arcadi, Adam Clark, 121
"Are Hyenas Primates?" (Frank), 343
"Are Primates Smarter than Birds?"

(Marler), 342
artificial symbol systems, 132–35, 249
Asquith, P. J., 324
associative learning, 347
asymmetric marker, 11–12
Avital, Eytan, 291, 299, 320

baboons, 116
  "carrying dance" and, 56
  curiosity and, 325–26
  elder pass on culture to, 248
  innovations by, 212
  male-female relations and, 261–62
  meeting for lunch, 136–37
  memory and, 328
  nonvocal communication, tail-drinking and, 26
  oblique learning and, 15
  social intelligence and, 331
backward conditioning, 10–11
bait-fishing, 215–16
Balda, Russell, 318
Baptista, Luis, 98, 100, 104
Barash, David, 107
Barry, Dave, 197n
bats
  communication and, 92, 105–6
  imprinting and, 63
  learning to fly, 295–96
  learning predation skills, 159
  midwifery and, 286–87
bears
  adoption of cubs by, 283

humans and, 199, 260
learning to climb, 37
learning predation skills, 162–63
learning what to eat, 148
moose learns to fear, 197–98
playing and, 47–48
teach young to hunt, 296
Beau Geste hypothesis, 110
beavers, 201
Bekoff, Marc, 47, 52–53, 54
Berger, Joel, 197, 198
Berry, Robert, 78–79
"best-of-N" strategy, 275, 276
Biben, Maxeen, 54
birds. *See also* specific types
dialects and, 100
fear of bird flying overhead, 185–86
female singing in, 99
imitation and, 102–3, 108
imprinting and, 60
intelligence of, 342–43
language and, vs. apes, 135
learn to fly, 41–44
learn which hawks are scariest, 190–91
mobbing by, 189–90
nest-building by, 240–43
nest parasitism and, 80–81
new neurons and songs, 100
parenting by, and age, 290–91
recognize own young, 81–82
rescue centers for, 41
song learning, 92, 95–99
territorial mimicry in, 110
vocal communication and, 92
vocal imitation and, 18
birth, 285–87
Black, Jeffrey, 275
blackbird, 117, 189–90, 216
Blaisdell, Frank, 45
Blumstein, Daniel, 194
Boccia, Maria, 336–37
Boesch, Christophe, 120–21, 136, 224, 247n, 272, 298–99, 350
Boesch, Hedwige, 120
Boinski, researcher, 162
Bond, Alan, 49–50, 53, 154
bonobo
artificial symbols and, 132–34
communications and, 122

gestures and, 138–39
learning through being told, 29–30
lets others borrow baby, 287
mirror test and, 336, 337–38
nest-building by, 239–40
playing with dolls, 56
pointing and, 138
social intelligence and, 330–32
stone tools made by, 227–28
tool use by, 219–20
weaning young, 289
Borgia, Gerald, 251
Born Free (Adamson and Adamson), 199
Boughman, Janette Wenrick, 105–6
Bouvier, Jérome, 301
bowerbird, 250–52, 343
Box, Hilary, 15
Boysen, Sarah, 322–23, 323
brain size, 328–29
Breland, Keller, 9–10
Breland, Marian, 9–10
Brindamour, Binti, 130
Brown, Eleanor, 102
Bshary, Redouan, 344, 345
budgies. *See* parakeets
buffalo, 193, 262–64
buntings, indigo, 325
Burger, Joanna, 12
Burghardt, Gordon, 57–58
Burton, Maurice, 14, 117, 296–97
bush-shrike, 107
bustards, 195–96
butterfly fishes, 192
Byers, John, 47, 343–44
Byrne, Jennifer, 178
Byrne, Richard, 17, 18, 178, 218, 303n, 314, 315, 329, 330

Caldwell, D. K., 117
Caldwell, M. C., 117
Cambefort, J. P., 212
Campbell, A.G., 109
Canadian Journal of Zoology, 111, 112
canaries, 96, 100
Cant, John, 39, 40
cardinals, 99
Caro, T. M., 293
Carter, Janis, 38–39, 145, 148, 158
catbird, 189
categorization, 317

catfish, 6, 198–99

cats, 17, 282
  climbing down trees, 40
  eating grass, 152
  imitation and, 18–19
  mirror and, 336
  predation skills, 165–68
  prey selection and, 169–70
  social play and, 52
  vocal communication and, 115–16

Chantek (orangutan), 23, 52, 57, 129–30, 137–38, 221–22, 271, 338

Chapais, Bernard, 333

Chapman, Graeme, 66–69

Chappuis, Claude, 106

Charcot, Jean, 59

cheetah, 76, 166–68, 192–93, 293

Cheney, Dorothy, 116, 318, 332

chick
  fear of predators in, 185
  learning to forage, 2
  learning about water, 32
  mother's voice and, 93
  pecking and aim in, 13–14
  running with large food, 14

chickadees. *See* tits

chicken
  trained to dance, 9, 10, 11
  social intelligence of, 343

chimpanzees
  aggression level of, 257–58
  American Sign Language and, 127–29, 131–32
  artificial symbols and, 132–35
  clinging by, 35–36
  coalitions of female, 257–58
  communication and, 91
  comparing cultures of, 266–68
  conditions favoring culture in, 248–49
  crawling, 37
  dialects and, 101
  disciplining of young by, 297
  drumming and, 120–121
  enculturation and, 271–72
  enriched environment and intelligence in, 341–42
  fear of snakes in, 187–88
  gestures and, 138–39
  imitation and, 23
  innovations by, 212

  language and, in wild, 136
  learning as adult, 244
  learning to climb, 38–39
  learning to walk, 36
  learning what to eat, 148
  learning where to find food, 158
  manipulation of others by, 331–32
  mate selection and age and, 291–92
  medicinal plants and, 152, 153
  mirrors and, 336, 337
  nest-building by, 239–40
  numerical skills of, 322–23
  play with dolls, 56
  personality and, 324
  poor mothering skills of, in wild, 309–10
  rehabilitation of, 145, 201, 267–68
  safety of young and, 289–90
  sex lives of, 279–80
  shoes used by, 226
  smiling and play-faces in, 35
  speech and, 123–24
  tools and, 222–27
  weaning young, 288–89

Choudhury, Sharmila, 275

chough, 55, 161

choz-choz, 47

cichlids, 234

Clarence (house sparrow), 15, 102–3, 243–44, 308

cleaner fish, 345

Clever as a Fox (Yoerg), 318

closed program, 3–5

coaching, 293, 294

cobra, 188–89

cockatiels, 277

cockatoo, 66–70, 232–33

Cody (orangutan), 33, 36–37, 60, 124–25

cognition theory, 270, 344

cognitive map, 324–25, 329

"cognitive structures," 249

Collet, Anne, 200

Collias, Elsie, 241, 243

Collias, Nicholas, 241, 243

communication, 91–140
  apes and speech, 123–25
  artificial symbol systems and, 132–35
  gestures, 138–39
  nonvocal, 120–22
  parrots and speech, 126–27

pointing, 137–38
sign language and, 127–32
translation and, 140
vocal, 91–119
competitors, 198–99
concept learning, 11–12, 317
conditioning, 6–7, 347
imitating vs., 18–19
operant, 8–10
Pavlovian, 6–8
condors, California, 31, 73–74, 145–46,
158, 268–69, 280–81
conservative learning, 14
cooperative hunting, 245–46, 342–45
copycatting, 18–19
Corbett, Jim, 210
Corbett, Laurie, 196
Corbo, Margarete Sigl, 41, 107
cormorant, double-crested, 342
correlation-learning device, 6
Coss, Richard, 186
cowbird, 80, 81, 83–84
Coyote, Martha, 114
coyotes, 54
crake, 276–77
crane
sandhill, 73, 147
whooping, 42, 73, 264–65, 267, 290
crawling, 36–37
critical or sensitive periods, 61, 95–97, 199
crocodile, 326
cross-fostering, 66–69, 79–80, 85–88
crows, 101–2, 110, 235–37, 311–12
Cuckoos, Cowbirds and Other Cheats
(Davies), 84–85
cuckoos, 80–81, 84–85
culture
baboons and male-female relations and,
261–62
bowerbird and, 250–52
buffalo voting posture and, 262–64
chimpanzee and, 257–58, 266–68
conditions favoring, 248–49
condors and, 268–69
debate over animals and, 246–47
defined, 247
elders pass on, to group, 247–48
enculturation and, 271–72
fads and, 254–55
landscape information and, 262

learning and, 349
learning as essence of, 246–47
learning new social behavior and,
255–56
lions and, 260–61
macaque reconciliation and, 256–57
macaque stone handling and, 252–54
male macaques caring for young and,
258–59
orangutans and rehabilitation and,
269–71
reindeer and caribou migration and,
265–66
social gaffs and, 259–60
special calls and, 249–50
whooping crane migration and, 264–65
in wild, 272
Curio, Eberhard, 189
curiosity, 317, 325–27
curlews, 143

damselfish, 94, 192, 198
Davies, N. B., 84, 160
deception, 335
deer, 34–35, 136
degu, 47
dialects, 100–102
Diamond, Adele, 321
Diamond, Jared, 248, 251–52
Diamond, Judy, 49–50, 53, 154
dingoes, 54, 156–57, 196
dippers, 159–60
disciplining of young, 289, 296–97
Distance Calls, 96
distraction displays, 203–5
dogs, 1–2
attacking snake, 188–89
barking, as superstition, 314
distraction displays and, 204–5
eating grass, 152
emulation and, 27
herding by, 304, 305
imitation and, 19
imprinting and community and
sheepdogs, 64–65
mirror and, 336
narcotics sniffing training, 25
pointing and, 137
problem solving and, 317
social play and, 52

training of, by humans, 304–5
understanding of humans by, 139–40
dolls, 56–57
dolphins
  brains of, 328
  bubbles, playing with, 50–51
  conditioning and, 29
  cooperation of, with fishermen,
    245–46
  cross-foster pilot whale, 80
  deception and, 335
  disciplining of young by, 297
  fads of, 254
  humans and, 201–2
  imitation and, 20–21
  intelligence of, 343
  learns to train trainer, 307
  "lucky corners" and, 314
  learn predation skills, 163
  learn to breathe at surface, 45
  midwifery and, 286
  mirror test and, 340
  mobbing of predator by, 192
  plays keep-away, 55
  pointing and, 139
  signature whistles of, 117–19
  taught to invent tricks, 29
  tools used by, 51, 233–34
  trained to herd sharks, 305–6
  vocal communication and, 92, 104–5
domains, 318–20
double-imprinting, 70
Douglas-Hamilton, Dudu, 85, 86
Douglas-Hamilton, Saba, 85, 86
drumming, 120, 232–33
Dublin, Holly, 153
ducks, 48, 62, 204, 282–83
duets, 113–14
Dunbar, Robin, 330
Dunn, Euan, 159
Durden, Ed, 44, 87–89
Durden, Kent, 44, 87–89

eagles, 1, 342
  adoption by, 283–84
  adopts goose, 87–89
  learning to fly, 44
  learns where to find food, 158
  mobbing of, by mongoose, 192
  object play and, 48

throwing things, 238
vervet alarm call for, 116
eating, learning about.
    See also predatory skills
  chimpanzee rehabilitation and, 268
  favorite and specialty foods, 155–57
  food caching and, 181–82
  foraging skills and, 177–80
  medicinal plants and, 152–154
  orangutan rehabilitation and, 270–71
  predation skills and, 158–82
  trying everything you see and, 154–55
  what not to eat, 149–52, 162
  what to eat, 147–49
  where to find food, 157–58
Eaton, Randall, 166–67
Economist, The, 269
eel, moray, 61, 192, 344–45
elephants
  elders pass on culture to, 248
  exploring by, 46
  learning about eating and, 15
  learn to move trunk, 37
  memory and, 327–28
  midwifery and, 286
  mirror test and, 336
  plants to induce labor and, 153
  tools used by, 234–35
Eloff, F. C., 164
Emlen, Steve, 325
emotional intelligence, 317
emulation, 27, 28
enculturation, 129, 271–72
enemies
  learning how to act in face of, 202–5
  learning to spot, 60
  recognition of competitors and, 198–99
  recognition of predators and, 184–98
  throwing things at, 231
environment
  enriched, intelligence and, 340–42
  innate behaviors and, 5
  social behavior and, 256
Espmark, Yngve, 266
Evans, Christopher, 194
"exceptional learning," 126
exploring, 46
eyes, fear of, 186
fads, 254–55
Fagen, Robert, 46

falcons, 41, 191, 342
"fear" stage, 184
Fenton, Neville, 108
ferrets, 145, 196–97, 278
Ficken, Millicent, 104
filial imprinting, 61–62
finches, 70, 79, 96–97, 184, 217
fish. *See also* specific types
    cleaning stations and Machiavellian
        strategy, 345
    hearing experiments, 6
    innovations by, 212–13
    inspection of predators by, 193–94
    intelligence of, 344–45
    landscape information and, 262
    learn to recognize competitors, 198–99
    mobbing of predators by, 192
    tools used by, 234
FitzGibbon, Clare, 192–93
"fixed threshold" strategy, 275
flycatcher, 160–61
flying, 41–44, 295–96
flying fox, 63
follow the leader, 55, 56
food. *See* eating; foraging; predation skills
food calls, 115
food preferences, 155
foot-drumming, 119
foraging, 177–80, 270–71, 300, 329
Fouts, Deborah, 128
Fouts, Roger, 128
Fox, E. A., 220
foxes, 1
    beginning to see and eyes opening,
        33, 35
    distraction displays and, 204
    exploring, 46
    grasping and, 33
    human structures and, 146
    learning about ice, 32–33
    learning to cache food, 181–82
    learning to forage, 300
    learning to recognize friends, 199
    learn what to eat, 149
    modeling behavior for young, 300
    rehabilitation, 200
    teaching animals to fear, 194–96
    trapping, 205–6
Fragaszy, Dorothy, 162, 213
Frank, Lawrence, 343

Freeberg, researcher, 98
French, Barbara, 159
Freud, Sigmund, 65
Fricke, Hans, 163, 344, 345
friends, learning to recognize, 199
"fright" stage, 184
Frisch, Karl von, 6
Fritz, Jo, 341
frog, 3
fruit fly larvae, 6–7
Fruth, Barbara, 239–40

Galdikas, Biruté, 23, 39, 120, 130, 157,
        221, 259, 270, 271, 288
Gallup, Gordon, Jr., 336
games, 55–56
Garcia, John, 150
Gardner, Allen, 127, 128
Gardner, Beatrix, 127–28
Gauthier, Carole, 333
gazelles, 192–93
geese
    cross-fostered by eagle, 87–89
    disciplining young, 297
    filial imprinting and, 62
    learning to fly, 41, 43–44
    mate selection and parenting skills,
        275–76
    migratory vs. sedentary populations of,
        264
    nesting on islands and, 313
    no fear of, in young birds, 185
    parenting skills, age and, 291
    social position and, 333
    stimulus enhancement and, 16
genes, 4, 5, 31–32
gestures, 138–39
Ghost with Trembling Wings, The
        (Weidensaul), 303n
gibbons, 113–14, 338–40
Gibson, Kathleen, 15, 153, 218, 230
Glickman, Stephen, 325
goats, 65
gobies, 325
Goodall, Jane, 101n, 212, 222–24, 289,
        291–92, 309–10
Goodwin, Derek, 110–11
Gopnik, Alison, 335
gorillas
    chest-beating and, 120

crawling by, 37
enculturated, 271
enriched environment and intelligence
    and, 341
hand-eye coordination and, 34
imitation and, 23
intelligence of, 312
learning in, vs. human babies, 322
learn to forage, 177–78
learn to reach, 321–22
mirror test and, 336, 338
nest-building by, 239, 240
parenting and, 273–74
parenting skills learned by, 308, 309
social distance learned by, 259–60
tool use and, 218–19
Gould, Carol Grant, 48, 215–16, 344
Gould, James L., 11–12, 48, 215–16, 344
Grey, Hunt, 237
Grey, Russell, 237
Griffin, Andrea, 194
grooming theory of language, 330
groupers, 344–45
grouse, 203–4
grunt, French, 262
Guinet, Christophe, 301
gulls, 12, 284–85, 291
guppies, 202, 212–13

Haas, Femke den, 149
habituation, 1, 65–66, 73–74
Hamerstrom, Frances, 191
hamsters, 64
Hancock, David and Lyn, 42–43
hand clap, 138
hand control, 33–34
Hansell, Michael, 217
Hauber, Mark, 82–83
Hausberger, Martine, 136
Hauser, M. D., 293
Hauser, Marc, 158, 320, 321, 325
hawks
    cooperative hunting and, 343
    fear of, in young birds, 185–86
    imprinting and, 78–79
    learning what to eat, 147
    learning which are scariest, 190–91
    mate selection and, 291
    mobbing of, 191
    object play and, 48

Hayes, Cathy, 123–24, 187–88
Hayes, Keith, 123–24, 187
hearing, 32
Hedren, Tippi, 169–70
Heinrich, Bernd, 81–82, 107, 188, 191,
    299, 315–16
Henry, J. David, 32–33
Henty, C. J., 216
herons, 158–59, 215–16
Hess, Eckhard, 13–14
Heyes, Cecilia, 214
Higuchi, Hiroyoshi, 215
hippos, 47
Hiraiwa, Mariko, 224
Hohmann, Gottfried, 239–40
honeybees, 11–12, 48, 344
honeyeater, 189, 190n
horses, 34–35, 122–23, 285, 290
Hsu, Minna, 267, 268
Huber, Ludwig, 19
Huffman, Michael, 253
human children
    babbling and, 97–98
    handling things, 34
    imitation and, 24
    inhibiting action in favor of another,
        321–22
    learn to blow nose, 32
    learn to stand and walk, 34–35
    learn to walk and talk, 294
    pointing and, 137
    smiling and, 35
    social play and "theory of mind," 53
    sticking out tongues, 334–35
humans
    animals' habituation to, and
        rehabilitation, 65–66, 145–46,
        200–202
    animal intelligence vs., 312–13
    animals learning to recognize as friend
        or foe, 199–202
    animals raised by, and enculturation,
        271–72
    apprenticeship and, 297–98
    brain of, 328
    cultures of, 247
    fear of snakes in, 186
    individual personality and, 324
    lion encounters with, as learned
        behavior, 260–61

mirror test and, 336
nature vs. nurture and, 4–5
nest-building by, 239
taste aversion in, 150
tigers eating, 210–11
tigers learn to associate with danger, 348
training of animals by, 304–8
vervet alarm call for, 116
hummingbirds, 48, 92, 104, 318–19
Humphrey, Nicholas, 329
Hunt, Gavin, 237
Huntingford, F. A., 202
hyenas, 343
hypothesis testing, 317

Imanishi, Kinji, 247
imitation, 17–24. *See also* mimicry
chimpanzees and, 271–72
emulation vs., 27
intelligence and, 314
sticking out tongue and, 334–35
tiger and, 347–48
imprinting, 60–61
critical or sensitive periods and, 61
cross-fostering and, 66–77
environment and, 5
filial, 61–62
nest parasites and, 82–84
sexual, 65–77
to learn community, 64–65
to recognize siblings, 63–64
recognizing children and, 80–82
incest taboos, 63–64
Ingmanson, Ellen, 122, 219, 220
inhibiting action in favor of another,
321–22
innate (inborn, instinctive) behavior, 2,
5–6, 31, 166–67, 182
innovation, 211–16
intelligence, 311–46
aspects of, 317
birds vs. primates and, 342–43
brain size and, 328
cleaner fish and, 345
cognitive maps and, 324–25
curiosity and, 325–27
deception and, 335
defining, 314–15
dolphins and whales and, 343
domain-specific, 318–20

enriched environment and, 340–42
extractive foraging and, 329
fish and, 344–45
geese nesting and, 313
gorilla and IQ test, 312
human vs. animal, 312–13
hyenas, octopuses, buffalo, bees and,
343–44
imitation and, 335
insight and, 315–16
learning and, 314–15
learning to reach for seen object and,
321–22
manipulation and, 334
memory and, 327–28
mirror test and, 335–40
number skills, 322–23
object permanence and, 320–21
personality and, 324
problem solving and, 317
reliability and, 332–33
social intelligence and, 333
social manipulation and, 329–32
superstition and, 314
theory of mind and, 334–35
tigers and, 313
"isolate song," 98
isolation, 98, 349
"Is There Teaching in Nonhuman
Animals?" (Caro and Hauser), 293
Itani, Junichiro, 258–59, 324

Jablonka, Eva, 291, 299, 320
jackdaw, 191–92
Jacobs, Dr. Catherine, 242
jaguars, 186
James, William, 18
Janik, researcher, 118
jays
learn to fly, 42
memory and, 318
mimicry and, 110–11
mirror and, 336
nest location and, 281–82
search image and, 157
tool use by, 209–10
young help parents with new brood, 292
Johanowicz, Denise, 256
Johnston, Robert, 64
Joines, Steven, 274

Jolly, Alison, 332
junco, 158
Ju/wasi people, 260–61

Kachanoff, Kim, 64
Kamil, Alan, 318
kangaroo, grey, 46
Kanzi (bonobo), 29–30, 56, 133–34, 135, 138, 140, 227–28, 287, 289, 330–31, 332, 338
Kawamura, Shunzo, 247
Kea, Bird of Paradox (Diamond and Bond), 49
keas (New Zealand parrots), 28
    eating at dump, 154–55
    social facilitation and, 28
    social play and, 53
    trapping, 207
    trial-and-error in play and, 49–50
kestrel, 282
Kilham, Benjamin, 37, 148
killdeer, 203n
killer whales, 1
    dialects of, 101
    fads and, 254
    imitation and, 21
    mirror test and, 336
    nursing calves and, 288
    synchronization of movements by, 21
    teaching beaching, 301–2
King, researcher, 98
kingbird, 191
kingfishers, 215
King Solomon's Ring (Lorenz), 62
Kipling, Rudyard, 234
Kipps, Clare, 15, 102–3, 143, 243–44, 308
Kleiman, Devra, 47
Knoxville, Tennessee, zoo, 21
Koko (gorilla), 128–29, 271, 304, 312, 338
Krüger, Kobie, 16, 140, 173–75, 188–89, 193
Krüger, Kobus, 173
Kruuk, Hans, 149
kudu, 193
Kuhl, Patricia, 335
Kummer, Hans, 5, 56, 137, 211–12, 248, 261–62
Laidler, Keith, 33, 36, 124–25

Laland, Kevin, 212
landscape, 262–66
language
    bonobos and, 134, 249
    chimpanzees and enculturation, 271–72
    evolution of tool use and, 218
    grooming theory of, 330
    learning methods, 349
    teaching parenting skills to gorilla through, 274
latent learning, 11
Laurance, William, 63, 206–7
Leakey, Louis, 222, 223
learning. *See also* learning methods; *and specific skills*
    as adaptation of tremendous power, xi
    baby tigers and, 347–49
    basics, 31–58
    to climb, 37–40
    to communicate, 91–140
    to crawl, 33–34
    conclusions about, 349–50
    culture and, 245–72
    defined, ix–x
    to distinguish between enemies, friends, and noncombatants, 184, 199
    to explore, 46
    to fly, 41–44
    hand and body control, 31–32
    how not to be eaten, 183–208
    how to eat and find food, 141–82
    "how to learn," 12
    innate behavior modified by, 4–6
    intelligence and, 311–46
    invention and tools and, 209–44
    maturation vs., 13–14
    to parent, 273–310
    playing and, 46–58
    preparation for, 2–3
    reasons for, 2–3
    selective advantage of, 4
    social, 14–27
    to swim, 44–45
    to walk and run, 34–36
    who parent is, and imprinting, 5
    wildlife rehabilitation and, x–xi
    your species, 59–89
learning methods, 1–30
    being told, 29–30

emulation, 27
imitation, 17–24
mixture of, 22
observational, stimulus and local
enhancement, 15–17
operant conditioning, 8–10
Pavlovian conditioning, 6–8
practice, 12–13
social, 25
social enhancement, 25–27
social facilitation, 15
social, horizontal vs. vertical, 14–15
trial and error and, 12
lemurs, 25–27, 325, 333
Leo (lion), 140, 173–75, 193
leopard
fear of, in young animals, 186
learning jungle, 142
man-eating, 210
predation skills and, 166, 173
sexual imprinting and, 76, 77
vervet alarm call for, 116
Lewin, Aroger, 135
Leyhausen, Paul, 115, 166
Limongelli, Luca, 230, 231
lions
adopt oryx calf, 85–86
culture and, 260–61
fads and, 255
inspection of, by prey, 193
learn to recognize humans as foes,
199–200
man-eating, 210, 211
predation skills and, 164, 166, 169–71,
173–75
sexual imprinting and, 76–77
wildebeest fetus and, 86–87
llamas, 77, 93
lobtail, 217
local enhancement, 16, 17
locality imprinting, 143
Locke, Lt. Colonel, 7, 348
Lockley, R. M., 79
Lopez, Diana, 302
Lopez, Juan Carlos, 302
Lord, Jeanne, 200
Lorenz, Konrad, 43, 44, 62, 72–73, 191,
333
Lore of the Lyrebird, The (Pratt), 108
Losey, George, 198

lovebirds, 5–6, 278
Lucy (chimpanzee), 23, 36, 38–39, 53, 91,
145, 148, 222
lyrebirds, 108–9

macaques
fear of leopard spots in, 186
grasping and hand coordination, 33
individual personality and, 324
innovations by, 212, 213–15
mirror test and, 336–37
"paternal" caring for young and,
258–59
play-faces in, 35
reconciliation and, 256–57
respect and expertise and, 221
social manipulation and, 334
social rank, 333
stone-handling culture and, 252–54
teach baby to crawl, 295
teach young independence, 295
tube-and-stick problem and, 232
macaws
learn to fly, 42–43
mimicry and duets, 114
McCarthy, Sarah, 34, 285
McCowan, Brenda, 118
Macdonald, David, 146, 149, 181, 199,
199, 206, 300–301
McGowan, Kevin, 191
McGrew, W. C., 247
Machiavellian Intelligence (Byrne and
Whiten), 329, 330
McKeever, Kay, 78, 306–7
Maestripieri, Dario, 295
"Mafia hypothesis," 84–85
magnificent frigate birds, 56
magpies, 84–85, 191
manipulation, 330–32, 334
Mann, Janet, 45
Mannu, Massimo, 229
Maoris, 154
Marcström, V., 204
Marino, Lori, 340
Marler, Peter, 342
marmots, 107
Marzluff, John, 281–282
Matata (bonobo), 132–33, 287, 289, 330,
337–38
Mateo, Jill, 64

mates and mating. *See also* nesting behavior; sexual behavior
  baboons and marriage, 261–62
  baboons and social intelligence and, 331
  barnacle geese and, 274–76
  black robins and, 277–78
  bowerbirds and, 250–52
  California condors and, 281
  chimpanzees and older, 291–92
  cockatiels and marriage, 277
  cross-fostered cockatoos and, 69
  discerning real from dummy mate and, 276–77
  geese and, 275–76
  imprinting and, 60, 74–75
  imprinting and incest taboos, 63–64
  inspection of predators and, 194
  learning to recognize, 3
  lovebirds and, 278
  mimicry and calling home, 110
  parenting skills and choice of, 274–76
  sexual imprinting and, 65–66
Matsuzawa, Tetsuro, 134–35, 225
maturation, 13–14
Mayr, Ernst, 3, 4
medicinal plants, 152–54
Meinertzhagen, Col. Richard, 302–3, 303n
Meltzoff, Andrew, 334–35
memory, 317, 318, 327–28
mice, 61, 63–64
midwifery, 286–87
migration, 264–66, 325
Miles, H. Lyn, 129, 130, 221–22, 271
Miles, Hugh, 45
Miller, Olive Thorne, 41–42
Millham, Cheryl, 283
Milton, Katharine, 255, 329
mimicry, 102–3, 106–13. *See also* imitation
mind, theory of, 334–35
Mind of the Raven (Heinrich), 107
minnows, 193–94, 243
mirror test, 335–41
"Misbehavior of Organisms, The" (Breland and Breland), 9
moas, 154, 155
mobbing, of predator, 189–92
modeling behavior, 300–301

model-rival method, 126, 130, 133, 303
modular abilities, 318
mongoose, 192, 282
monkeys, 17. *See also* macaques
  alarm calls and, 12, 116–17, 294, 318
  babbling and, 97
  coaching and, 294
  curiosity and, 325
  fear of snakes and, 187
  foraging of, and intelligence, 329
  hand-eye coordination and, 34
  imitation and, 19
  innovations by, 212, 213
  learn foraging skills, 161–62
  learn to be social, 255–56
  learn to climb, 37–38
  learn to react to alarm calls, 12
  learn to recognize predators, 183
  learn what to eat, 147
  millipedes used to repel mosquitoes by, 153
  Muppets on television and, 57
  nest-building by, 239
  reaching for what they see, 321
  reliability and, 332–33
  social play and, 53, 54
  teaching children to lead, while foraging, 295
  tools and, 228–32
  vocal learning and, 119–20
Montgomery, Sy, 211
Moore, Bruce, 18
moose, 197–98
Morgan, Kathleen, 258
Morton, Alexandra, 1, 21, 288
mosquito repellents, 153
muskox, 55
Myberg, Arthur, Jr., 94
mynah, 92, 200–201
Myowa, Masako, 335

Nature, 65
neighborhood, learning, 142–46
neighbors, learning to recognize, 94–95
nesting behavior
  adoption of eggs, 282–83
  building nest, 239–43
  changing, 282
  experience improves, 292
  parasitism and, 80–85

in peach-faced vs. Fischer's lovebirds, 5–6
selecting location for, 281–82
New Scientist, 217
New Yorker, 111
Nice, Margaret, 184–85
Nicolai, Jürgen, 4
Niff (red fox), 181–82, 199, 199
Nim Chimpsky (chimpanzee), 131–32
Nishida, Toshisada, 224, 288–89
Noad, Michael, 254
nonvocal communication, 136–39
    baboons and, 136–137
    dogs understand human, 139–40
    gestures and, 138–39
    pointing, 137–38, 139
    sound signals, primates and, 120–21
Nottebohm, Fernando, 100, 101
Nottebohm, Marta, 101
numerical skills, 317, 322–24, 341–42
nursing baby, 287–88, 309
nut cracking
    capuchin monkeys and, 228–29
    chimpanzees and, 225, 266–67, 272,
        298–99
    crows and, 235–36

Oatley, Terry, 111
object permanence, 320–21
object play, 48–51
oblique learning methods, 15
octopuses, 51, 324, 343
onespot fringehead, 156
"one-step-decision" strategy, 275, 276
open program, 3, 4
operant conditioning, 6, 8–10
opossums, 3
opportunity teaching, 293, 349
orangutans
    attempt to teach to speak, 124–25
    culture and rehabilitation and, 269–71
    imitation and, 23–24, 272
    intelligence of, 342
    learning hand control, 33
    learning sign language, 129–30
    learning to sit and crawl, 36–37
    learning to swing in "pole trees," 39–40
    learning what to eat, 148–49
    learning where to find food, 157–58
    mirror test and, 336, 338
    nest-building by, 239

nursing young, 287–88
play with dolls, 57
pointing and, 137–38
raised by humans, don't recognize own
    species, 60
rehabilitation of, 143, 148–49
social and cognitive intelligence of, and
        reaching
for food, 323–24
social distance learned by, 259, 267
social play and, 52
tools and, 220–22
vocal communication and, 120
wean young, 288
orcas, 51
ordination, 322
oriole, 107
oryx calf, adopted by lioness, 85
osprey, 302–3, 303n
Osten, Wilhelm von, 122–23
otters
    discipline young, 297
    distraction displays and, 204
    favorite foods and, 155
    hide-and-seek and, 55–56
    learn to find octopuses in cans, 155–56
    learn to swim, 44–45
    learn to forage, 180
    learn to spot enemies, 60
    operant conditioning and, 10
    see first snow, 32
    tools used by, 233
Ottoni, Eduardo, 229
owls, 1, 3, 32, 77–78, 184–85, 188,
    306–7

pandas, 21–22, 47, 55, 74–76
parakeets (budgies), 19, 55
parents and parenting, 273–310
    active teaching by, 298–99
    adoption and, 282–85
    apprenticeship as teaching by, 297–98
    baby learns to recognize, 60
    disciplining young and, 296–97
    egg tending and, 280–281
    giving birth for first time, 285–86
    imprinting and, 61–62
    improving skills with age, 289–91
    learning skills of, 273–74, 308–9
    mate selection and, 274–76

modeling behavior and, 300–302
nest location and, 281–82
not letting others borrow baby, 287
nursing, 287–88
open vs. closed programs and, 3
protecting safety of young, 289–290
providing time and confidence to learn
     in safety and, 299–300
teach young to eat, 306
teach young to hunt, 160–61, 296,
     299–303, 306
teach young to walk, crawl and fly,
     294–96
teenagers learn, by helping with new
     brood, 292–93
types of teaching to young, 293–94
weaning by, 288–289
Parker, Sue Taylor, 218, 248, 314
parrots, 106–7. See also cockatoo; keas;
     macaws
dialects of, 101
fear of predators, 185–86
learn to eat pine nuts, 178
mimicry by, 106–7
mirror test and, 336
object permanence and, 320–21
preening and, 68–69
rehabilitation and, 144
teaching, to speak, 125–27
teach young, 303–4
vocal communication and, 92–94
vocal imitation, 18
"partner-hold" tactic, 276
partridges, rock, 185
Passion (chimpanzee), 36
password, 82–83
"paternal care" phenomenon, 258–59
Patterson, Gareth, 76–77
Patterson, Penny, 128, 338
Pavlovian conditioning, 6–8
peacocks, 72–73
pelicans, 234, 342
penguins, 48, 59, 93–94, 291
Pepperberg, Irene, 9, 125, 126–27, 303–4,
     320
personality, 324, 349
pestle pounding, 225–26
Pfungst, Oskar, 123
Piaget, Jean, 320
pigeons, 29, 312–13, 331

pigs, 326–27
pike, 193–94, 202–3
pilotbirds, 108
pinnipeds, 92
Pitt, Frances, 204
play, 46–58
adult animals and, 57–58
benefits of, 47
defined, 46
dolls and, 56–57
games and, 55–56
learning about individuals through, 54
object, 48–51
as practicing, 12–13, 46–47
risks of, 47
social, 52–57
tiger cubs and, 347
toys and, 57, 234
trial-and-error learning in, 49–50
plovers, 205
pointing, 137–40
poisonous food, 162
polar bears, 47–48, 162–63, 199, 283
porpoise, 29
Poti, Patricia, 314
Povinelli, Daniel, 39, 40
Powell, David, 55, 76
practicing, 1, 12–13, 46–47
Prairie Night, 197n
Pratt, Ambrose, 108–9
predation skills
inborn, 165–67
learning, 158–77, 181–82, 300
learning to hunt, 141, 296, 299–303,
     348
teaching owls, by rehabilitators, 306–7
teaching young, 296, 300–303
tiger cub learning, 348
predators
alarm calls to warn of, 116–117
distraction displays to, 203–5
innate fear of spots and, 186
innate fear of overhead bird and,
     185–86
innate fear of snakes and, 186–89
innate vs. learned fear of, in jackdaws,
     191–92
inspection of, 192–94
learning to avoid, and rehabilitation, 145
learning to fear, from mother, 197–98

learning to recognize, 183, 184–85
learning what to do in face of, 202–5
learning which are most dangerous, 190–92
mobbing of, 189–90, 192
protecting children from, 290
showing, it's been spotted, 203
teaching animals to fear, in rehabilitation, 194–97
prey
learning what is good, 147
specialties, tiger cub and, 348
primates. *See also* specific types
as extractive foragers, and brains, 329
born with open eyes, 35
clinging by, 35–37
fear of leopard spots, 186
fear of snakes, 186–87
imitation and, 22–24
innovation by, 211–12
learning to grasp, 33–34
social intelligence of, and manipulation, 329–30
teaching young to crawl and walk, 34, 295
vocal communication and, 92, 119–20
Prince-Hughes, Dawn, 219
Prins, Herbert, 262–63
problem solving, 317
program-level imitation, 27
prosimians, 25–26
Pryor, Karen, 10, 11, 19, 29, 40, 114
Przewalski's horses, 143
ptarmigans, 79, 185, 203–5, 291
puffin, 79

quokkas, 195

rabbits, 2, 60, 117, 196
raccoons, 9, 10, 324
Ramakrishnan, Uma, 186
Ran, Arum, 211
"random mating" strategy, 275
rats, 17
discipline young, 296
enriched environment and exploring by, 46
intelligence, 341
latent learning and mazes, 11
learned taste aversion in, 150

learn to eat pine nuts, 178–80
operant conditionings, 8
social play and, 54
traps and, 207
ravens, 2, 299
dialect and, 102n
food calls, 115
intelligence and, 315–16
king of the castle, 55
moose detects predators with, 198
pelt invaders with rocks, 237–38
recognize own chicks, 82
taste aversion and, 151–52
vocal communication and, 92, 107
where to find food and, 158
Reader, Simon, 212
reconciliation, 256–57
Redshaw, Margaret, 321–22
Reid, Les, 146
reindeer, 265–66
Reiss, Diana, 118, 307
reliability, 332–33
Riedman, Marianne, 155, 180
Riggio, Robert, 94
Ristau, Carolyn, 205
rosella, green, 108
robins
imprinting and, 60, 70–72
mating and, 277–78
mobbing predator, 189–90
robin-chats, 110, 111, 113
Robinson, Norman, 108
rock throwing, 237–39
Ron (tiger), 164, 165, 348
Rood, Jon, 192
Rowell, Thelma, 136
Rowley, Ian, 66–69
Rumbaugh, Duane, 249
Running with the Fox (Macdonald), 206
Rusch, Kathryn, 104
Russo, Stefani, 83
Russon, Anne, 23, 24, 39, 248, 270, 271, 342

safety of young, 289–90, 299–300
salmon, 61
Salmoni, Dave, 164, 165
salt desert cavy, 31, 47
Savage-Rumbaugh, Sue, 30, 132–35, 138, 228, 249, 289, 330

Schaik, Carel van, 220
Schaller, George, 87, 168, 170, 193, 255
Schevill, Bill, 114
Schuchmann, Karl-Ludwig, 104
Science, 311
Scientist in the Crib, The (Meltzoff, Gopnik, and Kuhl), 335
sea bass, 16
seal
    killer whales catch, 301–2
    learning to fear humans, rehabilitation and, 200
    mimicry and, 111–12
    playing by, 47
    vocal communication and, 92, 93
sea lions, 92
search image, 154
seeing, 32–33
self-conception, 39–40, 328, 336
self-handicapping, 54
Seligman, Martin, 150
sensory input, 32
"sequential comparison" strategy, 275, 276
Sevcik, Rose, 249
sexual behavior
    bonobos and, 138–39
    chimpanzees and, 279–80
    ferrets and, 278
    social play and, 55
    tigers and, 278–79
    vocal communication and, 92
sexual imprinting, 65–66
    cross-fostered birds and, 66–74
    giant cats and, 76–77
    giant pandas and, 74–75
    goshawks and, 78–79
    owls and, 77–78
    zebra finches and, 79
Seyfarth, Robert, 116, 318, 332
Shah, Anup, 136
Shah, Manoj, 136
Shapiro, Gary, 130
sharks, 192, 305–6
shearwaters, 48, 79
sheep (lamb), 47, 52, 65
Sheldrick, Daphne, 189
Sherman, Paul, 83
Sherman (chimpanzee), 132, 138, 337
Shettleworth, Sara, 318
Shillito, Daniel, 336

shrike, 111
Shumaker, Robert, 24, 270, 323, 342
siblings, 60, 63–64, 93
signature sounds, 117–19
Singh, Arjan, 77, 142, 173, 176, 177
site tenacity, 143
size constancy, 320
Skinner, B. F., 8, 9
Slater, researcher, 118
smell, imprinting and, 63–64
Smithsonian, 270
Smits, Willie, 270
Smolker, Rachel, 118, 163, 201–2, 317
snakes, 116, 186–89, 231
Snowdon, Charles, 136
Snyder, Helen, 74, 281
Snyder, Noel, 74, 281
Snyder, Phil, 283
social-cognitive differences, 323–324
social conditioning, 2
social cues, 332
social facilitation, 15, 28
social learning, 14–27
    birds and, 343
    bonding, grooming, and language, 330
    culture and, 255–56
    defined, 14–15
    emulation as, 27
    horizontal, 14–15
    manipulation and, 329–32, 334
    mixed forms, 28
    oblique, 15
    other forms, 25–27
    rank and, 333–34
    recognize members of troop and, 60
    theory of mind and, 334–35
    vertical, 14
social play, 46
    dolls and, 56–57
    games and, 55–56
    learning about individuals through, 54
    learning social skills through, 52–54
    learning submissive behavior through, 54
    learning to take turns and self-handicapping, 54
    morality, fair play and, 54
    sexual behaviors and, 55
Soler, Manuel, 84
Sonerud, Geir, 204

songbirds, 92, 95–100
song templates, 95–96
sparrows
 dialects, 100
 learning by adult, 243–44
 music appreciation of, 102–3
 rehabilitation of, 143
 social facilitation and, 15
 song learning, 99–100
 stimulus enhancement and, 17
 subsong, 97
 threat recognition experiment and,
  184–85
 training to do tricks, 308
spatial memory, 318
Species Survival Plan, 279
spider, hunting by, 319–20
Sprague, Dan, 265
Squier, Leslie, 327
squirrels, 22, 46, 192, 317
Sroges, Richard, 325
stalking, 167, 347
Stander, P. E., 170
starling, 41, 96, 107, 116–17
stickleback, three-spined, 202–3
sticks, as tools, 220–21, 223–24, 226,
 230–33, 237
stimulus enhancement, 15–17, 23,
 25–28
Stirling, Ian, 162–63
stoats, 190
stone handling, 252–54
stone tools, 227–29
storks, 285
Strier, Karen, 183
subitization, 322
submissive behavior, 54
subsong, 97, 98, 102
Summers-Smith, D., 103
Sunquist, Fiona, 169, 172, 181
Sunquist, Mel, 169, 172, 181
superstition, 314
Swallow, Alice, 112
Swallow, George, 112
swans, 290–91
swimming, 44–45
symmetric marker, 11

takahe, 190
tamarin, golden lion, 144, 145, 150–51

Tarou, Lori, 55
teaching, by animals
 active by chimpanzees, 298–99
 apprenticeship and, 297–98
 debate over, 293–94
 defined, 293
 most effective methods, 349
 young to hunt, 296, 299–303
 young to walk, crawl, and fly, 294–96
Temerlin, Maurice, 53, 91, 148
Temple, Stanley, 282
Terkel, Joseph, 178
terns
 "contact dipping" learning, 13
 learning predation skills, 159
 practicing hunting, 12–13
 teaching young to hunt, 302
 training ravens not to eat, 151–52
ter Pelkwyk, Joost, 184–85
Terrace, Herbert, 131
territorial mimicry, 110
territory, 60, 198–199
Thalar, Ellen, 185
Thapar, Valmik, 170, 172
Ibis (journal), 302
Thinking Ape, The (Byrne), 303n
Thomas, Elizabeth Marshall, 60, 86–87,
 260, 317
Thomsett, Simon, 283–84
Thorndike, Edward, 18
thrushes, 107, 143, 216
tigers
 associate light with cattle, 7–8
 imprinting and, 76
 learn about sex, 278–79
 learn predation skills, 164–66, 168–72,
  176–77, 347, 348
 learn to drag carcass, 181
 learn to hunt, 141–42
 learn to open carcass, 181
 learn to see, 2
 learn specialty foods, 156
 man-eating, 210–11
 rehabilitation and, 267
 social environment and, 348–49
 social play and, 52
 what is needed to learn to survive,
  347–49
Tilson, Ron, 141, 279
titmouse, 316, 342

tits (chickadees), 316
  black robins raised by, 71, 72
  learning by young, 300
  peck open milk bottles, 16–17
  vocal communication and, 92, 100
toads, 63
Todd, John, 198–99
Todt, Dietmar, 126
Tolman, E. C., 11
Tomasello, Michael, 247, 271–72
tools, 217–44
  birds vs. primates and, 342
  blue jays and, 209–10
  bonobos and, 219–20
  capuchin monkeys and, 228–32
  changing views of animals and,
    217–18
  chimpanzees and, 212, 222–27, 249
  cichlids and, 234
  cockatoos and, 232–233
  crows and, 235–37, 311–12
  dolphins and, 233–34
  eagles and, 238
  elephants and, 234–35
  gorillas and, 218–219
  intelligent, 218
  ladder made by chimpanzees, 226–227
  language and, 218
  macaques and, 221, 232
  nests and burrows and, 239–43
  object play and, 51
  orangutans and, 220–21
  otters and, 233
  ravens and, 237–38
  social play and, 53–54
  stone, made by bonobos, 227–28
  thrushes and, 216
  vultures and, 238–39
tortoises, 73
towhee (chewink), 189
training of animals, by humans, 304–9.
    See also specific animals and skills
trap happiness, 205–7
trial and error, 2, 30
  defined, 12
  hunting and, 319–20
  imitation vs., 22
  play and, 49–50
triggerfish, 163
Tulley, J. J., 202

turns, taking, 54
turtles, 57–58, 61
Tyack, Peter, 118

Varty, John, 164, 165
Victoria (tiger), 76, 347
Viki (chimpanzee), 23, 124, 187–88
Visalberghi, Elisabetta, 213, 229, 230, 231
vocal communication
  birds and dialects, 100–101
  birds and new songs, 100
  birds and wrong songs, 100
  codes and signals and, 115–17
  duets and, 113–14
  female birdsong and, 99
  isolation and birdsong and, 98
  mammals and birds learn, 92
  mimicry and, 106–13
  signature whistles and, 117–19
  songbird templates and, 95–97
  sound of mother's voice and, 93
  sound of neighbors and, 94–95
  sound of siblings and, 93–94
  subsong and babbling, 97–98
  teaching apes to speak, 123–125
  teaching parrot to speak, 125–26
vocal learning
  birds and humans amd, 136
  birds vs. primates and, 343
  dolphins and whales, 343
Voelkl, Bernhard, 19
voles, 292–93, 296–97
"voting posture," 263–64
vultures
  fear of, by birds, 191
  learning where to find food, 158
  tool use by, 217, 238–39

Waal, Frans de, 5, 12, 17, 214, 247, 256,
    272
walking, 34–36, 294–95
wallabies, 46, 194–95
Wallace, Mike, 269
walrus, 92
warblers
  black robin eggs and, 71
  nest parasitism and, 80
  object play and, 49
  vocal communication and, 92
  young help parents with next clutch, 292

Washington Post, 197n

Washoe (chimpanzee), 128, 332

wasp, 217

Wasser, Samuel, 52

water, learning about, 32–33

waterbuck, 193

wattle-eye, 107

Wayre, Philip, 44

weaning, 288–89

weaver, village, 241–43

Weidensaul, Scott, 303n

Weisbord, Merrily, 64

West, researcher, 98

Western Birds (journal), 235

whales

  bubbles used as tools by, 51

  cross-fostered by dolphins, 80

  deception and 335

  duets, 114

  elders and, 248

  hunting by, 216–17

  intelligence of, 343

  mimicry and, 112–13

  object play and, 49

  song fads and, 254–55

  vocal communication and, 92

"What's So Special about Using Tools?"
  (Hansell), 217

whipbirds, 108

White, Jan, 283

Whitehead, Hal, 45, 49

Whiten, Andrew, 329, 330

Wickler, Wolfgang, 113, 344, 345

wildebeest, 87, 193

Wilderness Family, The (Krüger), 16, 173

wildlife rehabilitation, xi

  California condors and, 268–69

  chimpanzees and, 267–68

  culture and, as obstacle, 267

  imprinting and, 65–66, 77–78

  learning environment and, 143–45

  learning to fear humans and, 200–202

  orangutan and, 269–71

  owls and learning what to hunt, 306–7

  social gaffs and, 259–60

  teaching animals to spot predators and,
    194–97

Wilkinson, Mrs., 108

Wilson, Susan, 47

"win-shift" vs. "win-stay" strategy, 318–19

Winters, Patricia, 295–96

Wolfie (dog), 140, 174, 175, 193

wolves

  adoption by, 284

  moose teaches young to fear, 197–98

  pointing and, 140

  rehabilitation and, 145

  social play and, 52

  vocal communication and, 94–95

woodpeckers, 343

Wrangham, Richard, 56

wrens, striped-back, 249–50

Wylie, Philip, 317

Yamakoshi, Gen, 225

Year of the Greylag Goose, The (Lorenz), 44

Yoerg, Sonja, 8, 160, 317, 318

Zann, Richard, 79, 96